Nonlinear Dynamics of the Lithosphere
and Earthquake Prediction

T0142085

Springer
Berlin
Heidelberg
New York
Hong Kong
London
Milan
Paris
Tokyo

Physics and Astronomy

http://www.springer.de/phys/

Springer Series in Synergetics

http://www.springer.de/phys/books/sssyn

SSSyn – An Interdisciplinary Series on Complex Systems

The success of the Springer Series in Synergetics has been made possible by the contributions of outstanding authors who presented their quite often pioneering results to the science community well beyond the borders of a special discipline. Indeed, interdisciplinarity is one of the main features of this series. But interdisciplinarity is not enough: The main goal is the search for common features of self-organizing systems in a great variety of seemingly quite different systems, or, still more precisely speaking, the search for general principles underlying the spontaneous formation of spatial, temporal or functional structures. The topics treated may be as diverse as lasers and fluids in physics, pattern formation in chemistry, morphogenesis in biology, brain functions in neurology or self-organization in a city. As is witnessed by several volumes, great attention is being paid to the pivotal interplay between deterministic and stochastic processes, as well as to the dialogue between theoreticians and experimentalists. All this has contributed to a remarkable cross-fertilization between disciplines and to a deeper understanding of complex systems. The timeliness and potential of such an approach are also mirrored – among other indicators – by numerous interdisciplinary workshops and conferences all over the world.

Vladimir I. Keilis-Borok
Alexandre A. Soloviev (Eds.)

Nonlinear Dynamics of the Lithosphere and Earthquake Prediction

With 133 Figures and 51 Tables

 Springer

Professor Dr. Vladimir I. Keilis-Borok
Professor Dr. Alexandre A. Soloviev
Russian Academy of Sciences
International Institute of Earthquake Prediction
Theory and Mathematical Geophysics
Warshavskoye sh., 79, kor. 2
117556 Moscow, Russia

Library of Congress Cataloging-in-Publication Data
Nonlinear dynamics of the lithosphere and earthquake prediction/Vladimir I. Keilis-Borok, Alexandre A. Soloviev (eds.).
p.cm.– (Springer series in synergetics, ISSN 0172-7389)
Includes biblographical references.
1. Earthquake prediction. 2. Geodynamics–Mathematical models. I. Keilis-Borok, Vladimir Isaakovich. II. Soloviev, Alexandre A., 1947- III. Springer series in synergetics (Unnumbered)
QE538.8 .N66 2002   551.22–dc21   2002030442

ISSN 0172-7389

ISBN 978-3-642-07806-4

Springer-Verlag Berlin Heidelberg New York
a member of Springer Science+Business Media

http://www.springer.de

© Springer-Verlag Berlin Heidelberg 2010
Printed in Germany

The use of general descriptive names, registered names, trademarks, etc. in this publication does not imply, even in the absence of a specific statement, that such names are exempt from the relevant protective laws and regulations and therefore free for general use.

Dataconversion and production by LE-TeX Jelonek, Schmidt & Vöckler GbR, Leipzig
Cover design: *design & production*, Heidelberg
Printed on acid-free paper

Preface

The vulnerability of our civilization to earthquakes is rapidly growing, raising earthquakes to the ranks of major threats faced by humankind. Earthquake prediction is necessary to reduce that threat by undertaking disaster–preparedness measures. This is one of the critically urgent problems whose solution requires fundamental research. At the same time, prediction is a major tool of basic science, a source of heuristic constraints and the final test of theories.

This volume summarizes the state-of-the-art in earthquake prediction. Its following aspects are considered:

– Existing prediction algorithms and the quality of predictions they provide.

– Application of such predictions for damage reduction, given their current accuracy, so far limited.

– Fundamental understanding of the lithosphere gained in earthquake prediction research.

– Emerging possibilities for major improvements of earthquake prediction methods.

– Potential implications for predicting other disasters, besides earthquakes.

Methodologies. At the heart of the research described here is the integration of three methodologies: *phenomenological analysis* of observations; *"universal" models of complex systems* such as those considered in statistical physics and nonlinear dynamics; and *Earth-specific models* of tectonic fault networks. In addition, *the theory of optimal control* is used to link earthquake prediction with earthquake preparedness.

Focus. This scope, broad as it is, covers a specific part of the much wider field of earthquake prediction, which is intrinsically connected with most of the solid Earth sciences, as well as with many branches of other natural sciences and mathematics. Specifically, we review the research aimed at *unambiguously defined algorithms and their validation by advance prediction*. That focus is central both for a fundamental understanding of the process expressed in seismicity and for preventing damage from earthquakes, for a scholar in quest of a theory and a decision-maker with responsibility for escalating or relaxing disaster preparedness. Both are in dire need of hard facts, which only prediction can establish.

Consecutive approximations. The studies presented here regard the seismically active lithosphere as a nonlinear (chaotic or complex) dissipative system with strong earthquakes for critical transitions. Such systems may be predictable, up to a limit, only after averaging (coarse-graining). Accordingly, we consider prediction based on a holistic approach, "from the whole to details." The problem of prediction is posed then as a successive, step-by-step, narrowing of the time interval, territory, and magnitude range where a strong earthquake can be expected. Such division into successive approximations is dictated by similar step-by-step development of critical transitions. At the same time, this division corresponds to the needs of disaster preparedness.

Most of the findings described here concern intermediate-term prediction (with alarms lasting years) based on premonitory seismicity patterns. There are compelling reasons to expect that these findings are applicable to other data and other stages of prediction. We also consider the background stage of prediction the identification of areas where epicenters of strong earthquakes can be located.

Content. This volume consists of six chapters.

Chapter 1 outlines the fundamentals of earthquake prediction: (i) Hierarchical structure of fault networks. (ii) Origin of the complexity of the lithosphere that is a multitude of mechanisms destabilizing the *stress–strength* field. The strength field is particularly unstable, so analysis of the stress field *per se* might not always be relevant. (iii) General scheme of prediction, using the pattern recognition approach. (iv) Four paradigms of earthquake prediction research concerning basic types of premonitory phenomena, their common features (long-range correlations, scaling, and similarity), and their dual nature, partly "universal" and partly Earth-specific.

Chapter 2 explores seismicity generated by hierarchical lattice models with dynamic self-organized criticality. Modeled seismicity shows the typical behavior of self-similar systems in a near-critical state; at the same time, it exhibits major features of observed seismicity, premonitory seismicity patterns included. The heterogeneity of the strength distribution introduced in the models leads to the discovery of three types of criticality. The predictability of the models varies with time, raising the problem of the prediction of predictability, and, on a longer timescale, the prediction of the switching of a seismic regime.

Chapter 3 describes the model of a block-and-fault system; it consists of rigid blocks connected by thin viscoelastic layers ("faults"). The model is Earth-specific: it allows us to set up concrete driving tectonic forces, the geometry of blocks, and the rheology of fault zones. The model generates stick-slip movement of blocks comprising seismicity and slow movements. Such models provide a very straightforward tool for a broad range of problems: (i) the connection between seismicity and geodynamics; (ii) the dependence of seismicity on the general properties of fault networks, i.e. the fragmentation of structures, the rotation of blocks, the direction of the driving forces, etc.; (iii) obviously, direct modeling of earthquake prediction.

Chapter 4 describes a family of earthquake prediction algorithms and their applications worldwide. Several algorithms are put to the test, unprecedented in rigor and scale. By and large, about 80% of earthquakes are anticipated by alarms, and alarms occupy 10 to 30% of the time–space considered. Particularly successful is the advance prediction of the largest earthquakes of magnitude 8 or more. Recently, advance predictions have been posted on web sites, along with accumulating scores of their outcomes, successes, and failures alike: see http://www.mitp.ru/predictions.html and http://www.phys.ualberta.ca/mirrors/mitp/predictions.html

Chapter 5 connects earthquake prediction with earthquake preparedness. The general strategy of the response to predictions consists of escalation or deescalation of safety measures, depending on the expected losses and the accuracy of the prediction. The mathematical solution of that problem is based on the theory of optimal control. Much can be done by applying this strategy on a qualitative level.

Chapter 6 concerns background prediction: the recognition of still unknown areas, where epicenters of strong earthquakes may be situated, i.e. where strong earthquakes can nucleate. These are densely fragmented structures, *nodes*, formed about fault intersections. Recognition is based on geological and geophysical data, satellite observations included. Maps of such areas have been published since the early 1970s for numerous regions of the world, including such well-studied ones as California and the Circumpacific. Subsequent seismic history confirmed these maps: 90% of the new earthquakes (61 out of 68) occurred within predicted areas; in 19 of these areas, such earthquakes had been previously unknown. This method is among the best validated and less widely known, illustrating an awareness gap in earthquake prediction studies.

Collaboration. The findings reviewed here were obtained because of broad cooperation comprising about 20 institutions in 12 countries and several international projects. The authors have been privileged to have permanent collaboration with the Abdus Salam International Center for Theoretical Physics, the Universities of Rome ("La Sapienza") and Trieste (Italy), the Institute of the Physics of the Earth, Paris, and the Observatory of Nice (France), Cornell and Purdue Universities, the University of California, Los Angeles, and the United States Geological Survey (USA). The authors are deeply grateful to our colleagues: C.J. Allegre, B. Cheng, V. Courtillot, J.W. Dewey, J. Filson, U. Frisch, A.M. Gabrielov, I.M. Gelfand, M. Ghil, A. Giesecke, J.H. Healy, L.V. Kantorovich, L. Knopoff, I.V. Kuznetsov, J.-L. Le Mouel, B.M. Naimark, W. Newman, E. Nyland, Yu.S. Osipov, G.F. Panza, L. Pietronero, V.F. Pisarenko, F. Press, A.G. Prozorov, I.M. Rotwain, D.V. Rundqvist, M.A. Sadovsky, D. Sornette, D.L. Turcotte, S. Uyeda, I.A. Vorobieva, I.V. Zaliapin and A. Zelevinsky.

We worked in the fascinating environment of the International Institute of Earthquake Prediction Theory and Mathematical Geophysics, the Russian Academy of Sciences, and can hardly describe our eternal debt to its faculty and staff.

Acknowledgements. Considerable part of the work was done under the auspices of the International Decade of Natural Disasters Reduction (ICSU Project "Non-linear Dynamics of the Lithosphere and Intermediate-term Earthquake Prediction"). We received invaluable support from the James S. McDonnell Foundation (the 21st Century Collaborative Activity Award for Studying Complex Systems); the International Science and Technology Center (projects 1293 and 1538); the US Civilian Research & Development Foundation for the Independent States of the Former Soviet Union (projects RM0-1246 and RG2-2237); the US National Science Foundation (grants EAR-9804859 and EAR-9423818); the Russian Foundation for Basic Research (grant 00-15-98507); the NATO Science for Peace Program (project 972266); UNESCO (UNESCO-IGCP project 414); and the International Association for the Promotion of Cooperation with Scientists from the Independent States of the Former Soviet Union (projects INTAS/RFFI-97-1914, INTAS-94-232, INTAS-93-457 and INTAS-93-809).

The studies described in the volume were intensely discussed at the Workshops on Nonlinear Dynamics and Earthquake Prediction organized by the Abdus Salam International Center for Theoretical Physics; the last one convened in October 2001, right before this volume went to Springer-Verlag; it was supported by the European Commission (Contract HPCFCT-2000-00007).

Moscow   *V.I. Keilis-Borok*
May 2002   *A.A. Soloviev*

Contents

List of Contributors

E. Blanter,
blanter@mitp.ru
A. Gorshkov,
gorshkov@mitp.ru
A. Ismail-Zadeh,
aismail@mitp.ru
V. Keilis-Borok[1],
vkborok@mitp.ru
V. Kossobokov,
volodya@mitp.ru
G. Molchan,
molchan@mitp.ru
P. Shebalin,
shebalin@mitp.ru
M. Shnirman,
shnir@mitp.ru
A. Soloviev
soloviev@mitp.ru

International Institute
of Earthquake Prediction Theory
and Mathematical Geophysics,
Russian Academy of Sciences
Warshavskoye shosse, 79, kor. 2
117556, Moscow-556, Russia

[1]also
Institute of Geophysics and
Planetary Physics and Department
of Earth and Space Sciences,
University of California,
Los Angeles,
405 Hilgard av., IGPP
Los Angeles, CA 90095-1567, USA
vkb@ess.ucla.edu

1 Fundamentals of Earthquake Prediction: Four Paradigms

V.I. Keilis-Borok

1.1 Introduction

About a million earthquakes of magnitude 2 or more are registered each year worldwide. About a hundred of them cause serious damage and once or twice in a decade, a catastrophic earthquake occurs. The vulnerability of our world to earthquakes is rapidly growing due to well-known global trends: proliferation of high-risk construction, such as nuclear power plants, high dams, radioactive waste disposals, lifelines, etc.; deterioration of the ground and destabilization of engineering infrastructures in megacities; destabilization of the environment; population growth; and other factors, including the escalating socioeconomic volatility of the global village. Today a single earthquake with subsequent ripple effects may take up to a million of lives; cause material damage up to 10^{12}; destroy a megacity; trigger a global economic depression (e.g. if it occurs in Tokyo); trigger ecological catastrophe rendering a large territory inhabitable; and destabilize the military balance in a region (e.g., the Middle East). Regions of low seismicity became highly vulnerable; among them are the European and Indian platforms and central and eastern *United States*. These regions harbor scores of vulnerable megacities such as New York, Moscow and Rome.

As a result, earthquakes joined the ranks of the major disasters that, in the words of J. Wiesner, became "a threat to civilization survival, as great as was ever posed by Hitler, Stalin or the atom bomb." Earthquake prediction at any stage would open the possibility of reducing the damage by undertaking disaster-preparedness measures.

The problem. The problem of earthquake prediction consists of consecutive, step-by-step, narrowing of the time interval, space, and magnitude ranges where a strong earthquake should be expected. Five stages of prediction are usually distinguished. The background stage provides maps with the territorial distribution of the maximum possible magnitude and recurrence time of destructive earthquakes of different magnitudes. Four subsequent stages, fuzzily divided, include the time-prediction; they differ in the characteristic time interval covered by an alarm. These stages are as follows:

- *long-term* (10^1 years),
- *intermediate-term* (years),
- *short-term* (10^{-1} to 10^{-2} years), and
- *immediate* (10^{-3} years or less).

Such division into stages is dictated by the character of the process that leads to a strong earthquake and by the needs of earthquake preparedness; the latter comprises an arsenal of safety measures for each stage of prediction, as in preparedness for war.

Prehistory (20–20 hindsight). New fundamental understanding of the earthquake prediction problem was formed during the last 40 or so years, triggering entirely new lines of research. In hindsight, this understanding stems from the following unrelated developments in the early 1960s.

– F. Press initiated the installation of a state-of-the-art global seismological network augmented by some regional and local ones. Thus, a uniform database began to accumulate.

– E. Lorenz discovered deterministic chaos in an ordinary natural process, thermal convection in the atmosphere [Lor63]. This triggered the recognition of deterministic chaos in a multitude of natural and socioeconomic processes; however, the turn of seismicity and geodynamics in general came about a quarter of a century later [Kei90a,BCT92,Tur97,NGT94]. The phenomenon of deterministic chaos was eventually generalized by a wider concept of complexity [CKO+80,Hol95,HSSS98,Gel94].

– I. Gelfand and J. Tukey, working independently, created a new culture of exploratory data analysis that allows us to overcome the complexity of a process considered. Among the essential elements of this culture is a very robust representation of information and exhaustive numerical tests validating the results of analysis [GGK+76,Tuk77]. Specifically, pattern recognition of infrequent events developed by the school of I. Gelfand is widely used in the studies reviewed here.

– L. Knopoff and B. Burridge demonstrated that a simple system of interacting elements may reproduce a realistically complex seismicity, fitting many basic heuristic constraints [BK67]. That extended to seismology the abstract models of interacting elements developed in statistical physics.

– L. Malinovskaya found a premonitory seismicity pattern reflecting the rise of seismic activity [KM64]. This is the first reported earthquake precursor formally defined and featuring long-range correlations and worldwide similarity.

With broader authorship,

– plate tectonics established the connection between seismicity and large-scale dynamics of the lithosphere;

– research in experimental mineralogy and mechanics of rocks revealed a multitude of mechanisms that may destabilize the strength in fault zones.

Four paradigms. In the wake of these developments, the following four paradigms have been established at the crossroad between exploratory data analysis, statistical physics, and the dynamics of fault networks [Kei94, Kei96a].

  I. *Basic types of premonitory phenomena* comprising the variation in relevant observable fields.

  II. *Long-range correlations in fault system dynamics.* Premonitory phenomena are formed not only in the vicinity of the incipient source but also within a much wider area.

  III. *Partial similarity of premonitory phenomena* in diverse conditions, from fracturing in laboratory samples to major earthquakes worldwide and possibly even to starquakes.

  IV. *The dual nature of premonitory phenomena.* Some of them are "universal," common for complex nonlinear systems of different origin; others are Earth-specific.

Holistic approach to prediction. Complex systems are not predictable with absolute precision. However, after a coarse-graining (on not too detailed a scale), premonitory phenomena emerge and a system becomes predictable, up to the limits [FS87, MFZ+90, Kra93, Gel94, Hol95, Kad76]. Accordingly, prediction of complex systems requires a holistic approach, "from the whole to details" in consecutive approximations, starting with the most robust coarse-graining of the processes considered. Table 1.1 compares the holistic approach with the complementary (but not necessarily contradictory) reductionistic approach, "from the details to the whole."

  The studies reviewed here are based on the holistic approach. It makes it possible to overcome the complexity itself and the chronic imperfection of observations as well. This is achieved at an unavoidable price: the accuracy of prediction is limited.

"With four exponents I can fit the elephant" (E. Fermi). Earthquake prediction algorithms include adjustable parameters and other elements that have to be data-fitted retrospectively to "predict" past earthquakes. The designer of the algorithm does not know whether it will also predict future earthquakes and has at least to make sure that predictions are not sensitive to slight variations of adjustable elements. Such sensitivity analysis takes most of the effort in prediction research. It is based on the *error diagram*, a staple of that research (Sect. 1.4), and the link of prediction with disaster-preparedness.

The only final test of an algorithm is advance prediction. A series of experiments in advance prediction of strong earthquakes in numerous regions worldwide has been launched (see Chap. 4 and [KS99, MDRD90]).

  By and large the algorithms predict 80–90% of strong earthquakes, and alarms occupy 10–30% of the time–space considered. The major drawbacks are the rate of false alarms and the limited probability gain, between 3 and 10 for different algorithms.

Table 1.1. Two complementary approaches to earthquake prediction

"REDUCTIONISM" (from details to the whole)	"HOLISM" (from the whole to details)

Premonitory phenomena preceding an earthquake with linear source dimension L are formed

near the incipient source	**in a network of faults** of linear size **on a timescale**

$- 10^2 L$ tens of years
$- 10L$ years
$- L$ years to months
$-$ possibly, fractions of L, i.e., in the vicinity of the hypocenter, on a smaller timescale

Premonitory phenomena

are specific to mechanisms controlling the strength, e.g. friction, rock–fluid interaction, stress corrosion, buckling, etc.	**are divided into** $-$ "universal" ones common to many chaotic systems $-$ those depending on the geometry of fault network $-$ mechanism-specific ones.

Premonitory phenomena in different regions and energy ranges

are different	**are to a considerable extent similar**

Constitutive equations

are local	**are nonlocal**

Triggering of earthquakes is controlled

by a strength–stress difference in the incipient source	**also by geometric incompatibility near the fault junctions** that may supersede a strength–stress criterion

Indispensable for further research is the unique uniform collection of errors and correct predictions accumulated during these experiments.

This chapter outlines the fundamentals of earthquake prediction, as a common background for the subsequent chapters. It includes: the physical origin of complexity in the seismically active lithosphere (Sect. 1.2); the general scheme of prediction (1.3); the evaluation of prediction algorithms by error diagrams (1.4); the abovementioned paradigms and classification of premonitory seismicity patterns (1.5); the link between earthquake prediction and disaster–preparedness (1.6); and emerging possibilities of developing the next generation of earthquake prediction methods (1.7).

1.2   Lithosphere as a Complex Hierarchical System

Origin of complexity. Two major factors cause complexity of the lithosphere [Kei90a, Tur97]: (i) *a hierarchical structure* extending from tectonic plates to the grains of rocks; (ii) *instability* caused by a multitude of nonlinear mechanisms controlling the *strength–stress* field. On a timescale relevant to earthquake prediction, 10^2 years or less, these factors, by an inevitable conjecture, turn the lithosphere into a hierarchical dissipative complex system.

Critical transitions. A prominent feature of complex systems is the persistent reoccurrence of abrupt overall changes called *critical transitions* or *critical phenomena.*

Strong earthquakes may be regarded as critical phenomena in the lithosphere. Note that *an earthquake may be a critical phenomenon in a certain volume of the lithosphere and part of the background seismicity in a larger volume.*

1.2.1   Hierarchy

Blocks. The structure of the lithosphere presents a hierarchy of volumes, or blocks, that move relative to each other. The largest blocks are the major tectonic plates, of continental size, $10^3 - 10^4$ km in linear dimension. They are divided into smaller blocks, such as shields or mountain belts. After $15 - 20$ consecutive divisions, we come to about 10^{25} grains of rocks of millimeter size.

Boundary zones. Blocks are separated by less rigid "boundary zones," whose width is $10 - 100$ times smaller than the characteristic size of the blocks they separate. Boundary zones are named differently, depending on size. They are called fault zones, high in the hierarchy; then faults; sliding surfaces; and, finally, interfaces between grains of rock. Except at the lowest level of the hierarchy, a boundary zone presents a similar hierarchical structure with more dense division: it consists of blocks, divided by boundary zones, etc.

Nodes. Even more densely fractured mosaic structures, called *nodes*, are formed in the vicinity of the intersections and junctions of faults. Their origin is due, roughly speaking, to the collision of the corners of blocks [GKJ96, Kin83, Kin86]. The formalized definition of nodes is given in [AGG+77]. Nodes play a singular role in the dynamics of the lithosphere.

– *A special type of instability is concentrated within nodes* (Sect. 1.2).

– *Strong earthquakes nucleate in nodes.* As demonstrated in a series of studies, the epicenters of strong earthquakes worldwide are located within nodes, more precisely, within some nodes that can be identified by pattern recognition ([GGK+76], Chap. 6).

Nodes are well known in structural geology and geomorphology and play a prominent textbook role in geological prospecting. However, their connection with earthquakes is sometimes overlooked in earthquake studies.

Is the division "blocks ⇒ faults ⇒ nodes" always complete? We have stipulated above the division of a tectonic region into blocks separated by closed contours of faults. Such a division has developed throughout geological history and may be not complete, particularly in tectonically young regions. For example, some faults comprise a bundle of small ruptures that are not (or not yet) evolved into a hierarchical network; the boundary of a block may be a flexure not yet ruptured, etc.

Fault networks. Systems of boundary zones and nodes are called here fault networks; this term sounds more familiar, though it is less precise.

Fault network, a stockpile of instability. Boundary zones of different rank, from the Circumpacific seismic belt, with giant triple junctions for nodes, to interfaces between rock grains, with the corners of grains for nodes, their great diversity notwithstanding, play a similar role in lithosphere dynamics. Specifically, although tectonic energy is stored in the whole volume of the lithosphere and well beneath, energy release is to a large extent controlled by the processes in relatively thin fault networks. This contrast is due to the following reasons.

First, the strength of a fault network is smaller than the strength of the blocks it separates: fault networks are weakened by denser fragmentation and higher permeability to fluids. For that reason, tectonic deformations are concentrated in fault networks, whereas blocks move essentially "as a whole," with a relatively smaller rate of internal deformations. In other words, on the timescale directly relevant to earthquake prediction, tens of years or less, the major part of lithosphere dynamics is realized through deformation of fault networks and the relative movement of blocks.

Second, the strength of a fault network is not only smaller, but also highly unstable, sensitive to many processes there. This instability, central for understanding seismicity, is discussed below.

Physical and geometric instabilities. We term as "physical" the instability originated by a physical or chemical mechanism at the elementary (micro) level, and as "geometric," the instability controlled by the geometry of the fault network on a global (macro) level. These instabilities largely control the dynamics of seismicity, including the occurrence of strong earthquakes.

1.2.2   "Physical" Instability [Kei90a]

As in any solid body, deformation and fracturing in the lithosphere are controlled by the strength–stress field. Strength is in turn controlled by a great multitude of interdependent mechanisms concentrated in the fault network. We describe, for illustration, several such mechanisms starting with the impact of fluids.

Rehbinder effect, or stress corrosion [GK83, Tra85].

Mechanism. Many solid substances lose their strength when they come in contact with certain surface-active liquids. The liquid diminishes the surface tension μ and consequently the strength, which is proportional to $\sqrt{\mu}$ by Griffits criteria. When the strength drops, cracks may emerge under small stress, even gravity might suffice. This triggers expansion of fatigue: liquid penetrates cracks, they grow, and drops of liquid propel forward, until they dissipate. This mechanism requiring very little energy to generate fracturing was first discovered in metals and ceramics. Then such combinations of solid substances and surface-active liquids were recognized among common ingredients of the lithosphere, e.g., basalt and sulfur solution. When they meet, the basalt is permeated by a grid of cracks, and the efficient strength may instantly drop by a factor of 10 or more due to this mechanism alone.

Geometry of weakened areas. Orientation of cracks at each point depends on the stress field; it is normal to the main tensile stress. The stress field in the lithosphere may be exceedingly diverse. Strictly limited, however, is the geometry of weakened areas where cracks concentrate; such areas may be of only a few types, determined by the theory of singularities. Some examples are shown in Fig. 1.1, where thin lines show the trajectories or cracks. Each separatrix (a heavy line) separates the areas with different patterns of trajectories.

When the source of a liquid appears in a place such as shown in Fig. 1.1 by arrows, the liquid that penetrates the cracks concentrates in the shaded area, and its strength plummets. A slight displacement of the source across the separatrix may lead to a strong change in the geometry of such fatigue; it may be diverted to quite a different place and take quite a different shape, although not an arbitrary one.

A new dimension is brought into this picture by the evolution of the stress field, which is changing all the time for many reasons, including the feedback from this very effect. Such evolution may change the type of a singularity, make it disappear or create a new one, and the geometry of fatigue will follow suit.

Sensitivity to chemical composition. The Rehbinder effect is highly sensitive to the chemical ingredients of the fluid, even in microconcentrations. For example, gabbro and dolerite are affected only in the presence of iron oxides; Kamchatka ultrabasic rocks are affected by andesite lava liquids only in the presence of copper oxide, etc.

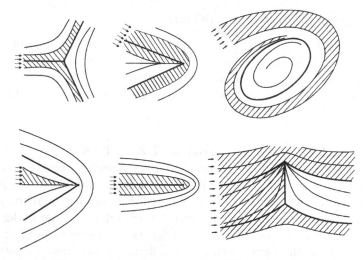

Fig. 1.1. Instability caused by stress corrosion. The geometry of weakened areas depends on the type of singularity and the place where the chemically active fluid comes in. After [GK83]

Self-excitation. The impact of chemically active fluids increases with stress in rocks, thus becoming self-exciting, because stress is always concentrated near the corners of rock grains.

Summing up, the Rehbinder effect brings a strong and specific instability into the dynamics of the lithosphere. This instability is controlled by the stress field and by the geochemistry of fluids. The migration of fluids is accompanied by the observable variations of the "fluids regime" and of electromagnetic and geochemical fields.

This effect might explain many premonitory seismic patterns. Such an explanation, however, has at least two limitations.

(i) The basic configurations of fatigue, as shown in Fig. 1.1, might be realized only in small areas. The inhomogeneity of stress and strength fields and the dissipation of fluids may destroy the formation of such configurations on the scale of the observed premonitory patterns, which is tens to hundreds of kilometers. More likely, these configurations are the elements composing a more complicated infrastructure of fatigue.

(ii) The Rehbinder effect is not a single major mechanism by which fault zones control the dynamics of the lithosphere. Even fluids alone may generate other equally strong mechanisms.

Nonlinear filtration [BKM83].

Mechanism. One of the competing mechanisms is the more conventional filtration of fluids through fault zones. This process is modeled in [BKM83] as the relative movement of impermeable blocks separated by a porous layer. The latter is connected with a source of fluid migrating along the gradient of pressure. The fluid acts as a lubricant that reduces the friction and triggers episodes of fast slip.

Further development brings in strong instability illustrated by Fig. 1.2. When the porosity is subcritical (below a certain threshold), the slip, once started, causes an increase in friction and self-decelerates. At most, the fluid will trigger vacillating creep or a slow earthquake. However, when the porosity exceeds a critical threshold, the slip causes a decrease in friction and the incessantly forming microcracks start to self-accelerate, grow, and merge at an escalating rate. The porosity can be raised above the critical threshold by infiltration of a fluid itself; this will increase the tension, and the pores will expand.

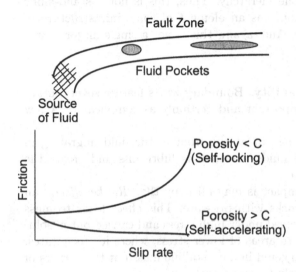

Fig. 1.2. Instability caused by the infiltration of a lubricating fluid. A change of porosity causes an abrupt change in the slip rate. After [BKM83]

Destabilization wave. The propagation of critical porosity is described by the equation

$$\partial\varphi/\partial t = V^2 \Delta\varphi^\alpha,$$
$$\varphi = \mu\rho; \quad V^2 = k_0 p_0/(\mu_0 \eta_0), \quad \alpha > 2,$$

where μ is porosity; ρ is the density of the fluid; and k_0, p_0, μ_0, and η_0 are the characteristic values of permeability, pressure, porosity, and viscosity, respectively. Δ is the Laplace Operator.

This is the famous nonlinear parabolic equation studied by Ya. Zeldovich and I. Barenblatt [Bar96]. In the specific case considered here, nonlinearity reflects the change of porosity and permeability from pressure.

Due to the nonlinearity, perturbations of $\varphi(x, t)$ propagate at a final velocity proportional to V. The values of V computed for realistic parameters of Earth's crust include the range 10–10^2 kilometers per year, the same as for the observed migration of seismicity along fault zones.

A model of an earthquake source. The source is modeled as a residual pocket of a fluid, where high background pressure raises the velocity of filtration. The filtration front may quickly cross and destabilize such a pocket, turning it into an earthquake source.

Summing up, instability caused by nonlinear filtration also explains many features of real seismicity, e.g., its migration, seismic cycle, and certain earthquake precursors, such as the rise of seismic activity and earthquake clustering. It also suggests some premonitory changes in the fluid regime and the electromagnetic field.

However, one can see the same limitations as in stress corrosion. First, such instabilities may rise simultaneously within boundary zones of different rank and interact along the hierarchy. Thus, this is not a stand-alone model but, like stress corrosion, has an element of some infrastructure of filtration-generated instability. And, again, this is not a single major source of instability.

Other mechanisms of instability. Boundary zones feature several other mechanisms, potentially as important and certainly as complicated. A few more examples follow.

"Fingers of fluids" springing out at the front of the fluid migrating in a porous media [Bar96]. The fluids may act as lubricants and create the destabilization described above.

Dissolution of rocks. Its impact is magnified by the *"Riecke effect,"* an increase in the solubility of rocks with pressure. This effect leads to mass transfer. Solid material is dissolved under high stress and carried out in solution along the stress gradient to areas of lower stress, where it precipitates. The Riecke effect might be triggered in a crystalline massif at the corners of rock grains, where stress is likely to concentrate.

Petrochemical transitions. Some of them tie up or release fluids, as in the formation or decomposition of serpentines. Other transitions cause a rapid drop in density, such as in the transformation of calcite into aragonite. (This would create a vacuum and unlock the fault; the vacuum will be closed at once by hydrostatic pressure, but the rupture may be triggered.)

Sensitivity of dynamic friction to local physical environment [Lom91].

Mechanical processes, such as multiple fracturing, buckling, viscous flow, etc.

The impact of pressure and temperature on most of the above mechanisms.

This list, by no means complete, illustrates the diversity of mechanisms that cause the physical instability.

1.2.3   "Geometric" Instability [GKJ96]

The geometry of fault networks might be, and often is, incompatible with tectonic movements, including earthquakes. This leads to stress accumulation, deformation, fracturing, and a change in fault geometry, jointly destabilizing the fault network. Two integral measures of that instability have been found by A. Gabrielov et al. [GKJ96]: geometric incompatibility concentrated within nodes and kinematic incompatibility spread across the fault network.

Each measure estimates the integrated effect of tectonic movements on a wide range of timescales from seismicity to geodetic movements (e.g., recorded by GPS) to neotectonics.

Geometric incompatibility

A simple example. The nature of geometric incompatibility is illustrated in Fig. 1.3 that shows the intersection of two strike-slip faults separating moving blocks. If the movements indicated by arrows in Fig. 1.3a could go on, the corners A and C would penetrate each other, and an intersection point would split into a parallelogram (Fig. 1.3b). In the more general case of a finite number of faults, their intersection point would split into a polygon.

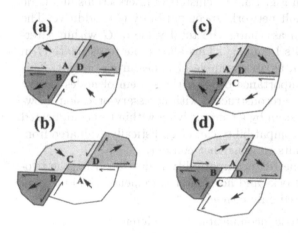

Fig. 1.3. Geometric incompatibility near a single intersection of faults. Initial position of the blocks (**a, c**); extrapolation of initial movement (**b, d**); the locked node: movement is physically unrealizable without fracturing or a change in the fault geometry (**a, b**); the unlocked node (**c, d**). After [GKJ96]

Such splitting is not possible in reality; instead, the collision of the corners triggers the accumulation of stress and deformations near the intersection followed by fracturing and changes in the fault geometry. The intensity of this process is characterized by the expansion of that unrealizable polygon with time,

$$S(t) = Gt^2/2,$$

where S is the area of the polygon, t is the time elapsed since the collision, and G is a measure of geometric incompatibility. The following manifestations of this phenomenon illustrate its impact on the dynamics of seismicity.

Locked and unlocked intersections. When the corners of blocks tend to overlap (Fig. 1.3a,b), the ensuing compression locks up the intersection turning it into asperity. When all corners diverge (Fig. 1.3c,d), the ensuing tension unlocks the intersection turning it into "a weak link" in the fault network.

Formation of nodes. The phenomenon illustrated above was first described by McKenzie and Morgan [MM69] for a triple junction. They found a condition under which a single junction "can retain its geometry as the plates move," so that stress will not accumulate. G. King [Kin83, Kin86] suggests that in the general case, when that condition is not satisfied, the ensuing fracturing would not dissolve the stress accumulation but only redistribute it among newly formed corners, thus triggering a chain:

corners of blocks at the fault junction collide ⇒
⇒ stress accumulates ⇒
⇒ smaller faults appear and form new intersections ⇒
⇒ corners of the blocks at the new intersections collide ⇒ etc.

As a result, a hierarchy of progressively smaller and smaller faults is formed about an initial intersection; this is a node, recognizable by its densely mosaic structure and probably has self-similar fractal geometry [Kin83].

Geometric incompatibility of a fault network. In reality, we encounter not a single intersection, as in Fig. 1.3, but clusters of intersections in a node and interacting nodes in a fault network. Incompatibility G is additive. The analog of the Stokes theorem associates the total value of G within a territory with observations on its boundary. This allows one to estimate G in a complicated structure, such as an ensemble of nodes, from outside. This is of considerable practical importance because the system of nodes is very complicated and hardly can be reconstructed with necessary precision. However, one can surround the system by a contour lying within less complicated areas. Then, the geometric incompatibility can be realistically evaluated from the movements of the few faults crossing that contour.

So far, the theory of geometric incompatibility is developed for the two-dimensional case with rigid blocks and horizontal movements; the impact of strong earthquakes is estimated only coarsely.

Interplay of nodes. Geometric incompatibility in different nodes is interdependent because they are connected through the movements of blocks and on faults. A strong earthquake in a node might change its incompatibility, thus affecting the occurrence of earthquakes in other nodes. Observations indicating interaction of nodes have been described by A. Prozorov [Pro75, PS90]. These studies suggest that a strong earthquake is followed by "long-range aftershocks," a rise of seismic activity in the area, where the next strong earthquake is going to occur within about 10 years.

Kinematic incompatibility

Description. Let us apply the well-known Saint-Venant condition of kinematic compatibility [MJCB84] to the lithosphere; its discrete analog suitable for a fault network was introduced by McKenzie and Parker [MP67]. That condition ensures that the relative movements on faults can be realized through the absolute movements of blocks separated by these faults. In the simplest case shown in Fig. 1.3, this condition is $K = \sum v_i = 0$, where v_i are slip rates on the faults meeting at the intersection. The value of K is the measure of deviation from kinematic compatibility. Naturally, it is named kinematic incompatibility. A simple illustration of that phenomenon is the movement of a rectangular block between two pairs of parallel faults. The movement of the block as a whole has to be compensated for by relative movements on all faults surrounding it: if, for example, the movement takes place on only one fault, the stress will accumulate at other faults and in the block itself, thus creating kinematic incompatibility. Numerous manifestations of that phenomenon are described in Chapter 3.

Origin. Estimates of K obtained by analyzing observations may be different from zero for the following reasons:

– The errors in the observed slip rates or in mapping the fault network; estimates of K are widely used in tectonic reconstructions to identify such errors [Bir98].

– An unaccounted stress and deformation in blocks.

It is not always easy to separate these explanations.

Kinematic incompatibility of a fault network. It has some basic features in common with geometric incompatibility:

– Additivity: K may be also summed up for different parts of the network.

– *An analog of Stokes theorem* linking the value of K for a region with observations on its boundary.

A conjecture, how to use estimates of G and K for earthquake prediction, is discussed in Sect. 1.4.

1.2.4   Generalization: Complexity and Critical Phenomena

Summing up, the dynamics of the lithosphere is controlled by a wide variety of mutually dependent mechanisms concentrated predominantly within fault networks and interacting across and along the hierarchy. Each mechanism creates strong instability of the strength–stress field, particularly of the strength. Except for very special circumstances, none of these mechanisms alone prevails in the sense that the others can be neglected.

Even the primary element of the lithosphere, a grain of rock, may act simultaneously as a material point; a viscoelastic body; an aggregate of crystals; a source or absorber of energy, fluids, or volume, with its body and surface involved in different processes.

Assembling the set of governing equations is unrealistic and may be misleading as well. A well-known maxim in nonlinear dynamics says that "one cannot understand a chaotic system by breaking it apart" [CFPS86]. One may rather hope for a generalized theory (or at least a model) that directly represents the gross integrated behavior of the lithosphere. That brings us to the concept that *on the timescale relevant to the earthquake prediction problem, 10^2 years and less, the mechanisms destabilizing the strength* *of fault networks turn the lithosphere into a nonlinear hierarchical dissipative system, where strong earthquakes are the critical phenomena.* Upon the emergence of that concept, the lithosphere was called *a chaotic* system [Kei90a, NGT94, Tur97]; the more general term *complex system* is probably more adequate [Gel94, MaS76, Hol95, RTK00].

Since "criticality" and "universality" are currently used in rather different senses, from metaphoric to precise, we add here the introductory explanation of these concepts, given by D. Sornette [Sor00]:

"The idea of 'universality' (is rooted in) ... the theory of critical phenomena in the natural sciences (which) describes the peculiar organizational changes that can occur in fluids, magnets and many other condensed-matter systems.

Interactions between constituents (of a system) favour order, while "noise" or thermal fluctuations promote disorder. The referee of this fight between order and disorder is known as a "control parameter". Varying it can cause the fluid or magnet to go from an ordered to a disordered state. The transition may be "critical" in the sense that fluctuations of both competing states occur over all space and time scales ... and become intimately intertwined.

The transition leads to specific signatures in the form of "power-law" relationships between physical observables (such as density or magnetization) and the distance of the control parameter from its critical value. The concept of universality enters this picture from the remarkable empirical discovery – later understood within the framework of renormalization group theory – that the critical exponents of these power laws are universal. ... The exponents are independent of the system, be it made of atoms, molecules or magnetic spins."

Such a mind-set, we believe, is helpful through the rest of that book.

1.3   General Scheme of Prediction

Raw data. Typical of a complex system, the lithosphere exhibits permanent background activity, a mixture of interacting processes. It reflects the approach of a strong earthquake and, accordingly, provides the observations ("the raw data") for earthquake prediction.

Premonitory seismicity patterns. Prediction algorithms considered here use only a part of potentially relevant observations, the earthquake sequences. Prediction is based on the spatiotemporal patterns of seismicity that signal the approach of a strong earthquake; naturally, they are called *premonitory seismicity patterns.*

Scaling. Patterns preceding an earthquake of magnitude M are formed by earthquakes within an area and magnitude ranges depending on M.

Generalization. The essential features of prediction algorithms are transferable from seismicity to other relevant data.

1.3.1   Formulation of the Problem

The algorithms described here consider prediction as a pattern recognition problem:

> **Given** the dynamics of a relevant field in a certain area prior to some time t,
> **to predict** whether a strong earthquake will or will not occur within that area during the subsequent time interval $(t, t+\Delta)$.

In terms of pattern recognition, the "object of recognition" is the time t. The problem is to recognize whether or not it belongs to the time interval Δ preceding a strong earthquake. That interval is usually called the "*TIP*," an acronym for the "*time of increased probability*" of a strong earthquake.

Such prediction is aimed not at the whole dynamics of seismicity but only at rare extraordinary phenomena, strong earthquakes. In other words, the problem is to localize in time–space a specific singular trait of an earthquake sequence. This is different from prediction in a more traditional sense, amplitude extrapolation of a random field in a given time–space point. G. Molchan, who brought to attention this difference (Chap. 5), calls such predictions "horizontal" and "vertical," respectively.

Pattern recognition of infrequent events [BVG$^+$66, GGK$^+$76, KP80] proves to be very efficient in that approach to prediction. This methodology has been developed by the school of I. Gelfand for the study of rare phenomena of highly complex origin, a situation where classical statistical methods were inapplicable.

The probabilistic side of prediction is reflected in the rates of errors evaluated by error diagrams (see Sect. 1.4 and Chap. 5).

1.3.2   An Early Example [KM64]

The first premonitory seismicity pattern of the kind considered here was "*pattern Σ*" introduced in 1964. It comprises the premonitory increase in the total area of the ruptures in the earthquake sources in a medium magnitude range. The emergence of this pattern was captured by the function

$$\Sigma(t, s, B) = \sum_i 10^{Bm_i},$$

where m_i is the magnitude of the ith earthquake; the sum is taken over all earthquakes that occurred during the time interval $(t - s, t)$ within the region considered; $B \approx 1$. With this value of B, the summands are coarsely proportional to the source area (when $B = 0$ and $B = 3/2$, this sum would correspond to the number of earthquakes and their total energy, respectively).

First applications. The change of the function $\Sigma(t)$ in the time periods preceding 20 strong earthquakes worldwide was investigated in [KM64]. It was shown that the function $\Sigma(t, s, B)$ strongly increased 1 to 10 years prior to each of the earthquakes considered, indicating "*a direct connection between strong earthquakes and the very large scale features of the development of the whole Earth's crust.*" Figure 1.4 shows an example of the catastrophic Assam earthquake in India, 1950, M=8.6. The emergence of pattern Σ was captured by the condition $\Sigma(t) > C_\Sigma$, threshold C_Σ was determined uniformly for all regions.

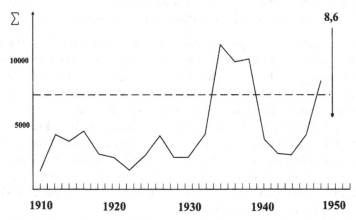

Fig. 1.4. Illustration of the premonitory seismicity pattern Σ: rise of the functional $\Sigma(t)$ before the Assam earthquake in India (1950, $M = 8.6$). The horizontal line shows the threshold C_Σ normalized by magnitude of target earthquake described in the text. After [KM64]

Pattern Σ was the first premonitory seismicity pattern that demonstrated the major features of patterns discovered later: long-range correlations and similarity. These features can be described as follows.

(i) *Long-range correlation:* The area of earthquake preparation can greatly exceed the source of the incipient earthquake. This is reflected in the large size of areas that had to be used for calculating of the function Σ (Fig. 1.5).

(ii) *Similarity:* Pattern Σ is self-adapting; it has a uniform definition for different magnitudes M of strong earthquakes targeted for prediction. Specifically,

– the area of preparation is a power-law function of M (Fig. 1.5), and

– the threshold for identifying pattern Σ is normalized by M, $C_\Sigma = 0.5 \times 10^{BM}$. This means that the area unlocked by medium magnitude earthquakes reached at least half of the area that will be unlocked by an incipient strong earthquake.

These features have been confirmed by subsequent studies and eventually evolved in the earthquake prediction paradigms (Sect. 1.5).

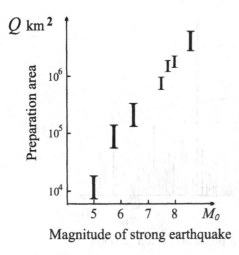

Fig. 1.5. Long-range correlation in the formation of premonitory seismicity patterns. $Q(M)$ is the area, where pattern Σ was formed prior to an earthquake of magnitude M. After [KM64]

Premonitory phenomenon $\Rightarrow$ precursor $\Rightarrow$ function. Pattern Σ illustrates the consecutive stages in the search for earthquake prediction algorithms.

– It started with the hypothetical *premonitory phenomenon*, the rise of seismic activity, which is one of the basic characteristics of seismicity.

– That phenomenon was captured by a specific *precursor*: a large area unlocked by earthquakes in a medium magnitude range. The same phenomenon is also captured by other precursors, for example, by the number of earthquakes not weighted by magnitude.

– Finally, that precursor was formally defined by the function $\Sigma(t)$.

Pattern Σ illustrates the general scheme of prediction described below.

1.3.3 Data Analysis

It comprises the following four steps:

(i) **A sequence of earthquakes is robustly described** by the functions $F_k(t)$, $k = 1, 2, ...$, each depicting a certain premonitory seismicity pattern (Fig. 1.6). With a few exceptions, the functions are defined in a sliding time window $(t - s, t)$; note that the value of a functional is attributed to the end of the window.

(ii) **The emergence of a premonitory seismicity pattern** is defined by the condition

$$F_k(t) \geq C_k.$$

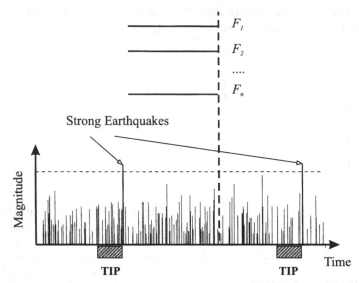

Fig. 1.6. General scheme of prediction. After [Kei90a]

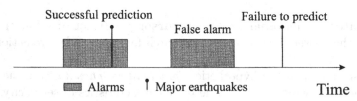

Fig. 1.7. Possible outcomes of prediction

The threshold C_k is usually defined as a certain percentile of the function F_k.

(iii) **An alarm** is triggered when a single pattern or a certain combination of patterns emerges; different patterns are used in different algorithms (see Chap. 4). An alarm lasts for a certain time period; in some algorithms, it is terminated if and when a strong earthquake occurs. The possible outcomes of prediction are illustrated in Fig. 1.7.

(iv) **The reliability of such a prediction algorithm** is evaluated by the error diagram summarizing the outcomes of a series of predictions (Chap. 4).

This scheme is open for the use of other data, not necessarily seismological [Kei96b].

The key element in the development of such an algorithm consists obviously of determining functions $F_k(t)$ that provide good predictions. The next section discusses how to evaluate an algorithm.

1.4   Error Diagrams

The behavior of a complex system cannot be predicted with absolute precision; one may reduce the rate of errors, but not eliminate them. The performance of a prediction algorithm is quantitatively characterized by three measures: (i) the rate of false alarms, (ii) the rate of failures to predict, and (iii) the relative space–time occupied by all alarms together.

Error diagrams showing the trade-off between these measures are pivotal in developing and validating of prediction methods, as well as in using predictions for enhancing earthquake preparedness (see Chaps. 4 and 5).

The danger of data fitting. Earthquake prediction algorithms inevitably include some adjustable elements, e.g., the values of numerical parameters, the observations used for prediction, the definition of precursors, and the selection of the magnitude scale. In lieu of an adequate theory, many such elements cannot be uniquely determined a priori. They have to be chosen retrospectively; we design an algorithm that performed well in the past.

That creates a danger illustrated by Fig. 1.8 (after [GDK$^+$86]). It shows the "prediction" of random numbers t_i (the vertical lines) by independent random numbers P. The figure shows the retrospective "prediction" of t_i by the values of P; large dots are the "large" values of P that trigger alarms shown by horizontal segments.

An apparently good success-to-failure score is obtained by retrospective adjustment of only two parameters: the threshold for declaration of alarm ($P > 70$) and the duration of a single alarm (5 time units). These parameters

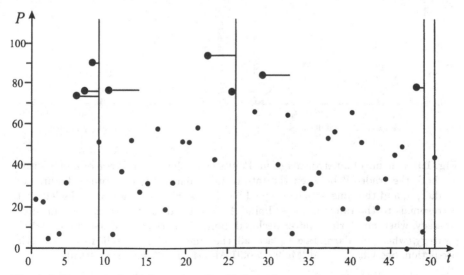

Fig. 1.8. Dangers of self-deception. Prediction of one series of random numbers by another. After [Kei96b]

control the trade-off between errors of different kinds. For example, raising the threshold to, say, 85, we eliminate one false alarm at the cost of an extra failure to predict.

Stability tests. To validate an algorithm under these circumstances, an exhaustive set of numerical tests is designed; they take a lion's share of efforts in the design of an algorithm [GGK$^+$76, GZNK00]. The results of such tests are summed up by the error diagrams.

Definition. Consider the basic test of a prediction algorithm by the error diagram:
 – The algorithm is applied to a certain territory during the time period T.
 – A certain number of alarms A has been declared, and A_f of them happened to be false.
 – N strong earthquakes have occurred, and N_m of them have been missed by alarms.
 – Altogether, the alarms cover the time D.
The *performance* of an algorithm in that test is characterized by three dimensionless parameters:
 – the relative duration of alarms, $\tau = D/T$;
 – the rate of failures to predict, $n = N_m/N$;
 – the rate of false alarms, $f = A_f/A$.

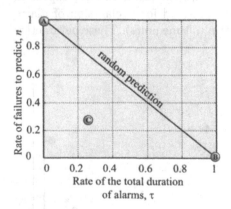
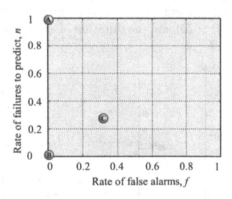

Fig. 1.9. Scheme of an error diagram. Points show the performance of a prediction method: the trade-off between the rate of false alarms, f, the rate of failures to predict, n, and the time–space occupied by alarms, τ. The diagonal in the left plot corresponds to the random guess. Point A corresponds to the trivial "optimistic" strategy, when an alarm is never declared; point B marks the trivial "pessimistic" strategy, when an alarm takes place all the time; point C indicates a realistic prediction. See Chap. 4 and [Mol90, Mol91, Mol94, Mol97] for more details

These three parameters are necessary in any test of a prediction algorithm regardless of the particular methodology.

Error diagram. This test is repeated for different combinations of adjustable elements of the algorithm. The results are summed up in the error diagram schematically illustrated in Fig. 1.9. Different points correspond to different combinations of adjustable elements.

Validation of prediction methods. The results of the numerical experiments are summed up by error diagrams. A prediction algorithm makes sense only if its performance is

(i) sufficiently better than a random guess, and

(ii) not too sensitive to variation of adjustable elements.

An error diagram is so far the only and a powerful tool for checking these conditions.

1.5 Four Paradigms

The paradigms discussed here have been found in the quest for premonitory seismicity patterns in observed and modeled seismicity. There are compelling reasons to apply them also to premonitory phenomena in other relevant fields.

1.5.1 First Paradigm: Basic Types of Premonitory Phenomena

The approach of a strong earthquake is indicated by the following changes in the basic characteristics of seismicity:

(i)   *Rise of seismic activity*

(ii) *Rise of earthquake clustering in space and time*

(iii) *Rise of the earthquake correlation range*

(iv) *Transformation of magnitude distribution (Fig. 1.10)*

(v)   *Rise of irregularity in space and time*

(vi) *Reversal of territorial distribution of seismicity*

(vii) *Rise of correlation between different components (decrease of dimensionality)*

(viii)*Rise of response to excitation*

Other relevant processes exhibit premonitory phenomena of the same types.

Patterns of the first two types, (i) and (ii), were found first in observations [GDK$^+$86, Kei90b, KS99] and then in models (see Chaps. 2 and 3 and [GKZN00]); patterns of the next three types, (iii)–(v), were found in the reverse order, first in models (Chap. 2 and [GZNK00, NTG95, Sor00]) and then in observations (see Chap. 4 and [SSS96]); reversal of territorial distribution of seismicity (vi) was found in observations and not explored yet through modeling (Chap. 4); the last two phenomena remain purely hypothetical so far.

Validation. Patterns of the first two types, rise of intensity and clustering, have been validated by statistically significant predictions of real earthquakes (see Chap. 4 and [MDRD90]); other patterns are undergoing different stages of testing.

Reminiscence of theoretical physics. The premonitory phenomena listed above bear a resemblance to the asymptotic behavior of a nonlinear system near the point of phase transition of the second kind. However, our problem is unusual for statistical physics: We consider not the equilibrium state, but the *growing disequilibrium* culminated by a critical transition.

Seismicity patterns. Premonitory phenomena of each type are depicted by different seismicity patterns. Systematically explored are the intermediate-term patterns, with characteristic duration of alarms years. A few examples follow.

– *Measures of seismic activity:* total area of ruptures in earthquake sources (Sect 1.3.2); accumulated strain release [Var89, BV93, BOS⁺98]; the number of earthquakes in a certain magnitude range (see Chap. 4 and [KLKM96, WH88]); the time period when a given number of earthquakes occurs (lower time obviously indicates higher activity) [SJ90], etc.

– *Measures of earthquake clustering:* the number of aftershocks closely following a medium magnitude main shock ("bursts of aftershocks," see Chap. 4 and [MDRD90]); swarms of main shocks having medium magnitudes [CGK⁺77, KKR80, KLJM82]; swarms of relatively small earthquakes [CCG⁺83] etc.

– *Measures of earthquake correlation range:* the distance between nearly simultaneous earthquakes and the number of faults with a nearly simultaneous rise of activity (patterns "ROC" and "Accord" defined in Chap. 4); and the distribution of link lengths in a single link cluster connecting earthquakes that occurred in a given time-space [ZHK01].

– *The measures of irregularity:* the variation in magnitudes or strain release [SS95, NTG95, Sor00].

– *The measures of premonitory transformation of magnitude distribution* (Gutenberg–Richter relation). That transformation is schematically illustrated in Fig. 1.10. One of its measures is the slope of the magnitude distribution in a relatively high magnitude range (pattern "Upward Bend;" see Chap. 2); another measure is the difference between its slopes for lower and higher magnitudes ("pattern γ," [RKB97]). The question mark in the figure indicates that "pattern γ," the reversal of curvature of the distribution, is yet less tested than the "Upward Bend."

Why are different measures used for the same premonitory phenomenon? These measures are certainly correlated, even by definition. Several of them are used instead of an "optimal" one for the following reasons.

– A premonitory phenomenon may have different manifestations on different timescales, spatial scales, and magnitude ranges.

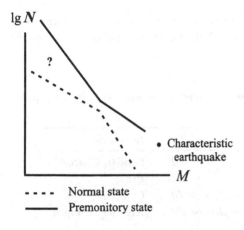

Fig. 1.10. Scheme of premonitory transformation of the Gutenberg–Richter relation $N(M)$. N is the number of earthquakes of magnitude M or larger. *Dashed and solid lines* correspond to time intervals far from a strong earthquake and close to it ("TIPs" in Fig. 1.6), respectively

– A set of measures is more reliable than a single one, due to the complexity of the processes considered and unavoidable noise.

– In lieu of an adequate theory, premonitory seismicity patterns have been found by heuristic analysis, and more compact definitions might just be overlooked so far.

1.5.2   Second Paradigm: Long-Range Correlations

The generation of an earthquake is not localized about its future source. A flow of earthquakes is generated by a fault network, rather than each earthquake by a segment of a single fault. Accordingly, the signals of an approaching earthquake come not from a narrow vicinity of the source but from a much wider area.

Size of areas where premonitory phenomena are formed. Let M and $L(M)$ be the earthquake magnitude and the characteristic length of its source, respectively. In the intermediate-term stage of prediction (on a timescale of years) that size may reach $10L(M)$; it might be reduced to $3L$–L in a second approximation (Chap. 4). In the long-term stage, on a timescale of tens of years, that size reaches about $100L$. For example, according to [PA95], the Parkfield (California) earthquake with M about 6 and $L \approx 10$ km "... *is not likely to occur until activity picks up in the Great Basin or the Gulf of California*" about 800 km away.

Historical perspective. An early and probably the first estimate of the area where premonitory patterns are formed was obtained for pattern Σ (Fig. 1.5). It is noteworthy that Charles Richter, who was generally skeptical about the feasibility of earthquake prediction, made exception to that pattern, specifically because it was based on long-range correlations. He wrote [Ric64]: "... *It is important that (the authors) confirm the necessity of considering a very extensive region including the center of the approaching* *event. It is very rarely true that the major event is preceded by increasing activity in its immediate vicinity.*" Table 1.2 shows similar estimates for other intermediate-term patterns.

Table 1.2. Estimations of the size of the area where premonitory seismicity patterns emerge

Measure	Year	$Q(L)^b$	Reference
Area of fault breaks	1964	$\sim 10L$	[KM64], Chap. 4
Distant aftershocks	1975	$\sim 10L$	[Pro75]
Earthquake swarms	1977	$\sim 5L - \sim 10L$	[CGK$^+$77]
Bursts of aftershocks, area of fault breaks, swarms	1980	$\sim 5L - \sim 10L$	[KKR80], Chap. 4
Algorithm **CN**^{a}	1983	$\sim 5L - \sim 10L$	[KR90], Chap. 4
Algorithm **M8**^{a}	1985	$\sim 5L - \sim 10L$	[KK90], Chap. 4
Benioff strain release	1989	$\sim 5L$	[Var89, BOS$^+$98, JS99]
Algorithm **SSE**	1992	$\sim 5L$	[LV92], Chap. 4
Activation of distant areas	1995	$\sim 100L$	[PA95]
Number of earthquakes	1996	$\sim 5L$	[KLKM96]
Correlation length via single link cluster	2001	$\sim 5L$	[ZHK01]
Near-simultaneous pairs of earthquakes	2001	$\sim 3L$	[SZK00]
Simultaneous activation of fault branches	2002	$\sim 10L$	[ZKA02]

^{a}Reference is not given to the original publication, but to the latest comprehensive reviews
^{b}L is the linear dimension of the source of the approaching strong earthquake. Premonitory phenomena have been observed on a timescale of years with one exception, tens of years in the 1995 entry

At the same time, long-range correlations have been often regarded as counterintuitive in earthquake prediction research on the grounds that redistribution of stress and strain after an earthquake in simple elastic models would be confined to the vicinity of its source ("Saint-Venant principle"). Sometimes that prompted the objection: *"earthquakes cannot trigger each other at such distances."* The answer is that earthquakes involved in long-range correlation do not trigger each other but reflect the underlying large-scale dynamics of the lithosphere. More specific explanations follow.

Mechanisms of long-range correlations. Correlation in earthquake occurrence at long distances greatly exceeding earthquake source dimensions is the prominent feature of seismicity dynamics; it is not confined to earthquake precursors. Among the manifestations of that correlation are the following phenomena: simultaneous changes of seismic activity within large regions [Rom93], migration of earthquakes along fault zones [VS83, KK97, MFZ+90, Mog68], and alternate rise of seismicity in distant areas [PA95] and even in distant tectonic plates [Rom93]. Global correlations have also been found between major earthquakes and other geophysical phenomena, such as Chandler wobble, variations of magnetic field, and the velocity of Earth's rotation [PB75, KP80]. Several mechanisms (not mutually exclusive) have been suggested to explain the long-range correlations. They may be divided into two groups.

(i) Some explanations attribute long-range correlations to a large-scale process controlling stress and strength in the lithosphere. Among such processes are the following.

- Microrotation of tectonic plates [PA95] and crustal blocks (see Chap. 3, [SV99b]); microfluctuations in the direction of mantle currents (Chap. 3). Each of them creates redistribution of normal and tangential stress and, consequently, redistribution of strength through a large part of the fault network.
- Migration of pore fluids in fault systems (see Sect. 1.2.2 and [BKM83]) affects lithosphere strength in the following ways: lubrication; stress corrosion and destabilization waves (see Sect. 1.2.2); and redistribution of hydrostatic pressure between the solid and fluid components of the fault zone.
- Hydrodynamic waves in the upper mantle [PBR98] that propagate through thousands of kilometers during decades and may trigger strong earthquakes connecting seismicity across the globe.
- Activity of creep fractures in the ductile part of the lithosphere. Deformation in the ductile part increases the stress in the brittle part thus triggering earthquakes [Aki96].
- Inelasticity and inhomogeneity of the lithosphere [Bar93]. Due to either of the mechanisms, the redistribution of stress after fracture extends to much greater distances than in a homogeneous elastic media.

Such mechanisms act under different circumstances, separately or jointly. Being rather common, they make long-range correlations inevitable.

(ii) In another approach, the lithosphere is regarded as a complex system; then the long-range correlations are again inevitable, as a general feature of such systems in a near-critical state [TNG00, BOS+98, SS95].

1.5.3   Third Paradigm: Similarity

Premonitory phenomena are similar (identical after normalization) in extremely diverse environments and in a broad energy range. The similarity is not unlimited, however, and regional variations of premonitory phenomena do emerge.

Normalization. Earthquake sequences used in a prediction algorithm are normalized to ensure that the algorithm is self-adapting, i.e., it can be applied without readaptation in regions with different seismic regimes.

The area to which the sequence belongs is normalized by $L(M)$, as described in Table 1.1.

The magnitude range is readapted by changing the minimum magnitude considered (see Chap. 4.)

The timescale in most prediction algorithms does not depend on M although according to the Gutenberg–Richter relation, earthquakes of smaller magnitudes occur more frequently. This is not a contradiction because the Gutenberg–Richter relation refers to a given region, the same for all magnitudes, whereas prediction is made for an area of a size proportional to $L(M)$.

This difference has a consequence not always recognized. Let $T_r(M)$ and $T_a(M)$ be the average return times of an earthquake of magnitude M in the whole region and in a smaller area of linear size $L(M)$, respectively. According to the well-known relations, $T_r(M) \sim 10^{bM}$, $L(M) \sim 10^{cM}$, and $T_a(M) \sim 10^{(b-cv)M}$. Here b is the slope of the Gutenberg–Richter relation, c determines the connection between the magnitude and the source dimension, and v is the fractal dimension of the cloud of epicenters. The existing estimates of parameters b (about 1), c (between 0.5 and 1), and v (between 1.2 and 2) do not contradict the hypothesis that the expression $(b - cv)$ is close to zero; accordingly, *earthquakes of different magnitudes might have about the same recurrence time in their own cells.*

Applications. Prediction algorithms, thus normalized, retain their predictive power in many cases: microfractures in laboratory samples; induced seismicity; earthquakes in subduction zones, major strike-slip fault zones, rift zones, and platforms (see Chap.4 and [GKR+97, KS99]). The corresponding seismic energy release ranges from a few ergs to 10^{25} ergs. However, the performance of prediction algorithms does vary from case to case.

Frontiers of similarity, the neutron star [KKC00]. An opportunity to explore, albeit qualitatively, the frontiers of similarity is provided by the registration of 111 flashes of energy radiated from the neutron star with celestial coordinates 1806-20 in the frequency band of soft γ-rays. These flashes were probably originated by "starquakes," i.e. fractures in the crust of the neutron star. Environments where starquakes and earthquakes originate are compared in Table 1.3.

Figure 1.11 compares the emergence of premonitory patterns before a major starquake and the strong earthquake in the Aquaba gulf, 11 Dec. 1995, $M = 7.3$. The functions capturing these patterns are taken from earthquake prediction algorithms (Chap. 4): Functions Σ, N, and Z capture the intensity of earthquake flow, L captures its deviation from the long-term trend, and B is the measure of earthquake clustering. The patterns work for Earth and the star, their fantastic difference notwithstanding. Only one parameter, the timescale for starquakes, had to be readjusted a posteriori.

Table 1.3. Environments of starquakes and earthquakes

Characteristics	Neutron star 1806–20	Earth
Composition of the crust	Lattice of heavy nuclei	Grains of rock
Radius	10 km	6371 km
Thickness of the crust	1 km	About 33 km
Density	10^{14} g/m^3	5.5×10^6 g/m^3
Energy release	Up to 10^{46} erg	Up to 10^{25} erg
Driving forces	Magnetic field	Convection

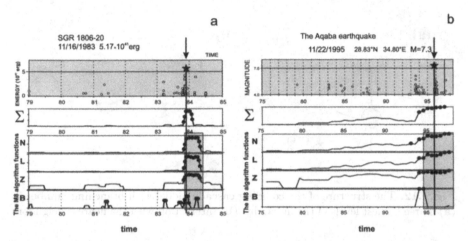

Fig. 1.11. Frontier of similarity of premonitory seismicity patterns. (a) Starquake registered on 16 Nov. 1983, a coarse equivalent of its magnitude is about 20. (b) Aqaba earthquake, 22 Nov. 1995, $M = 7.3$. The panels show the sequence of star-quakes (*left*) and earthquakes (*right*). *Stars* indicate major events that are targeted for prediction. Other panels show the functions capturing premonitory seismicity patterns defined in Chap. 4. B is the measure of clustering; other functions are different measures of seismic activity. Dots indicate the emergence of a premonitory pattern. After [KKC00]

Limitations. The performance of some prediction algorithms still does vary from region to region (see Chap. 5 and [KS99, Ghi94]). It is not yet clear whether this is due to imperfect normalization or to Earth-specific limitations on similarity itself.

1.5.4 Fourth Paradigm: Dual Nature of Premonitory Phenomena

Some premonitory phenomena are "universal," common to hierarchical complex nonlinear systems of different origins; others are specific to the geometry of fault networks or to a certain physical mechanism controlling the strength–stress field in the lithosphere.

A. "Universal" premonitory phenomena

The meaning of "universality." The known premonitory seismicity patterns prove to be universal in the sense that one can reproduce them in the lattice models, specific not only to Earth.[1] Such models are developed in statistical physics and nonlinear dynamics (Chaps. 2 and 4).

An illustration is given in Figs. 1.12–1.14 (after [ZKG01b]). The model reproduces the first four types of the premonitory patterns listed in the first paradigm (Sect. 1.5.1). So far, this is the widest set of patterns reproduced in a model. This was done with a recently introduced model of "colliding cascades" [GKZN00, ZKG01a].

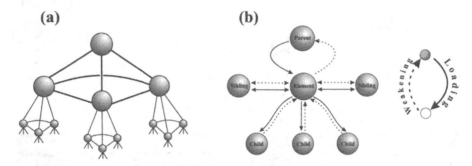

Fig. 1.12. The structure of the colliding cascade model with branching number 3. (a) Three highest levels of the hierarchy, (b) interaction with the nearest neighbors

The model. It describes the generation of critical phenomena by colliding cascades of loading and failures. Its major features are the following:

– The model has a ternary hierarchical structure (Fig. 1.12a).

– The load is applied at the top of the hierarchy and is transferred downward as a direct cascade.

– Failures are initiated at the lowest level of hierarchy and propagate upward as an inverse cascade.

– The cascades are interacting (Fig. 1.12b). Loading triggers failures and failures redistribute and release the load.

Eventually the failed elements "heal," thus ensuring permanent functioning of the system.

In applications to seismicity, the hierarchy imitates the structure of the lithosphere, loading imitates the impact of tectonic forces, and failures imitate earthquakes; a critical phenomenon, "a major earthquake," is failure of the top element.

---

[1] Obviously, seismicity and other fields relevant to earthquake prediction *are* Earth-specific. At the same time, these fields are subject to more general laws of nature. Among them, if our conclusions are correct, are the premonitory phenomena described here; under the same clause they would reflect the general laws of self-organization of complex systems.

Prediction of modeled seismicity. An example of an earthquake sequence generated by a colliding cascade model is shown in Fig. 1.13. The model generates earthquakes of discrete magnitudes. Prediction is targeted at the strongest synthetic earthquakes possible in the model, $M = 7$.

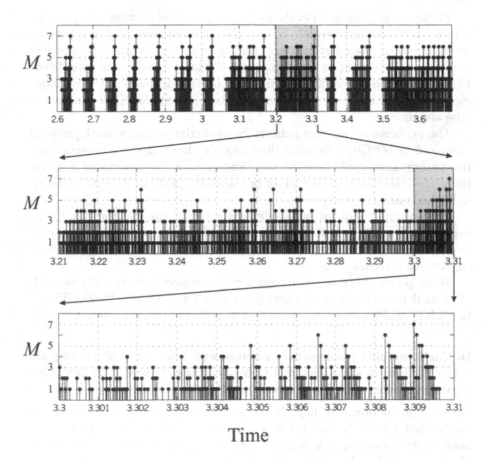

Fig. 1.13. Synthetic earthquake sequence consecutively zoomed. Shaded areas mark zoomed intervals. The model shows the rich variety of behavior on different timescales. Note that the ratio of timescales for the top and bottom panels is 10^2

Heuristic constraints. The modeled seismicity fits basic heuristic constraints, exhibiting, albeit coarsely, the major features of observed seismicity: the seismic cycle; intermittence of the seismic regime; the Gutenberg–Richter relation; aftershocks; long-range correlation; and the premonitory seismicity patterns indicated below.

Premonitory seismicity patterns. Central for this discussion is the fact that the model also reproduces the observed premonitory seismicity patterns of the first four types, (i) to (iv), listed in Sect. 1.5.1.

Moreover, the premonitory rise of the correlation range was found first in the colliding cascades model [GKZN00, GZNK00] and then in the observed seismicity of the Lesser Antilles [SZK00] and Southern California [ZKA02]. So far, other previously suggested patterns have not been explored with that model.

Performance. Figure 1.14a shows the emergence of premonitory patterns before major earthquakes ($M = 7$) in the modeled seismicity. For brevity, we show only 10 such earthquakes. Patterns of each type are defined in several magnitude ranges, so that altogether 23 patterns have been considered. Figure 1.14b shows a continuous time period that includes not only major earthquakes but also long time intervals between them; here, one can see the false alarms.

The performance of these patterns in predicting of major earthquakes is estimated in [ZKG01b] by error diagrams. In the analysis of observations, this performance would be quite satisfactory. As in the analysis of observations, the collective performance of premonitory patterns is better than an individual one.

The universality of patterns is demonstrated by the fact that the model has nothing specific only to generation of earthquakes, and it can be equally interpreted in more general terms, or, for example, in terms of economic or engineering systems.

Many premonitory seismicity patterns have also been reproduced with other models of interacting elements (see Chap. 2 and [PCS94,Tur99,HSSS98, KC95,NTG95,PSV97,YK92]), as well as with Earth-specific models (Chap. 3).

B. Earth-specific premonitory phenomena Phenomena of this kind are not yet defined in a clear-cut way to be incorporated directly in prediction algorithms. Here, we outline the evidence of what such phenomena might be.

Geometric incompatibility. Nodes are of obvious interest for prediction due to their role in the nucleation of strong earthquakes and in the development of the instability of a fault network. Generally speaking, the approach of a strong earthquake is reflected in the strength–stress field, which directly controls seismicity. However, it is hardly realistic to determine this field within a node, given its complexity. Monitoring the geometric incompatibility G seems to be highly promising:

– It is the only known control parameter that captures the state of a node relevant to seismicity; it shows whether the node is locked or unlocked; reflects accumulation of stress and fracturing, etc.

– It can be realistically monitored by observations from outside the nodes.

Reversals of geometric incompatibility might create or dissolve phenomena highly relevant to prediction [GKJ96]: asperities; relaxation barriers; weakest links; and alternation of seismicity and creep ("loud" and "silent" earthquakes).

(a)

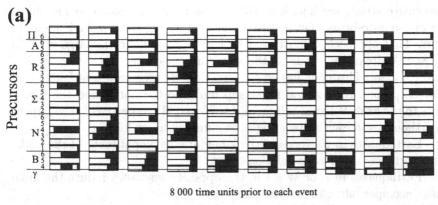

8 000 time units prior to each event

(b)

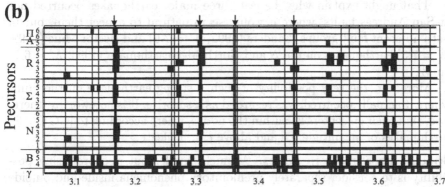

Fig. 1.14. Collective performance of premonitory seismicity patterns (PSPs) found in observed seismicity and reproduced by the colliding cascade model (after [ZKG01b]). The figure shows alarms generated by PSPs of the four types discussed in Sect. 1.5.1: the rise of earthquake correlation length (Π_m, A_m, R_m); the rise of seismic activity (Σ_m, N_m); the rise of clustering (B_m); and the transformations of the Gutenberg–Richter relation (γ). Each pattern is defined in different magnitude ranges indicated at the left side of the panel. *Shaded areas* show alarms triggered by the patterns. **(a)** Alarms preceding 10 major synthetic earthquakes ($M = 7$). Each box corresponds to a single major earthquake; the right edge of the box is its occurrence time. **(b)** Alarms during the continuous time interval. *Arrows and solid vertical lines* show the times of major earthquakes. *Dashed vertical lines* show the times of main shocks of magnitude $M = 6$ to illustrate their association with false alarms

If that inference is correct, these features would migrate from node to node at the large velocity typical of seismicity migration: tens to hundreds km/year [VS83].

Kinematic incompatibility K. This is a control parameter complementary to G. It is relevant to prediction because it captures the state of faults and the internal state of blocks within a given territory.

Where to expect the next earthquake? Both K and G change after consecutive strong earthquakes. This opens a new approach to identification of "soon-to-break" faults, where the next strong earthquake is going to occur. That approach is illustrated by estimates of G in the large node in the San Andreas fault zone at the southeastern corner of the famous Big Bend [GKJ96]. This node lies on the junction of three groups of faults: San Andreas proper, right-lateral strike-slip faults to the north, and the thrust faults to the west. We shall call them SA, N, and W, respectively. The impact of earthquakes in each system might be summed up as follows.

– Earthquakes in SA increase the kinematic incompatibility of the node, thus continuing stress accumulation there.

– Earthquakes in N or W act in the opposite way: they reduce the kinematic incompatibility, thus releasing stress.

That might explain why the last three major earthquakes occurred not at San Andreas faults, where it would seem natural to expect them, but in both adjacent fault groups, Landers (1992, $M = 7.6$) and Hector Mine (1999, $M = 7.3$) earthquakes on the N and Northridge earthquake (1994, $M = 6.7$) on W.

Moreover, the very possibility of such a strong earthquake on the fault, where it actually occurred, was regarded as a surprise. It is noteworthy, therefore, that their epicenters lie within the nodes, where according to [GGK+76], earthquakes of magnitude 6.5 and above can nucleate.

Specific precursors in nodes. A much simpler but also promising possibility is to consider separately premonitory phenomena inside and outside of nodes. The difference between these two cases was observed by Rundqvist and Rotwain [RR94] who applied the earthquake prediction algorithm CN in the Red Sea Rift zone. They found that during a correct alarm, the node was seismically silent, whereas during false alarms, it was active.

1.6   Earthquake Prediction
and Earthquake Preparedness

Given the currently limited accuracy of predictions, how can we use them for damage reduction? The key to this is to escalate or deescalate preparedness depending on the following:

– the content of the current alarm ("what and where is predicted");

– the probability of the false alarm; and

– the cost/benefit ratio of the disaster–preparedness measures.

Such is the standard practice in preparedness for all disasters, war included. The costly mistake that only a precise short-term prediction is practically useful (besides estimates of seismic hazard) sometimes emerges in seismological literature. Actually, as in defense, prediction might be useful if its accuracy is *known*, even if it is not high.

Diversity of damage. Earthquakes may hurt a population, an economy, and the environment in the very different ways listed below:
  – destruction of buildings, lifelines, etc;
  – triggering fires; release of toxic, radioactive and genetically active materials;
  – triggering of floods, avalanches, landslides, tsunamis, etc.
  Equally dangerous are the socioeconomic and political consequences of earthquakes:
  – disruption of vital services: supply, medical, financial, law enforcement, etc.;
  – epidemics;
  – drop of production, unemployment, and slowdown of an economy;
  – destabilization of a military balance;
  – disruptive anxiety of a population, profiteering, and crime.
  The socioeconomic consequences may also be inflicted by undue release of predictions.
  Different kinds of damage develop on different time and space scales, ranging from immediate damage to a chain reaction, lasting tens of years and spreading regionally if not worldwide.

Disaster–preparedness measures. Such diversity of damage requires a hierarchy of disaster–preparedness measures, from building code and insurance to mobilization of postdisaster services to a red alert. It takes different time, from decades to seconds, to undertake different measures; having different costs they can be maintained for different time periods, and they have to be spread over different territories, from selected sites to large regions. No single stage can replace another for damage reduction and, no single measure is sufficient alone.
  On the other hand, many important measures are inexpensive and do not require high accuracy of prediction. An example is the Northridge, California, earthquake, 1994, which caused economic damage of about $30 billion. Its prediction, published well in advance [LV92], is described in Chap. 4. Prediction was not precise. The alarm covered an area 400 km in diameter and a time period of 18 months. However, an inexpensive low key response to this prediction (e.g., out of turn safety inspection) would be well justified if even a few percent of the damage were prevented.

Response to predictions. The framework for the optimal choice of disaster-preparedness measures undertaken in response to an alarm was developed by G. Molchan (Chap. 5; see also the early papers by Kantorovich et al. [KKM74, KK91]). Optimization is based on error diagrams and the trade-off between the costs and benefits of different measures.

Error diagrams generalize the concept of prediction. Traditionally, a prediction is made with a single combination of adjustable elements chosen as the best in some sense. Disaster–preparedness measures would be more flexible and efficient if prediction is made in parallel with several such combinations, i.e., by several versions of an algorithm. So far, this has not been done.

1.7   A Turning Point: Emerging Possibilities yet Unexplored

The studies discussed in this volume suggest yet unexplored possibilities for developing the next generation of prediction algorithms. These possibilities stem from the following developments:

- abundance of relevant observations;
- models of dynamics of seismicity (Chaps. 2 and 3);
- unique uniform collection of failures to predict, false alarms, and successes accumulated in tests of prediction algorithms (Chap. 4); and
- advances in nonlinear dynamics and geodynamics, the two fields at the opposite ends of the earthquake prediction expanse.

1.7.1   The Near-at-Hand Research Lines

Expansion of the observational base. Earthquake precursors might exist in many observable fields: tectonic movements in the whole velocity range from seismicity to creep to geodetic movements; satellite topography; fluid regime; geochemistry; electric and magnetic fields, etc. It seems promising to explore these precursors with the *common scaling* and *common types* of premonitory phenomena, as have been found for premonitory seismicity patterns (Sect. 1.5).

Why should we expect a connection between precursors in different fields? All of them might reflect the same internal process in the fault network, namely, the development of instability. Some precursors might not be field-specific, but depend only on the geometry of the network, or might not be Earth-specific at all (Sect. 1.5.4).

Where should we look for precursors? Within a considerable distance from the earthquake targeted for prediction, up to $100L(M)$ and $10L(M)$ for long-term and intermediate-term precursors, respectively; here, M is the earthquake magnitude, and $L(M)$ is the size of its source.

What types of precursors should we look for in other fields? The basic types described in Sect. 1.5.1 (the first paradigm) deserve particular attention, at least for a start.

Could precursors be field-specific or region-specific? Possibly, yes. It might be easier to find them in the background of "universal" precursors.

Transition to short-term prediction. The following possibilities emerge:

  – *Identification of short-term precursors in the background of intermediate-term alarms.* The first observations on such precursors are discussed in Chap. 4. This would at least reduce the rate of false alarms, the stumbling block in the search for such precursors.

  – *Identification of foreshocks* by criteria formulated in [MD90, YK92]: the low slope of the Gutenberg–Richter relation and the rise of magnitudes in time.

  A broader goal would be to integrate all stages of prediction, from long-term to immediate in successive approximations. A singular success in predicting of the Haicheng earthquake in China, 1975 [MFZ+90], seems highly encouraging.

Using the geometry of fault networks. The following research goals seem feasible.

  – *Premonitory changes in geometric and kinematic incompatibilities* (see Sect. 1.5).

  – *Differences between precursors in fault zones and nodes*; possibly, also, in blocks, with their quite different dynamics.

  – *Region-specific precursors.* Their existence may be inferred from the different performance of prediction algorithms in different regions (see Chap. 4).

  A powerful tool for this research is provided by block models of seismicity (Chap. 3). The collection of prediction errors, by now accumulated, is highly relevant to the last research line because experiments in advance prediction have been made in more than 20 regions worldwide.

A simpler model? A rich variety of "universal" (nonlinear dynamics type) models reproduces about the same basic features inherent in the dynamics of seismicity. This suggests the possibility of designing a simpler model, which would reproduce, though coarsely, the bulk of these features.

"Prediction of predictability." The prominent feature of seismicity is the intermittence of "seismic regimes," strikingly different in frequency and irregularity in the occurrence of strong earthquakes. Typically, for complex systems, the regimes may abruptly switch from one to another in the same region and migrate from one territory to another on a regional or global scale. Premonitory phenomena and the degree of predictability might be different for different regimes. This raises the following problems (Chap. 2):

Given observations on seismicity (and/or other relevant fields),
• to identify the current regime;
• to predict its switching to another one.

  Promising approaches to these problems are provided by "universal" models of seismicity (Chap. 2, [ZKG01a]).

1.7.2   The Goals

If our conjectures are relevant to reality, the research outlined above converges on the following goals.

The next generation of prediction algorithms, with a
   – five- to tenfold increase in accuracy compared with existing algorithms,
   – adaptation to the needs of disaster–preparedness.

A unifying theory of critical phenomena in the lithosphere.

These are the parts of a broader issue:

   – the emergence of newly integrated dynamics of solid Earth,
   – extending from a fundamental concept succeeding plate tectonics to prediction and (with luck) control of geological and geotechnical disasters, and
   – linking the study of a wide class of critical phenomena in a hierarchical complex system formed, separately and jointly, by nature and society.

Finally, this discussion includes conjectures which might be wrong. What seems reasonably certain is that the emerging possibilities outlined above remain unexplored and deserve attention.

2 Hierarchical Models of Seismicity

M. Shnirman and E. Blanter

2.1 Introduction

The earthquake process is a very complicated natural phenomenon that may be investigated from different viewpoints by different tools and methods. This present chapter focuses on the scaling properties of seismicity and its relation to earthquake prediction. Different hierarchical models of seismicity are used to show the theoretical basis and the origin of conclusions drawn in this chapter.

2.1.1 Modeling and Hierarchy

Modeling of seismicity has several preferences: it is possible to obtain arbitrarily long sequences of events; the change in model parameters shows which of them are essential for effects observed; different experiments allow distinguishing random results from stable ones. We cannot present the description, let alone enumeration of seismic models together with their specific advantages; instead, we mention two of them that influence many others and has become classic: string-mass or slider-block model [CL82, BK67] and avalanche or sand-pile model [BT89]. These and many other models are described in remarkable books [NGT94, Tur97, Sor00, RTK00]. These books, considered in aggregate, give the overall picture of seismic modeling. We have no needs and possibility to compete with them. Thus we focus on the special field, hierarchical modeling. Models of this kind are well suited for studying multiscale phenomena like seismicity. It is generally believed that earthquakes occur as a result of the nonlinear dynamics of Earthlithosphere, which is actually considered a multiscale system with a hierarchical structure. Self-similarity of fault systems, of crack propagation, and of rocks structures was established in impressive amount of work (see, for example, [Ito92, Kin83, SGPS84]); therefore, hierarchical modeling has wide use in seismology. Although hierarchical models reflect the self-similarity of the lithosphere, they do not always present the same scaling properties as the observed seismicity.

2.1.2 Self-similarity of Seismicity

A widely known statistical feature of seismic processes described by the Gutenberg–Richter law is self-similarity. The Gutenberg–Richter law shows the power law relationship between the number of earthquakes and their energy for the worldwide seismicity [GR44]:

$$\log N = a + \beta \log E .$$

The commonly used form of the Gutenberg–Richter law is called also the magnitude–frequency relation. It represents earthquakes as a function of their magnitude:

$$\log N = A - bM . \tag{2.1}$$

It was established that the magnitude–frequency relation (2.1) is also linear for several seismoactive regions and the variability of seismic conditions in different regions is reflected in the b-value, the slope of the Gutenberg–Richter plot [Kro84]. The b-value is a very important characteristic of seismicity: as a mean value (estimated for a long time interval), it reflects the fractal properties of the seismic region at hand [BT94]; temporal variation of the b-value reflects the preparation of a strong earthquake and can be used for earthquake prediction (see, for example, [Smi81]). The linear magnitude–frequency relation is usually understood as evidence of criticality. Seismicity is then considered a critical system although the origin of this criticality is not yet clear. The discussion starts with the question of what influences the scaling properties of seismicity and how the self-similarity of seismic processes may be realized in hierarchical models.

2.1.3   Inverse Cascade Models

Different static and dynamic hierarchical models of seismicity are presented here. All of these models describe the development of defects from low to high scaling levels of the system at hand and therefore belong to inverse cascade models. The notion of an inverse cascade appears as opposed to a direct cascade of energy transfer from high to low scaling levels well known in studies of turbulence [Fri95]. Hierarchical seismic models may include mechanisms of both inverse and direct cascades [GKZN00], but two-cascade models are more complicated and are not considered here.

Static hierarchical models with an inverse cascade of defect development can be applied to describe experiments on rupture or to study the statistical properties of the seismic process averaged over a long time interval when seismicity may be considered a stationary process. Details of temporal seismicity evolution are completely neglected in static hierarchical models; the linear magnitude–frequency relation appears instantaneously because of the phase transition [ALM82], or it may be realized as a general kind of system behavior when space heterogeneity is taken into account [SB98,SB99]. The stable linearity of the magnitude–frequency relation is also presented in the static hierarchical model with nonmonotonic conditions of defect formation [BS97]. Static hierarchical models show how the characteristics of stress inhomogeneity may be reflected in the b-value of the Gutenberg–Richter law, although the description is rather abstract.

Dynamic hierarchical models describe the temporal evolution of the seismic process considered, and a synthetic catalog may be generated and compared with the observed data. The synthetic catalog contains emergence times and magnitudes of events. Although spatial features of the system at hand appear only as certain integrated parameters of modeling, the inhomogeneity of the stress field may be taken into account as in the static case [SB01]. The linear magnitude–frequency relation in dynamic hierarchical models may result from inhomogeneity or from the feedback relation between the evolution of stress and events emerging in the system [NS90]. Dynamic hierarchical models allow us to understand how the healing properties of a system influence the magnitude–frequency relation and the b-value. Inspecting the evolution of system parameters in the dynamic hierarchical model [BSLA97], it is possible to observe all basic features found in the temporal evolution of seismicity: the Gutenberg–Richter law; the seismic cycle; the foreshock and the aftershock activity; the Omori law; and the spatial and temporal variations in the predictability of strong events [BSL99].

We do not discuss here a very interesting but quite a different variant of dynamical hierarchical model well known as SOFT (Scaling Organization of Fracture Tectonics) [AMCN95].

2.1.4   Earthquake Prediction and Synthetic Seismicity

The problem of earthquake prediction attracts common interest because of its great practical importance. The most developed direction of the earthquake prediction theory considers earthquakes of high magnitude as the goal of prediction. If a catalog of earthquakes covers a time interval sufficiently long and there are enough strong earthquakes within it, then one can estimate the predictive quality by employing different measures of predictability [Mol90]. The statistical significance of results was checked for certain algorithms of earthquake prediction, such as the CN and M8 [KR90, KK90], as well as for some particular precursors of strong earthquakes [MDRD90]. Small numbers of strong earthquakes make estimation of the prediction quality quite uncertain when short time intervals or particular seismic regions are considered. The measure of predictability can differ when the same algorithm is applied to different regions, and results for the retrospective and forward prediction can also diverge [NR96]. So far, the discussion of the predictability of earthquakes appears to follow from the lack of stability in earthquake prediction. To understand the nature of the predictability of strong earthquakes, we turn to synthetic catalogs of seismicity capable of providing any desired number of strong events and allowing us to change seismic conditions.

The earthquake prediction problem may also be reformulated for synthetic events, and the relation between the predictability of high-scale events and the main time constants of system evolution may be established in several cases. An interesting example where predictability of strong events in synthetic catalogs is investigated for self-organized systems is presented in [PC94]. As to hierarchical models, the predictability variation obtained for synthetic events [BSL99] shows that for a given earthquake prediction algorithm, the quality of prediction may be different for different seismic regions and time periods even when the parameters of the algorithm are optimized. The relation between predictability and the parameters of system evolution allow us to recognize earthquake prediction not only as an important practical problem, but also as a possibility of characterizing regional seismicity by the predictive quality of its strong events. Thus, the predictability of strong events may be considered a physical characteristic of the seismic process at hand. This characteristic is studied below for different hierarchical models of seismicity, and its relation to other properties of system evolution is established in the simplest model case.

This chapter reviews basic results obtained for different hierarchical models and clarifies their physical meaning for real seismicity and earthquake prediction. The description focuses on the simplest models that allow us to draw rather clear and transparent conclusions. The explicit description of each model, its particular properties, and secondary effects may be found in the original papers cited in the text.

The chapter is organized as follows: static hierarchical models and the role of inhomogeneity in the scaling properties of seismicity are considered in Sect. 2.2; Sect. 2.3 is devoted to dynamic hierarchical models, basic time constant, their influence on the observed b-values, and the predictability of strong events; Sect. 2.4 briefly discusses what basic seismic features may be realized in the dynamic hierarchical model involving a feedback relation for time constants. Sect. 2.5 summarizes the conclusions and presents a discussion of possible applications and prospects.

2.2   Static Hierarchical Models

The first static hierarchical model with inverse cascade was developed by Allegre et al. [ALM82]. It describes crack propagation in homogeneous hierarchical media under uniform stress and indicates a phase transition when the applied stress reaches its critical value. This section demonstrates that the phase transition obtained in [ALM82] is a general case of the system behavior for the homogeneous monotonic static hierarchical model with an inverse cascade.

A simple but quite general construction of a static hierarchical model with an inverse cascade is described below, and the discussion is centered on the scaling properties of the system, their origin, and the conditions of their existence. *The monotonicity* and *homogeneity* of static hierarchical models are defined and it is shown how nonmonotonicity and heterogeneity may be introduced in a system. The heterogeneous case of monotonic static hierarchical models is considered in detail and the nonmonotonic case is only mentioned, one can see [BS97] for details.

2.2.1   General Description

We consider a hierarchical system of elements with branching number n (Fig. 2.1a). Each element at level $l + 1$ is linked to a group (its associate group) of n elements at the previous level l. Each element of the system may be in one of two possible states; it can be broken or unbroken (Fig. 2.1b). A broken element is called a defect.

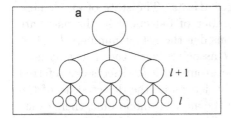 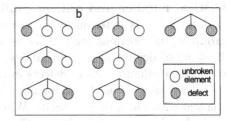

Fig. 2.1. Hierarchical system with branching number $n = 3$: (**a**) hierarchical tree of elements; (**b**) critical configurations with the critical number $k_0 = 1$

Inverse cascade. The state of an element at level $l + 1$ is determined by the number k of defects in its associate group of n elements at the lower level l. The rule defining the state of an element at level $l + 1$ through the number k of defects among the elements in its associate group is called the destruction conditions. These conditions are assumed to be independent of level l, which means that the model structure is self-similar. The system behavior is described by the concentrations of defects $p(l)$ at level l ($l = 1,...,L$). Destruction rules may depend not only on the number k of defects among the associate group, but also on the spatial configuration of defects in the group as assumed in [ALM82]. Such a rule does not change possible system behavior, but complicates the description; hence the discussion below treats the simplest case where the states of elements depend on the number k alone.

Heterogeneity. When the destruction condition is the same for all elements of the system, the model is called *homogeneous*. A *heterogeneous* model generally has different conditions of destruction for different elements. Thus, a homogeneous model is the degenerate case of a heterogeneous one.

Monotonicity. Assume that a defect of an element at level $l+1$ is induced by k_0 defects in its associate group at level l. The model has *monotonic* destruction conditions when a defect of an element at level $l+1$ is also induced by k defects in its associate group at level l for any $k > k_0$. When it is not true, destruction conditions are *nonmonotonic*.

This definition implies that homogeneous and monotonic conditions of destruction are defined by the lowest number k_0 of defects in an associate group of n elements at level l producing a defect at the higher level $l+1$. Consequently, there exist n different monotonic and heterogeneous rules of destruction for a system with branching number n. In nonmonotonic and homogeneous conditions, it is necessary to define a set of numbers k_i of defects in associate groups. Such a case was described in [BS97], where the conditions of destruction were homogeneous and nonmonotonic.

Critical number and critical configurations. The state of an element at level $l+1$ is determined by the number of defects k in the associate group at the previous level l. Let us consider the set of numbers k_i ($1 \leq k_i \leq n$) of defects in a group at level l associate with the defect at level $l+1$. All configurations of defects in the group with k_i defects are referred to as *critical configurations*, because they determine the defect state of the element with which they associate. When the model is monotonic, there exists a number $k_0 = \min\{k_i\}$ which is the minimum number of defects sufficient for a critical configuration. This number k_0 is referred to as a *critical number*. Any configuration with k_0 or more defects in a group of n elements is critical (Fig. 2.1b). The critical number is the same for all elements of a homogeneous hierarchical model.

Densities of defects. It is assumed that defects at the first level of the system are random and found from the Bernoulli distribution with parameter p_0. Let us denote the density of defects (the ratio of their number to the number of elements) at level l by $p(l)$. The density of defects at the first level $p(1) = p_0$ is a parameter of the model, the densities of defects at higher levels can be calculated when the density p and the critical number k_0 are fixed. The density of defects at level $l+1$ can be expressed from the density of defects at the previous level l as follows:

$$p(l+1) = F\left[p\left(l\right)\right] , \tag{2.2}$$

where $F(p)$ denotes the probability that a critical configuration of defects appears in a group of n elements if p is the probability of a defect. The density of configurations containing exactly k defects in a group of n elements at level l takes the form

$$W_k(p) = C_n^k p^k (1-p)^{n-k} ,$$

where $p = p(l)$ is the density of defects at level l, C_n^k denotes the binomial coefficients,and n is a branching number. The density of all configurations with k or more defects is as follows:

$$\Phi_k(p) = \sum_{j=k}^{n} W_j(p) . \tag{2.3}$$

The transition function for a homogeneous model with critical number k_0 is then defined by (2.3) with $k = k_0$; therefore,

$$F(p) = \sum_{j=k_0}^{n} C_n^j p^j (1-p)^{n-j} .$$

Events. Some defects at level l correspond to a defect at the upper level $l + 1$. In real observations, the defect entering a defect at the upper level usually is not detectable. Therefore events are defined as defects at level l which *do not enter* a defect at the upper level $l + 1$. The density of events at level l is denoted as $P(l)$. For the homogeneous model, the density of events at level $l < L$ is represented as follows:

$$P(l) = \sum_{j=1}^{k_0-1} C_{n-1}^{j-1} p^j(l)[1 - p(l)]^{n-j} . \tag{2.4}$$

All defects at the highest level are events:

$$P(L) = p(L). \tag{2.5}$$

Magnitude–frequency relation. The magnitude is actually used as a measure of earthquake energy. A relation between the magnitude M of an earthquake and the size S of its source is established in the form [US54]

$$\log_{10} S \approx M + \text{const} . \tag{2.6}$$

By definition, the size of an element in the model is the quantity increasing with level l as follows:

$$S(l) = S_0 n^l . \tag{2.7}$$

The number of elements decreases according to the law

$$N_e(l) = Cn^{L-l} , \tag{2.8}$$

where C denotes the number of elements at the highest level L of the system. In view of (2.6), the magnitude of defects at level l is defined as follows:

$$M(l) = l \cdot \log_{10} n + \text{const} . \tag{2.9}$$

Let us represent the average number of defects at level l in the form

$$N(l) = Cn^{L-l}p(l) , \tag{2.10}$$

and obtain from (2.10) and (2.9) the magnitude–frequency relation for the model which has analogy with (2.1) for seismicity and reads as follows:

$$\log_{10} N(l) = -M(l) + \log_{10} p(l) + \text{const} . \tag{2.11}$$

The magnitude–frequency relation for events is similar to (2.11):

$$\log_{10} \nu(l) = -M(l) + \log_{10} p(l) + \text{const} , \tag{2.12}$$

where $\nu(l) = Cn^{L-l}P(l)$ denotes the average number of events at level l. It follows from (2.11) that the form of the magnitude–frequency relation for defects is completely determined by their density $p(l)$ and by its variation with level l; similarly, the magnitude–frequency relation for events (2.12) is determined by the density of events $P(l)$.

2.2.2   Phase Transition in a Homogeneous Model

The transition function F, see (2.2), is monotonic for the monotonic model; therefore there exists a limit $0 \leq \lim_{l\to\infty} p(l) \leq 1$, and there are only three possibilities for the value of this limit listed below.

Stability. When l increases, the densities of defects $p(l)$ tend to zero. In this case, high levels of the system remain intact,and the perturbation does not reach high scales. It follows from (2.4) that the density of events $P(l)$ also tends to zero, similarly for $p(l)$. Therefore one cannot observe any strong event at high levels of the system if it is large enough. This kind of behavior is referred to as *stability*.

Scale invariance. When l increases, the densities of defects $p(l)$ tend to a constant value p_0 ($0 < p_0 < 1$). This means that the measure of the destruction at different levels is asymptotically the same and the probability of an event is independent of its scale. This kind of behavior is referred to as *scale invariance*. It is evident from (2.11) and (2.12) that the magnitude–frequency relation in this case is linear and the slope is equal to unity.

Catastrophe. When l increases, the densities of defects $p(l)$ tend to unity. High levels of the system are completely destroyed. This kind of behavior is referred to as a *catastrophe*. It follows from (2.4) that the density of events $P(l)$ tends to zero when l grows.

It is easy to show (see [BS97] for details) that the domain of stability always exists for a homogeneous system when $k_0 > 1$. When the density of defects at the lowest level of the system $p(1)$ is close to zero, it is stable,

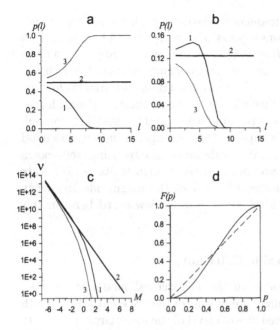

Fig. 2.2. Different kinds of behavior for a homogeneous hierarchical system with branching number $n = 3$ and critical number $k = 2$: stability (curves 1), scale invariance (curves 2), and catastrophe (curves 3). The following functions are plotted: (**a**) the density of defects; (**b**) the density of events; (**c**) magnitude–frequency relation; and (**d**) the transition function

and both densities $p(l)$ and $P(l)$ tend to zero (Fig. 2.2a and b, curve 1). The corresponding magnitude–frequency relation has a downward bend as in the exponent of an exponential function (Fig. 2.2c, curve 1). When $k_0 = n$, the system demonstrates stability for all values of parameter $p(1)$.

Similarly, the domain of catastrophe exists when $k_0 < n$. When the density of defects at the lowest level of the system $p(1)$ is close to unity, the system demonstrates catastrophic behavior, the density of defects $p(l)$ tends to unity, and the density of events $P(l)$ approaches zero (Fig. 2.2a and b, curve 3). The magnitude–frequency relation for events has a double exponential downward bend (Fig. 2.2c, curve 3). When $k_0 = 1$, there exists a catastrophe for all values of parameter $p(1)$.

Summarizing the two previous cases, it is seen that when $1 < k_0 < n$, there exist both stability and catastrophe. An example of the transition function $F(p)$ is presented in Fig. 2.2d. The map F has two stable fixed points ($p = 0$ and $p = 1$) and one unstable fixed point $p = p_{cr}$ [see the intersection of the curve $F(p)$ with the dashed diagonal line in Fig. 2.2d]. When p is less than p_{cr}, $F(p)$ is below the diagonal line [$F(p) < p$]; when p is greater than p_{cr}, $F(p)$ is above the diagonal line [$F(p) > p$]. Thus, all values $p(1) < p_{cr}$ imply stability, and all values $p(1) > p_{cr}$ imply catastrophe. The phase transition from stability to catastrophe occurs when the density of defects at the lowest level $p(1)$ reaches its critical value $p(1) = p_{cr}$ (Fig. 2.2d). Scale invariance is observed for $p(1) = p_{cr}$: densities of defects $p(l)$ and events $P(l)$ approach nontrivial constant values (Fig. 2.2a and b, curve 2), the corresponding magnitude–frequency relation is linear, the slope is equal to unity (Fig. 2.2c, curve 2).

Thus, the linear magnitude–frequency relation for a homogeneous monotone model can be realized at a single point $p(1) = p_{\mathrm{cr}}$ of the phase transition from stability to catastrophe. When $p(1) < p_{\mathrm{cr}}$, the system appears to be in the domain of stability, and the magnitude–frequency relation has a downward bend. However, if parameter $p(1)$ is close enough to its critical value, the magnitude–frequency relation is linear for an essential number of scale levels. Earthquakes have a finite number of possible scales, and the statistics for high levels is quite poor; therefore, it is impossible to distinguish the neighborhood of an unstable critical point from stable scale invariance by using well-known seismological observations. The proximity of the system to its critical state may be characterized by the downward bend of the magnitude–frequency relation or by the b-value, which reflects the same downward bend in this case.

2.2.3   Heterogeneity and Stable Criticality

Let us now introduce heterogeneity in the hierarchical model and choose different critical numbers k for different elements of the system. Elements with different critical numbers are mixed at each level of the hierarchy. The density of elements with the critical number k is denoted by a_k; densities a_k are called below the concentrations of the mixture. Favoring the self-similarity of system, the concentrations a_k are assumed to be the same for different levels of the system (a_k are independent of l). Let us note that this self-similarity condition does not mean the self-similarity of system behavior, as one can see in the homogeneous model. The transition function $F(p)$ determining the relation between the densities of defects at two successive levels [see Eq. (2.2)] is represented in the form

$$F(p) = \sum_{k=1}^{n} a_k \Phi_k(p) \, ,$$

where Φ_k is defined by (2.3). The density of events at level $l < L$ has the form

$$P(l) = \frac{1}{n} \sum_{k=1}^{n} k C_n^k p^k(l) [1 - p(l)]^{n-k} (1 - \sum_{i=1}^{k} a_i) \, . \qquad (2.13)$$

As above, all defects are also events at the highest level of the system: $P(L) = p(L)$.

Concentrations a_k control the heterogeneity of the system. Let us see now how its behavior depends on a_k for the simplest case of a hierarchical system with branching number $n = 3$.

Critical stability. Let us consider a system with branching number $n = 3$ where elements with critical numbers $k = 1$, $k = 2$, and $k = 3$ are mixed with concentrations a_1, a_2, and a_3, respectively. It is easy to show (see [SB98] for details) that densities of defects $p(l)$ approach zero when l increases for all $p(1)$ if the concentrations of the mixture a_k satisfy the following conditions:

$$0 < a_1 < 1/3 ,$$
$$a_3 < 1/3 .$$

The ratio between the densities of defects at two successive levels to the first order of small quantity $p(l)$ is expressed as follows:

$$\frac{p(l+1)}{p(l)} \sim 3a_1 . \tag{2.14}$$

It is seen from (2.14) and (2.13) that the density of events also approaches zero:

$$P(l) \sim (3a_1)^l .$$

It follows from the power form describing the decrease of densities $p(l)$ and $P(l)$ that the magnitude–frequency relation, for both defects and events, is linear and has a slope equal to $b = -\log a_1/\log 3$. This kind of behavior is referred to as *critical stability*; it is illustrated by curve 1 in Fig. 2.3a–c.

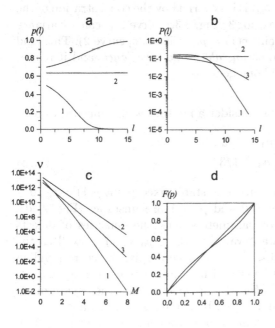

Fig. 2.3. Different kinds of behavior for a heterogeneous hierarchical system with branching number $n = 3$. Concentrations of the mixture are as follows: $a_1 = 0.1$, $a_2 = 0.7$, and $a_3 = 0.2$ for plots (**a**)–(**c**); $a_1 = a_3 = 0.5$, and $a_2 = 0$ for plot (**d**). Different values of $p(1)$ correspond to the following behavior: stability, $p(1) = 0.5$ (curves 1); scale invariance, $p(1) = 0.6364$ (curves 2), and catastrophe, $p(1) = 0.7$ (curves 3). The following characteristics are plotted: (**a**) the density of defects; (**b**) the density of events; (**c**) the magnitude–frequency relation; and (**d**) the transition function

Critical catastrophe. Catastrophic behavior in a heterogeneous model with $n = 3$ sets in for all values of parameter $p(1)$ when concentrations of the mixture a_k satisfy the following conditions:

$$1/3 < a_1 \,,$$
$$0 < a_3 < 1/3 \,.$$

In this case the density of defects $p(l)$ approaches unity, and the density of events $P(l)$ approaches zero (see [SB98] for details):

$$P(l) \sim (3a_3)^l.$$

The magnitude–frequency relation for events is linear with a slope $b = -\log a_3/\log 3$. The linear form of the magnitude–frequency relation means that the behavior of the system is critical; therefore this kind of behavior is referred to as *critical catastrophe*. It is illustrated by curve 3 in Fig. 2.3a–c.

Unstable scale invariance. When $0 < a_1 < 1/3$ and $0 < a_3 < 1/3$, the system demonstrates phase transition from critical stability to critical catastrophe. The magnitude–frequency relation remains linear for all values of parameter $p(1)$, but its slope changes in the neighborhood of the critical point $p(1) = p_{\mathrm{cr}}$. When $p(1) < p_{\mathrm{cr}}$, the slope of the magnitude–frequency relation is governed by the concentration of the most fragile elements, $b = -\log a_1/\log 3$ (Fig. 2.3c, curve 1); when $p(1) > p_{\mathrm{cr}}$, it is governed by the concentration of the most stable elements, $b = -\log a_3/\log 3$ (Fig. 2.3c, curve 3); scale invariance ($b = 1$) is observed for a single point $p(1) = p_{\mathrm{cr}}$ (Fig. 2.3c, curve 2). This kind of system behavior is referred to as *unstable scale invariance*, and it relates to the phase transition as in the homogeneous case.

Stable scale invariance. Let us consider a mixture with concentrations:

$$a_1 > 1/3 \,,$$
$$a_3 > 1/3 \,. \tag{2.15}$$

The transition function, $F(p)$ has only one stable fixed point $p(1) = p_0$ ($0 < p_0 < 1$); the fixed endpoints $p = 0$ and $p = 1$ are unstable (Fig. 2.3d). This means that for all values of parameters $p(1)$, the densities of defects $p(l)$ approach a nontrivial constant value p_0 ($0 < p_0 < 1$) as well as the densities of events $P(l) \to P_0$ ($0 < P_0 < 1$). Accordingly, the corresponding magnitude–frequency relation is linear with a slope equal to unity, $b = 1$. Thus the system demonstrates *stable scale invariance*, and the asymptotical behavior of the system may be illustrated by curve 2 in Fig. 2.3a–c.

Let us consider the phase diagram for mixture concentrations (Fig. 2.4). Conditions (2.15) mean that both the most fragile elements (described by concentration a_1) and the most stable ones (described by concentration a_3) are mixed with large weights ($a_1 > 1/3$; $a_3 > 1/3$). The system, far from homogeneous in this case, is strongly heterogeneous. Thus, stable scale invariance is the result of the strong heterogeneity of the system. From this point of view, it is not strange to see the unity slope in the Gutenberg–Richter law for global seismicity. Deviations from unity slope for local seismoactive regions show that their heterogeneity is not strong enough to produce scale invariance. Nevertheless, the critical behavior characterized by the linear magnitude–frequency relation is a general property of a heterogeneous system. The b-value reflects the heterogeneity of the system, it is governed by one of two concentrations a_1 or a_3 describing the density of the most fragile or the most stable elements, respectively. It is remarkable that the characteristics of heterogeneity have no effect on the b-value in the domain of stable scale invariance.

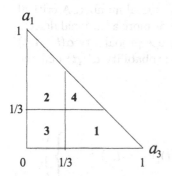

Fig. 2.4. Phase diagram for a heterogeneous static model with branching number $n = 3$. The mixture is parameterized by two free parameters: a_1 and a_3. There are four domains of system behavior: stability (domain 1); catastrophe (domain 2); unstable scale invariance (domain 3); and stable scale invariance (domain 4)

It should pointed out that the above results were obtained for the simplest case of $n = 3$; however, they are similar for any n. When n is arbitrary, the b-value is controlled by a_1 in the case of stability and by a_n in the case of catastrophe. The phase diagram becomes more complicated for larger n (see the example of $n = 5$ in [SB99]), but stable scale invariance is still associated with strong heterogeneity of the system.

2.3   Dynamic Hierarchical Models

The most important problem in extending a static model to a dynamic one consists of a proper definition of the process allowing the renormalization of a time step. The temporal evolution of the model can be organized in different ways (see for example [NSH$^+$00]); for the simplicity only one formalism is considered below. Extension of a homogeneous hierarchical static model described in Sect. 2.2.1 was discussed in [NS90]. This extension runs below in the form of a simple and rather general example of a dynamic hierarchical model.

2.3.1   General Description of the Dynamic Model

Consider a hierarchical system with branching number n and two possible states of elements: broken (defects) and unbroken state (Fig. 2.1), as in the static case (Sect. 2.2.1). The density of defects from level l at time t is denoted by $p(l,t)$. The evolution of the system is governed by the competition of two processes: the emergence of new defects and the healing of old defects.

Emergence of new defects. An unbroken element from level l becomes a defect at time t with the probability $\alpha(l,t)$ referred to as the emergence probability. The emergence probability at the first level of the system is a constant $\alpha(1,t) = \alpha_0$. The emergence probability at an upper level $l+1$ is determined by the conditional probability of obtaining a new critical configuration of defects at level l at time t. As in the static case, a configuration is critical when it has k or more defects; k is called a critical number. A critical configuration is considered new at time t when one or more additional defects emerge in this configuration at time t. The emergence probability $\alpha(l+1,t)$ at level $l+1$ is expressed through the emergence probability $\alpha(l,t)$ and the density of defects $p(l)$ at level l as follows:

$$
\begin{aligned}
&\alpha(l+1,t) \\
&= \left(\sum_{i=0}^{n-1} \left\{ C_n^i p^i(l,t)[1 - p(l,t)]^{n-i} \sum_{j=k-i}^{n-i} \alpha^j(l,t)[1 - \alpha(l,t)]^{n-i-j} \right\} \right) \\
&\quad \times \left\{ \sum_{i=0}^{n-i} C_n^i p^i(l,t)[1 - p(l,t)]^{n-i} \right\},
\end{aligned}
$$

where k denotes the critical number, which is assumed to be the same for all elements of the system.

Healing of old defects. A defect at level l that emerged at time t_0 can return to the unbroken state in any time $t > t_0$ with probability $\beta(l)$, called the healing probability. It is constant in time and depends on scale levels as follows:

$$
\beta(l) = \beta_0 c^l \tag{2.16}
$$

where c is the scaling parameter$(0 < c \leq 1)$, it reflects the increase of the mean lifetime of defects with scale level.

Kinetic equation. The density of defects evolves according to the kinetic equation:

$$
p(l,t+1) = [1 - p(l,t)]\alpha(l,t) + p(l,t)[1 - \beta(l)], \tag{2.17}
$$

where the first term on the right side is the density of new defects that emerged at time $t+1$ and the second term denotes the density of old defects not healed at time $t+1$. New defects are called events; the density of events from level l at time t is denoted by $q(l,t)$; it is expressed as follows:

$$q(l,t+1) = [1 - p(l,t)]\alpha(l,t) \ . \tag{2.18}$$

The magnitude–frequency relation. It is defined as in the static model and its scaling properties are determined by the density of events $q(l,t)$:

$$\log_{10} N(l,t) = -M(l) + \log_{10} q(l,t) + \text{const} \ ,$$

where $N(l,t)$ denotes the number of events from level l at time t and the magnitude $M(l)$ of events at level l is defined by (2.9).

Let us point out that the notions of events in dynamic and static models are different. Really, instead of new defects, we have to introduce events as new defects which are not relevant to any event of the superior level. Asymptotically, however, there is no difference between these two definitions, and therefore we use the simplest one.

Note also that the discrete time in the model can be easily rescaled to any smaller time step by multiplying the emergence probability at the lowest level α_0 and the healing parameter β_0 by the ratio of the new to the old time step. All formulas and elements of the model description remain the same.

2.3.2 Stationary Solution and Phase Transition

When time t tends to infinity, the model described above has the stationary solution:

$$\lim_{t\to\infty} \alpha(l,t) = \alpha(l) \ ,$$

$$\lim_{t\to\infty} p(l,t) = p(l) \ ,$$

$$\lim_{t\to\infty} q(l,t) = q(l) \ .$$

Let us describe the properties of this stationary solution. Substitute the stationary solution in the kinetic equation (2.17) and obtain:

$$p(l) = \frac{\alpha(l)}{\alpha(l) + \beta(l)} \ . \tag{2.19}$$

Divide by $\alpha(l)$ and arrive at

$$p(l) = \left[1 + \frac{\beta(l)}{\alpha(l)}\right]^{-1} \ . \tag{2.20}$$

It is seen that the behavior of the system is controlled by the ratio $\beta(l)/\alpha(l)$. It was assumed that $\beta(l) \sim c^l$; hence three cases of $\alpha(l)$ are possible.

When $\alpha(l)$ approaches zero more rapidly than c^l, the density of defects $p(l)$ goes to zero and the density of events $q(l)$ also tends to zero. This kind of behavior is called stability, as in the static case.

When $\alpha(l)$ tends to zero like c^l, the density of defects $p(l)$ approaches a nontrivial constant value, $p(l) \to p_0$. The magnitude–frequency relation calculated for defects is then linear with the slope equal to unity, so this is a case of scale invariance. In contrast to the static case, the density of events goes to zero like c^l. Thus the magnitude–frequency relation for events is also linear, but the slope depends on the scaling parameter of healing: $b = 1 - \log_{10} c / \log_{10} n$.

When $\alpha(l)$ approaches a nonzero constant value or goes to zero more slowly than c^l, the density of defects $p(l)$ approaches unity, and the density of events $q(l)$ tends to zero like c^l. The magnitude–frequency relation is then linear with the slope $b = 1 - \log_{10} c / \log_{10} n$. Thus, catastrophic behavior for a dynamic model is critical.

As in the static case, the stationary solution of the dynamic model demonstrates the phase transition from stability to catastrophe for all critical numbers $1 < k < n$. The critical point $\alpha_0 = \alpha_0^{cr}$ depends on the critical number k and the parameters of healing. The magnitude–frequency relation is linear when $\alpha_0 \geq \alpha_0^{cr}$ and has a downward bend otherwise. One observes at the critical point $\alpha_0 = \alpha_0^{cr}$ both the transition from stability to the catastrophe and from noncritical to critical behavior.

Let us point out that in contrast to the static case, the slope of the magnitude–frequency relation at the critical point is not equal to unity. The density of events $q(l)$ for $\alpha_0 \geq \alpha_0^{cr}$ decreases with level like c^l; therefore the slope of the magnitude–frequency relation is $b = 1 - \log c / \log n$. Scale invariance is then possible only when $c = 1$. It is seen that the scaling properties of a dynamic system depend on the scaling properties of the healing process and the scale invariance of events is impossible without the scale invariance of the healing process.

Another difference from a static model is observed in the special case $k = n$. For $k = n$, a static model displays stability behavior alone, in contrast to a dynamic model that allows a catastrophe for relatively large values of the emergence probability at the lowest level α_0. Thus, a catastrophe is always possible in a dynamic model when the emergence process is stronger than healing.

2.3.3   Heterogeneity in the Dynamic Model

Let us consider a dynamic heterogeneous model which is a generalization of both the static heterogeneous model described in Sect. 2.2.3 and the dynamic homogeneous model described in Sects.2.3.1–2.3.2. For simplicity, hierarchy with branching number $n = 3$ is assumed.

Heterogeneity. The critical number k in the homogeneous dynamic model considered above was assumed to be the same for all elements of the system. In line with the static heterogeneous model (Sect. 2.2.3), it is assumed that there are three kinds of elements in the system (recall that $n = 3$); a critical configuration for kth kind of elements contains k or more defects in the associate group at the lower level. Different kinds of elements are mixed with concentrations a_k, where $a_1 + a_2 + a_3 = 1$. The self-similarity of the mixture is assumed; therefore, concentrations a_k are the same for all levels of the system. The density of defects of the kth kind from level l at time t is denoted by $p_k(l, t)$; then the total density of defects is a weighted sum of p_k:

$$p(l,t) = a_1 p_1(l,t) + a_2 p_2(l,t) + a_3 p_3(l,t). \tag{2.21}$$

Healing probability. As above, the healing probability is defined by (2.16).

Kinetic equations. Unlike a homogeneous system, the present model has three kinds of emergence probability $\alpha_k(l, t)$ assigned to each kind of element; kinetic equations are set for different p_k:

$$p_k(l, t + 1) = p_k(l,t)[1 - \beta(l)] + [1 - p_k(l,t)]\alpha_k(l,t) . \tag{2.22}$$

The second term on the right-hand side of (2.22) stands for the density of new defects, referred to as events, so that the density of events of the kth kind $q_k(l, t)$ is represented as follows:

$$q_k(l, t + 1) = [1 - p_k(l,t)]\alpha_k(l,t) . \tag{2.23}$$

The density of all events from level l at time t is a weighted sum of $q_k(l, t)$:

$$q(l,t) = a_1 q_1(l,t) + a_2 q_2(l,t) + a_3 q_3(l,t) . \tag{2.24}$$

The kinetic equation for the density of all kinds of defects has the form

$$p(l, t + 1) = [1 - p(l,t)]\alpha(l,t) + p(l,t)[1 - \beta(l)] , \tag{2.25}$$

where $\alpha(l,t)$ is, by definition, the emergence probability of new defects. The density of all takes the form

$$q(l,t) = [1 - p(l,t)]\alpha(l,t) . \tag{2.26}$$

Emergence probabilities. The value of the emergence probability $\alpha(l,t)$ obtained by substituting of (2.21)–(2.22) in (2.25) satisfies the following equation:

$$\alpha(l,t)[1 - p(l,t)] = \sum_{k=1}^{3} a_k \alpha_k(l,t)[1 - p_k(l,t)] . \tag{2.27}$$

The lowest level of the system differs from the other levels because there are no differences between elements. The emergence intensity of new defects at the lowest level is assumed to be constant, $\alpha(1,t) = \alpha_0$. The emergence probabilities of new defects $\alpha_k(l+1,t)$ at level $l+1$ are determined by the conditional probability of obtaining a new critical configuration at the previous level l. The distributions of different kinds of elements at two successive levels are assumed to be independent; therefore emergence probabilities at an upper level $\alpha_k(l+1,t)$ are calculated from the emergence probabilities $\alpha(l,t)$ and the density of defects $p(l,t)$ at the lower level l:

$$
\begin{aligned}
&\alpha_1(l+1,t) \\
&= \frac{[1-p(l,t)]^3 \left\{\alpha^3(l,t) + 3\alpha^2(l,t)[1-\alpha(l,t)] + 3\alpha(l,t)[1-\alpha(l,t)]^2\right\}}{1-p^3(l,t)} \\
&\quad + \frac{3p(l,t)[1-p(l,t)]^2 \left\{\alpha^2(l,t) + 2\alpha(l,t)[1-\alpha(l,t)]\right\}}{1-p^3(l,t)} \\
&\quad + \frac{3\alpha(l,t)p^2(l,t)[1-p(l,t)]}{1-p^3(l,t)} .
\end{aligned}
\tag{2.28}
$$

$$
\begin{aligned}
\alpha_2(l+1,t) &= \frac{[1-p(l,t)]^3 \left\{\alpha^3(l,t) + 3\alpha^2(l,t)[1-\alpha(l,t)]\right\}}{1-p^3(l,t)} + \\
&\quad + \frac{3p(l,t)[1-p(l,t)]^2 \left\{\alpha^2(l,t) + 2\alpha(l,t)[1-\alpha(l,t)]\right\}}{1-p^3(l,t)} \\
&\quad + \frac{3\alpha(l,t)p^2(l,t)[1-p(l,t)]}{1-p^3(l,t)} .
\end{aligned}
\tag{2.29}
$$

$$
\begin{aligned}
\alpha_3(l+1,t) &= \frac{[1-p(l,t)]^3\alpha^3(l,t) + 3p(l,t)[1-p(l,t)]^2\alpha^2(l,t)}{1-p^3(l,t)} \\
&\quad + \frac{3\alpha(l,t)p^2(l,t)[1-p(l,t)]}{1-p^3(l,t)} .
\end{aligned}
\tag{2.30}
$$

When time t goes to infinity, the heterogeneous model has the stationary solution:

$$
\lim_{t\to\infty} \alpha_k(l,t) = \alpha_k(l) ,
$$

$$
\lim_{t\to\infty} \alpha(l,t) = \alpha(l) ,
$$

$$
\lim_{t\to\infty} p_k(l,t) = p_k(l) ,
$$

$$
\lim_{t\to\infty} p(l,t) = p(l) ,
$$

$$\lim_{t\to\infty} q_k(l,t) = q_k(l) \,,$$

$$\lim_{t\to\infty} q(l,t) = q(l) \,.$$

The scaling properties of the stationary solution are treated for different concentrations of the mixture a_k. Analogously to a homogeneous dynamic model, the heterogeneous model always demonstrates catastrophic behavior when the emergence process is stronger than healing; it was analytically proved for the heterogeneous model [SB01]. The critical point α_{cr} is the boundary for the catastrophe setting and depends on mixture concentrations a_k and the parameters of healing. Depending on the mixture concentrations, three kinds of behavior are realized when α_0 runs from zero to unity: phase transition from stability to catastrophe; phase transition from scale invariance to catastrophe, and the catastrophe itself. Now we consider domains of stability, scale invariance, and catastrophe to describe the scaling properties of events in each domain.

Domain of scale invariance. In the domain of scale invariance, the density of defects $p(l)$ approaches a constant value different from zero and unity, and the emergence intensity $\alpha(l)$ approaches zero like c^l. Then, (2.26) means that the density of events $q(l)$ also approaches zero like c^l. Substitute this approximation in the magnitude–frequency relation (2.11), and obtain the slope of the magnitude–frequency relation in the domain of scale invariance:

$$b = 1 - \frac{\log_{10} c}{\log_{10} 3} \,. \tag{2.31}$$

This slope is greater than unity when $c < 1$, less than unity when $c > 1$, and equal to unity only when $c = 1$. Thus the unity slope of the magnitude–frequency relation means that the scale invariance of destruction $[p(l) = \text{const}]$ is complemented by the scale invariance of healing $\beta(l) = \text{const}$.

The catastrophe domain. In the domain of catastrophe, the emergence intensity $\alpha(l)$ approaches a constant value when level l increases. It follows from (2.19) that the density of unbroken elements has the following form:

$$1 - p(l) = \frac{\beta(l)}{\alpha(l) + \beta(l)} \,.$$

The healing probability $\beta(l)$ approaches zero like c^l; therefore the density of unbroken elements $(1 - p(l))$ goes to zero like c^l. It is seen from (2.26) that the density of events $q(l)$ is a product of the emergence probability $\alpha(l)$ and the density of unbroken elements $(1 - p(l))$. The emergence probability $\alpha(l)$ approaches a constant value, the density of unbroken elements $(1 - p(l))$ goes to zero like c^l, and therefore, the density of events $q(l)$ also tends to zero like c^l. This is similar to the behavior in the domain of scale invariance; therefore, one obtains the same slope for the magnitude–frequency relation (2.31).

The domain of stability. Let us consider (2.28)–(2.30) for $t \to \infty$ under conditions $p \to 0$ and $\alpha \to 0$. Up to the first order of smallness, emergence intensities $\alpha_k(l)$ take the following form:

$$\alpha_1(l) = 3\alpha(l) ,$$
$$\alpha_2(l) = 0 , \qquad\qquad (2.32)$$
$$\alpha_3(l) = 0 .$$

Substitute $\alpha_k(l)$ in (2.27) and obtain

$$\alpha(l+1)[1 - p(l+1)] = 3a_1\alpha(l)[1 - p_1(l+1)] . \qquad (2.33)$$

The density of defects $p_1(l)$ can be expressed from $\alpha_1(l)$ and $\beta(l)$; then

$$1 - p_1(l+1) = \frac{\beta(l+1)}{\alpha_1(l+1) + \beta(l+1)} . \qquad (2.34)$$

Substitute (2.34) into (2.33), use (2.32), and obtain

$$\alpha(l+1)[1 - p(l+1)] = \frac{3a_1\alpha(l)[1 - p(l)]c[\alpha(l) + \beta(l)]}{3\alpha(l) + \beta(l)c} . \qquad (2.35)$$

The densities of events are defined by (2.23); hence $\alpha(l)(1 - p(l))$ can be replaced by $q(l)$. Substitute $q(l)$, divide by $\beta(l)$, and obtain from (2.35) the form

$$\frac{q(l+1)}{q(l)} = \frac{3a_1c[\varepsilon(l) + 1]}{3\varepsilon(l) + c} ,$$

where $\varepsilon = \alpha(l)/\beta(l)$ approaches zero in the domain of stability. This results in the approximation

$$\frac{q(l+1)}{q(l)} = 3a_1 .$$

Consequently, the density of events $q(l)$ increases with level as $(3a_1)^l$. Substitute this approximation in the magnitude–frequency relation (2.11) and obtain the slope:

$$b = -\frac{\log_{10} a_1}{\log_{10} 3} .$$

To compare this slope with the slope of the magnitude–frequency relation in the domain of scale invariance, it is necessary to estimate the limit con-

centration a_1 in the domain of stability. Let us return to (2.33). Substitute
the form of $1 - p_1(l + 1)$ given by (2.34) and the similar form of $1 - p(l + 1)$
in (2.33),and obtain the following equation after some transformations:

$$\frac{3a_1}{c} = \frac{\varepsilon(l + 1)}{\varepsilon(l)} \, .$$

The right-hand side is less than unity; thus, the desired condition for mixture
concentrations in the stability domain takes the form

$$a_1 < \frac{c}{3} \, . \tag{2.36}$$

Let us compare the slopes of the magnitude–frequency relation in the domains
of stability, scale invariance, and catastrophe. Owing to the condition (2.36),
the slope in the stability domain always exceeds the slope in the domains of
scale invariance and catastrophe. The same result was obtained for the static
heterogeneous model (see Sect. 2.2.3).

In terms of the magnitude–frequency relation, scale invariance is not
the same for the dynamic and static models; the b-value is not equal to
unity for the dynamic case and depends on the scaling parameter c of the
healing process. However, the magnitude–frequency relation in this domain
has a constant slope independent of concentrations a_k.The characteristics of
heterogeneity contribute to the b-value in the domain of stability and have no
influence on the scaling properties of events in the domains of scale invariance
and catastrophe. Similarly to the homogeneous case, the healing properties
of the system are always reflected in the slope of the magnitude–frequency
relation.

2.3.4   The Feedback Relation and the Evolution of Scaling Properties

The previous section treated the influence of heterogeneity on the generation
of stable criticality reflected in the linear form of the magnitude–frequency
relation. Another way to stabilize the critical behavior is by introducing
the feedback relation in the system. A dynamic hierarchical model with the
feedback relation was suggested in [NS90]. The model described below is
similar to that in [NS90]; it has several simplifications, but the features of its
behavior are qualitatively the same.

Let us consider the dynamic homogeneous model described in Sect. 2.3.1
with branching number $n = 3$ and critical number $k = 2$ and assume that
the emergence probability at the lowest level α_0 is not constant any longer,
but depends on the energy (or stress) function as follows:

$$\alpha_0(t) = 1 - \exp\left[-\kappa E(t)\right] \, .$$

The evolution of the energy function $E(t + 1)$ is controlled by the constant input flow ΔE and by the dissipation $R(t)$:

$$E(t + 1) = E(t) \exp \left[\frac{\Delta E - \lambda R(t)}{E(t)} \right] . \tag{2.37}$$

The dissipation is assumed to be the weighted sum of events at all levels of the system:

$$R(t) = \sum_{l=1}^{L} q(l, t) h^l , \tag{2.38}$$

where $h > 1$ is chosen to provide the weighted contribution of events at different scales in dissipation R. It is assumed that the dissipation from each event is determined by its volume $V(l)$. It follows from (2.7) that the volume of events at level l is determined by the branching number of the system as follows:

$$V(l) = n^{3l/2} . \tag{2.39}$$

The density of events $q(l, t)$ takes the form

$$q(l, t) = \frac{N(l, t)}{N_e(l)} ,$$

where $N(l, t)$ denotes the number of events at level l that emerged at time t and $N_e(l)$ is the total number of elements at level l. It follows from (2.10) that $N_e \sim n^l$. The dissipation from events at level l equals $N(l, t)V(l)$. It enters in (2.38) as $q(l, t)h^l$, and the result is as follows:

$$q(l, t)h^l = N(l, t)V(l) .$$

Substitute here the form of $q(l, t)$ and the asymptotics for $V(l)$ to obtain the estimation of h as $h = \sqrt{n}$. For the case considered here, $n = 3$, and obviously, $h = 1.73$.

Evidently, the kinetics of defect generation is essentially controlled by random fluctuations of density and intensity about their mean values. Thus, to perform modeling, it is necessary to randomize the initial equations by changing $\alpha(t)$ in (2.17) through random realization of the Poisson distribution with corresponding rates.

The number of levels $L = 20$ was chosen for modeling. Initial conditions were

$$p(l, 0) = 0 \quad \text{for all } l ,$$

and dimensionless time was measured in the number of steps.

When the parameters of the feedback relation, λ and ΔE, are properly chosen, the emergence probability α_0 determined by (2.37) increases in the domain of stability (no events are observed at high levels of the system) and decreases in the domain of catastrophe (events emerge at high levels of the system); therefore the system always returns to its critical state. This difference between energy dissipation inside the domain of stability and the domain of catastrophe is controlled by h, which ensures that the principal contribution to the dissipation is determined by high scale events. This model with the feedback relation is called an FB model below.

The magnitude–frequency relation averaged over a long time period T demonstrates linear behavior (Fig. 2.5a, curve 1). However, the magnitude–frequency relation estimated over a short time period can have a strong downward bend (Fig. 2.5a, curves 2 and 3) but can also be linear (Fig. 2.5a, curve 4). The average magnitude of events calculated in a sliding time window oscillates with time (Fig. 2.5b). Thus, the magnitude–frequency relation for the feedback model demonstrates strong temporal variation, which may be used for predicting of strong events in the model.

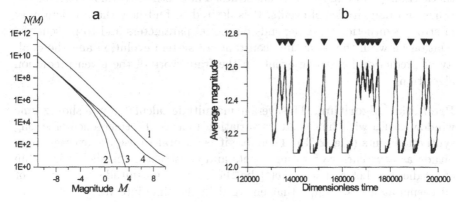

Fig. 2.5. magnitude–frequency relation for the dynamic model with the feedback relation ($n = 3$, $k_0 = 2$): (a) the magnitude–frequency relation for a long time interval $T_1 = 70000$ (curve 1) and short time intervals $T_i = 2000$ ($i = 2, 3, 4$) (curves 2–4); (b) temporal evolution of the average magnitude (solid line) and strong events of 18–20 levels (triangles)

For the homogeneous dynamic hierarchical model, the magnitude–frequency relation has a downward bend in the domain of stability (see Sect. 2.3.2). This bend decreases when the system approaches its critical state. When the system tends to the threshold of catastrophe ($\alpha_0 \rightarrow \alpha_0^{cr}$,) the emergence probability of high level events increases. On the other hand, dissipation begins to work only when the system reaches its critical state. These reasons allow us to construct the precursor of a strong event: When the downward bend of the magnitude–frequency relation decreases, the probability of a strong event increases. This precursor was formalized in the algorithm of prediction for the dynamic model with the feedback relation [NS90] and very good predictive quality for strong events was obtained in a particular case of model parameters [NS94]. Tests of this algorithm for real earthquakes in several seismoactive regions gives a predictability value very far from random prediction but certainly worse than the model result. It was observed that different seismoactive regions yield a different predictability of strong earthquakes for this algorithm of prediction [Nar]. The variation of the predictability of strong events in different regions was also observed when a more complicated algorithm of earthquake prediction was used [Rot]. The section below treats the variation in the predictability for strong events and analyzes its origin for the FB model.

2.3.5   Prediction and Predictability of Strong Events

This section describes a very simple algorithm for predicting strong events in the FB model considered above. In order to estimate the predictability of strong events, it is natural to randomize the equations and generate a synthetic catalog of events on different scales. The synthetic catalog contains the times and magnitudes of events. It is desired to find how the predictability of strong synthetic events depends on model parameters and to present the relation between the basic time constants of system evolution and the quality of prediction for strong events in the framework of the given prediction algorithm.

Prediction algorithm. The average magnitude calculated in a sliding time window varies with time, and its variation reflects the emergence of strong synthetic events (Fig. 2.5b). Therefore it is natural to use the average magnitude as a precursor of strong events and construct a threshold algorithm of prediction. Let a function $b(t)$ be the average magnitude calculated for all events at levels $l \geq l_{min}$ that emerged in the time interval $[t - \Delta, t - 1]$. If the functional $b(t)$ exceeds a threshold value η, then an *alarm* is declared for the time t (Fig. 2.6a). Events of several highest levels $L_0 \leq l \leq L$ are considered as *strong events* to be predicted. If an alarm is declared at the time of a strong event, this event is considered a *success*. If there is no alarm at the time of a strong event, this event is considered a *failure*.

Quality of prediction. Divide the number of failures by the total number of strong events, and denote the result by n_0. The ratio of the total alarm time to the total time period of prediction called the alarm time fraction is denoted by τ. The n_0 versus τ diagram (Fig. 2.6b) is used as a characteristic of prediction quality [Mol90]. For random prediction, this diagram is close to the diagonal line (the measure of successes is equal to the measure of alarms); a deviation from the diagonal line shows the reliability of prediction. Add the fraction of failures n_0 to the relative alarm time fraction τ and obtain the sum $\varphi = n_0 + \tau$ called the *total error* of prediction. It depends on the threshold

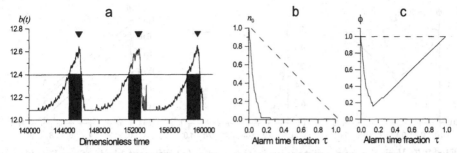

Fig. 2.6. The threshold algorithm for predicting strong events in the model: (**a**) declaring alarms (filled rectangles) for a given threshold (*dashed line*), filled triangles stand for strong events; (**b**) the diagram of n_0 versus τ, *dashed line* depicts random prediction; (**c**) the total error φ as a function of alarm time fraction τ, *dashed line* depicts random prediction

of prediction η. This threshold has a one-to-one correspondence with the alarm time fraction τ; therefore, τ is considered a probabilistic threshold of prediction. The total error in random prediction is close to unity for all thresholds, therefore the deviation of $\varphi(\tau)$ from the unity line reflects the quality of prediction (Fig. 2.6c). The threshold yielding the lowest total error is called the *optimal threshold*. The total error for the optimal threshold also shows the quality of prediction. The value φ_{min} is considered below as a predictability measure of strong events for different values of system parameters. The prediction quality for strong events is estimated as a function of model parameters by using the same algorithm of prediction. This approach allows us to recognize system characteristics that can influence the predictability of strong events. Let us now consider how different system parameters affect predictability in the model.

Variation of predictability. The predictability of strong events in the model changes with its parameters. It can be quite good (better than for any real seismicity, Fig. 2.7a, curve 1) or completely random ($\varphi \approx 1$ Fig. 2.7a, curve 3). The predictability of strong synthetic events is determined by the temporal evolution of the precursor (the average magnitude for our prediction algorithm); changes in the range of strong events (making L_0 smaller or larger) do not lead to significant changes in the result (Fig. 2.7b, curves 1 and 2). There is no evident correlation between the predictability of strong events and the slope of the magnitude–frequency relation calculated over the whole time interval; the same predictability value can correspond to different slopes, and the same slope in the magnitude–frequency relation may correspond to different predictability of strong events. Figure 2.7c illustrates the absence of a monotonic relation between the slope of the magnitude–frequency plot and the predictability of strong events.

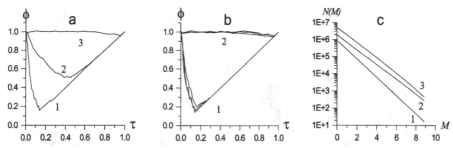

Fig. 2.7. Variation of predictability in the model: (**a**) different kinds of predictability (good, curve 1, moderate, curve 2, and poor, curve 3); (**b**) the influence of the range representing strong events, $L_0 = 17, 18, 19$, on the predictability for good (curve 1) and poor (curve 2) predictability; (**c**) the magnitude–frequency relation for different predictability (good, curve 1; moderate, curve 2; and poor, curve 3)

Predictability and the rate of energy. Let us consider the dynamics of predictability depending on the energy rate parameter ΔE when all other parameters are fixed. The minimum error φ depends on ΔE in a nonmonotonic way (as shown in Fig. 2.8a), indicating a range of parameter values with good predictability ($\varphi < 0.2$). When the energy rate is outside this range, the predictability becomes worse. Let us note that the domains of small and large ΔE values are not similar; when ΔE approaches ΔE_{max}, predictability rapidly becomes random and has relatively small oscillations; when ΔE decreases tending to ΔE_{min}, the prediction error φ slowly increases and has oscillations of very high amplitudes. This means that for small energy rates ΔE the estimation of predictability is less stable under the condition that the time interval for modeling remains the same.

Predictability and energy dissipation. Let us turn to changing the energy dissipation parameter λ (all other parameters are fixed). The nonmonotonic relation between the predictability of strong events and λ is inverted with respect to $\varphi(\Delta E)$ (Fig. 2.8b). Small values of energy dissipation and high values of energy rate have similar effects on the predictability of strong events.

Predictability and healing. Let us study the relation between the predictability and the healing properties of the system. Healing is determined by two parameters: β_0 and c. Let us fix all parameters of the system including the scaling c and inquire how the prediction quality changes when β_0 increases. There is a nonmonotonic relation between β_0 and the error of prediction φ for strong events (Fig. 2.8c) with two minima. Considering $\varphi(\Delta E)$ and $\varphi(\lambda)$ for different values of β_0 it is possible to conclude that minimum values $\min(\varphi(\Delta E))$ and $\min(\varphi(\lambda))$ depend on healing (Fig. 2.8d). It follows that the

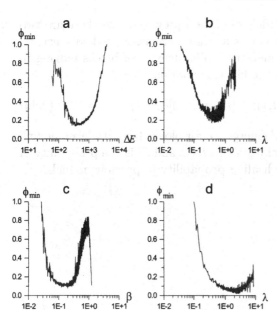

Fig. 2.8. The relation between the optimal total error and the basic parameters of modeling: (a) energy rate ΔE; (b) energy drop λ for $\beta_0 = 0.4$; (c) healing parameter β_0; and (d) energy drop λ for $\beta_0 = 0.2$

predictability of large synthetic events strongly depends on the parameters of the healing process.

2.4   Complex Hierarchical Model

It was shown above how the scaling properties of seismicity can be generated in simple hierarchical models. For simplicity, the description was restricted to dynamic models based on the idea of heterogeneity or the feedback relation. This simplified approach allows us to clarify the origin of features observed in the model. However, when it is desired to generate a synthetic catalog where several seismic properties are presented, it is possible to introduce both the heterogeneity and the feedback relation into one model. Such model constructed in [BSLA97] (referred to as a DSOC model) generates synthetic catalogs realizing different seismic properties. The basic principles of the model are described below together with results obtained from modeling. A detailed description of the model was presented in [BSLA97], and its simplified version was given in [BSL99, BSL98].

2.4.1   Description of the Model

Consider a hierarchical system of blocks having the following structure. Four blocks at a current level compose a block at the succeeding level. Each block can move in two orthogonal directions e_1 and e_2 (Fig. 2.9). The state of the system at each time is defined by the densities of blocks moving in two orthogonal directions, $p_1(l, t)$ and $p_2(l, t)$. Moving blocks in this model stand for defects in the hierarchical models described previously, so there are two kinds of defects; blocks starting to move at time t are associated with events, and two kinds of events are distinguished. The density of blocks moving in the ith direction is controlled by the kinetic equation

$$p_i(l, t+1) = p_i(l,t)[1 - \beta_i(l,t)] + [1 - p_i(l,t)]\alpha_i(l,t) , \quad i = 1, 2 , \quad (2.40)$$

where $\alpha_i(l,t)$ is the probability of emergence for blocks from lth level moving in direction e_i in the time interval from t to $t+1$ and β_i is the probability of damping, which is similar to the healing probability in previous models.

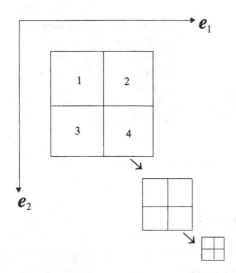

Fig. 2.9. Hierarchical system of blocks with branching number $n = 4$. Blocks can move along two orthogonal directions e_1 and e_2

The feedback relation for the emergence probabilities at the first level $\alpha_i(1,t)$ takes the form

$$\alpha_i(1,t) = 1 - \exp[-kE_i(t)] , \quad i = 1, 2 .$$

Two energy function, $E_1(t)$ and $E_2(t)$, enter this equation. Their evolution is controlled by two competing factors: dissipation resulting from all movements (defects) in the system and energy rate at time t:

$$E_1(t+1) = E_1(t) \exp\left[\frac{\Delta E_1 - \lambda_{11} R_1(t)}{E_1(t)}\right] \exp\left[\frac{-\lambda_{12} R_2(t)}{E_2(t)}\right] ,$$

$$E_2(t+1) = E_2(t) \exp\left[\frac{\Delta E_2 - \lambda_{22} R_2(t)}{E_2(t)}\right] \exp\left[\frac{-\lambda_{21} R_1(t)}{E_1(t)}\right] . \quad (2.41)$$

The second exponential term on the right-hand side of (2.41), which distinguishes it from (2.37), follows from the interaction of movements in orthogonal directions. This influence may be negative (when $\lambda_{ij} > 0$) or positive (when $\lambda_{ij} < 0$). The dissipation functions $R_i(t)$ have a form similar to (2.38):

$$R_i(t) = \sum_{i=1}^{L} \widetilde{p}_i(l,t)h^l \,,$$

where $\widetilde{p}_i(l,t)$ is the weighted sum of the densities of new movements (events) $q_i(l,t)$ and old movements, whose density is equal to $p_i(l,t) - q_i(l,t)$. The difference between the contributions of old and new defects is determined by parameter γ:

$$\widetilde{p}_i(l,t) = q_i(l,t) + \gamma[p_i(l,t) - q_i(l,t)] \,. \tag{2.42}$$

A block at level $l+1$ starts moving in direction i, when the critical number of blocks moving in direction i ($i = 1,2$) appears in the associate group at the lower level l. The rule determining the transfer of motion in direction i from low to high levels of the system (the inverse cascade) is expressed in a general form:

$$\alpha_i(l+1,t) = F_1[\alpha_i(l,t), p_i(l,t)]Q_1[\widetilde{p}_j(l,t)]$$
$$+F_2[\alpha_i(l,t), p_i(l,t)]Q_2[\widetilde{p}_j(l,t)],$$

where $i \neq j$, F_1 is determined by the probability of critical configurations with 3 and 4 moving blocks, and F_2 is determined by the probability of critical configurations with 1 and 2 moving blocks. Therefore, this model can be recognized as a particular case of the dynamic mixture model, where mixture concentrations a_k are not constant and depend on the densities of blocks at the same level moving in orthogonal directions. Two kinds of interaction between blocks moving in direction i and of mixture concentrations along direction j ($i \neq j$) are considered. First, the motion of a block in one of the directions can be prevented by neighboring blocks moving in the other direction. For example (Fig. 2.9), block 1 moving in direction e_1 presses block 2 and prevents its movement in direction e_2. Second, the interaction of motions along direction e_1 may also increase emergence probabilities in direction e_2. For example (Fig. 2.9), the movement of block 4 in direction e_1 reduces compression between blocks 3 and 4. Mixture concentrations in direction i at level l are determined as nonlinear functions of $\widetilde{p}_j(l,t)$ ($j \neq i$).

In contrast to previous models, damping (healing) probabilities β_i depend on time. The probability of damping is controlled by friction between blocks. Friction between a block moving along e_1 and its neighbors is enhanced by a motion along e_2 through a corresponding increase of compression. Friction damping the motion along e_1 is also likely to grow with the motion density along this same direction (due ,for example, to asperities in faults). Accordingly, $\beta_i(l,t)$ is supposed to depend on both $\widetilde{p}_i(l,t)$ and $\widetilde{p}_j(l,t)$ ($i \neq j$) which is defined by (2.42). As above, the damping probabilities are supposed to decrease with level l like c^l. The probability of damping at level l is thus defined as follows:

$$\beta_1(l,t) = [c_0 + c_1 \widetilde{p}_2(l,t) + c_2 \widetilde{p}_1(l,t)] \, c^l \,,$$
$$\beta_2(l,t) = [c_0 + c_1 \widetilde{p}_1(l,t) + c_2 \widetilde{p}_2(l,t)] \, c^l \,. \tag{2.43}$$

Let us note that (2.43) controls the temporal evolution of the healing probability and therefore allows us to obtain the variation of the predictability for strong events because it strongly depends on healing (see Sect. 2.3.5).

As in Sect. 2.3.4, it is necessary to randomize (2.40). The number of levels $L = 16$ was assumed in the modeling. Initial conditions $p(l,0) = 0$ were prescribed at all levels l. Dimensionless time was again measured in the number of steps.

2.4.2 Seismic Patterns in the Model

Seismic cycle. For a set of system parameters, the model demonstrates strong heterogeneity in the temporal distribution of events (Fig. 2.10a). Intervals of high and weak model activity alternate. Similar behavior was observed by Mogi [Mog74] for the seismicity of the Pacific belt as the alternation of seismic activation with seismic quiescence and was called a seismic cycle.

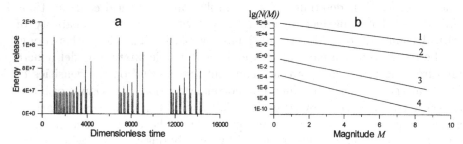

Fig. 2.10. Temporal variation of activity in the DSOC model: (a) seismic cycle in terms of Mogi, intervals with high and low activity alternate; (b) magnitude–frequency relation for different activity intervals: all events for a long time interval [2000,9000], the slope is 0.98 (curve 1); an interval after a strong event [4500,6300] yields different slopes for events corresponding to different directions, the slopes are 0.9 (curve 2) and 1.6 (curve 3); all events for a quiescence interval [5500,6300], the slope is 1.93 (curve 4)

Dynamic self-organized criticality (DSOC). When averaged over a long time interval, the magnitude–frequency relation is linear with a slope determined by the healing properties of the system. Short intervals of averaging also result in a linear magnitude–frequency relation, but its slope varies depending on the parameters of heterogeneity related to the synthetic activity (Fig. 2.10b).

Foreshock and aftershock activity. Some strong events in the model are preceded by a growth of energy release in the system, and some strong events are followed by exceptionally high activity decreasing with time. The energy release is determined by the dissipation function $R(t)$ representing the weighted sum of all events that emerged in the system at time t (see Eq. 2.38). The dissipation function decreases linearly at both sides of its extremum on a log/log plot (Fig. 2.11a and b). This behavior is reminiscent of what is observed for foreshock and aftershock activity. The slope of the fitting straight line observed in reality is 2 in the case of foreshocks and about 1 in the case of aftershocks; the corresponding slopes in the model (in Fig. 2.10, they take the values 2.47 and 0.9, respectively) are quite reasonable. As in real seismicity, the synthetic foreshock activity does not precede all events. The ratio of strong events preceded by foreshocks depends on system parameters.

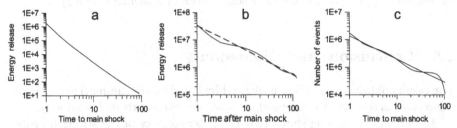

Fig. 2.11. Foreshock and aftershock activity in the model. The power law decrease of the energy release for the foreshock (**a**) and aftershock (**b**) activity; the Omori law for the decreasing number of aftershocks (**c**). Best fits are plotted by *dashed lines*. Slopes of the best fits are as follows: 2.47 (**a**), 0.9 (**b**), and 0.84 (**c**)

The Omori law. The Omori law for aftershocks was checked for synthetic events at the highest level L followed by aftershock activity. All events of levels $l < L$ that emerged after the main shock are called aftershocks, and N_a denotes their number. The decay of N_a is linear on a log/log plot (Fig. 2.11c). The slope of the fitting straight line varies, depending on model parameters, from 0.8 to 1.2. These values agree with those observed for the slope of the Omori law for aftershock activity after strong earthquakes [Uts61].

Variation of predictability. The predictability of strong events in the DSOC model varies depending on system parameters. A similar result is true for the FB model, and its origin is clear. However, the DSOC model also displays strong temporal variations in predictability for large events when the parameters of modeling and the algorithm of prediction are fixed (Fig. 2.12) [BSL99]. The range of temporal variability of the predictability for strong events depends on the healing parameters of the model and the rate of the energy functions.

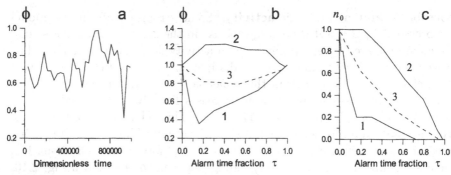

Fig. 2.12. Temporal variation of predictability in the DSOC model: (**a**) evolution of the minimum total error, the prediction is performed in a sliding time window; (**b**) the total error of prediction; and (**c**) the n_0 versus τ diagram. Plots are obtained for three intervals: the interval with the lowest $\varphi_{\min}$ (curves 1), the interval with the highest $\varphi_{\min}$ (curves 2), and the whole interval of modeling (curve 3)

2.5   Conclusions and Discussion

The results obtained for different hierarchical models are summarized to draw some conclusions and to discuss their possible relation to real seismicity.

Hierarchical modeling shows that the heterogeneity of the strength properties of the media can be a source of the linear magnitude–frequency relation. The slope of the Gutenberg–Richter plot reflects the parameters of heterogeneity when heterogeneity is not too strong; however, it becomes invariant in the domain of high heterogeneity. It is well known that the lithosphere of Earth must be considered only a heterogeneous medium, so the linear magnitude–frequency relation of seismicity can be naturally related to this feature. Scaling properties of seismicity that are reflected in the slope of the Gutenberg–Richter plot can be related to the parameters of heterogeneity, and this relation must be taken into account when comparing different seismic regions. The degree of heterogeneity can have a particular effect on the application of the same prediction algorithm for different seismic regions; the first attempt to describe the difference in seismic features depending on heterogeneity in the framework of earthquake prediction was undertaken in [BS99].

It is established that the predictability of strong events for hierarchical models with the feedback relation strongly depends on time constants (controlled in the FB model by the energy rate ΔE and the dissipation parameter λ). The whole spectrum of predictability is displayed when these time constants run over their defined domains. Good predictability exists only within some intervals of values $[\Delta E_{\min}, \Delta E_{\max}]$ and $[\lambda_{\min}, \lambda_{\max}]$ and decreases when the rate of energy and dissipation become too large or too small. A similar argument can explain the variation of earthquake predictability observed in different seismic regions. Predictability can be estimated when all evolutionary parameters for the system are known; however, the uncertainty of this estimation also depends on time constants and may be significant.

The healing process appears to be very important for the problem of predictability. The best possible predictability always depends on healing even when all other parameters are optimized. The change in the healing process with time allows us to obtain temporal variations of predictability. Thus, the estimation of healing constants is very important for assessing possible predictability and its reliability in the system. On the other hand, healing constants are reflected in the slope of the magnitude–frequency relation; analyzing by scaling properties of the seismic process, it is possible to obtain information on healing that strongly influences earthquake prediction.

The models described above have fixed evolutionary parameters in the majority of cases. The parameters of observed seismicity certainly vary depending on the seismic situation and the evolution of stress. The system dynamics for varying parameters appears to be much more complicated and presents special seismic features such as foreshock and aftershock activity, the Omori law, and temporal variations of predictability. In terms of earthquake prediction, the evolution of system parameters can change the efficiency of precursors and the best possible predictability of strong events. It follows that the problem of "prediction of predictability" for a given time interval deserves attention.

The next step in developing hierarchical models of seismicity consists of generating synthetic catalogs with events that have spatial coordinates in addition to temporal and magnitude characteristics. Then it will be possible to check predictability using spatial prediction algorithms, as done above for the average magnitude precursor. Considering hierarchical models, one can easily understand which properties of real seismicity are general for a given class of systems and which kinds of results are reliable, even if it is impossible to check them because of the small volume of observed data.

3 Models of Dynamics
of Block-and-Fault Systems

A. Soloviev and A. Ismail-Zadeh

A model of block-and-fault dynamics (block model for short) of the lithosphere was developed to analyze how the basic features of seismicity depend on the lithosphere structure and dynamics and to study the specific features of this dependence. A seismic region is modeled by a system of perfectly rigid blocks divided by infinitely thin plane faults.

We analyze major results obtained by numerical modeling of block structure dynamics and discuss the possibilities of reconstructing tectonic driving forces from the spatial distribution of seismicity, the clustering of earthquakes in the model, and the dependence of the occurrence of large earthquakes on fragmentation of the media and on rotation of blocks. These results show that modeling of block structure dynamics is a useful tool in studying relations that associate block movements and the geometry of faults with earthquake flow, including premonitory seismicity patterns, in testing existing earthquake prediction algorithms, and in developing new ones.

3.1 Introduction

Seismic observations show that features of seismic flow are different for different active regions (e.g., [Hat74, Kro84]). It is reasonable to suggest that this difference arises, among other factors, from contrasts in regional tectonic structures and in main tectonic movements that govern lithosphere dynamics in the regions. Laboratory studies show that this difference is controlled mainly by the rate of fracturing and the heterogeneity of the medium and also by the type of predominant tectonic movements [Mog62, SBV⁺80, SBB83].

It is difficult to detect the impact of a single factor on the features of a seismic flow by analyzing seismic observations because seismic flow is impacted by an assemblage of factors some of which could be more significant than that under consideration. It is also difficult if not impossible to single out the impact of an isolated factor by using seismic observations. This difficulty may be resolved by numerical modeling of processes that generate seismicity and by studying the synthetic earthquake catalogs obtained (e.g., [SCL97, GN94, AMCN95, NTG95, Tur97]).

One more reason to use models is that the statistical and phenomenological study of seismicity based on observed earthquake catalogs has the disadvantage that the reliable data cover, in general, a time interval of about 100 years or even less. This time interval is very short in comparison with the duration of tectonic processes responsible for seismic activity; therefore the patterns of earthquake occurrence identifiable in an observed catalog may be only apparent and may not repeat in the future. On the other hand, synthetic catalogs obtained by numerical modeling of the seismogenic process may cover very long time intervals thus allowing us to acquire a more reliable estimation of seismic flow parameters.

Mathematical models of lithosphere dynamics are also tools for studying earthquake preparation processes and are useful in earthquake prediction studies [GN94]. An adequate model should indicate the physical basis of premonitory patterns determined empirically prior to large events. Note that available data often do not constrain the statistical significance of premonitory patterns. The model can also be used to suggest new premonitory patterns that might exist in real catalogs.

Although there is no adequate theory of seismotectonic processes, various characteristics of the lithosphere, such as spatial heterogeneity, hierarchical block structure, different types of nonlinear rheology, gravitational and thermodynamic processes, physicochemical and phase transitions, fluid migration, and stress corrosion are probably relevant to the properties of earthquake sequences. The qualitative stability of these properties in different seismic regions suggests that the lithosphere can be modeled as a large dissipative system that does not essentially depend on the isolated details of specific processes active in a geological system.

The model of block structure dynamics exploits the hierarchical block structure of the lithosphere proposed by Alekseevskaya et al. [AGG+77]. The basic principles of the model were developed by Gabrielov et al. [GLR90]. In accordance with this model, the blocks of the lithosphere are separated by comparatively thin, weak, less consolidated fault zones, such as lineaments and tectonic faults. In seismotectonic processes, major deformation and most earthquakes occur in such fault zones.

A seismic region is modeled by a system of perfectly rigid blocks divided by infinitely thin plane faults. Displacements of all blocks is supposed to be infinitely small relative to their size. The blocks interact with each other and with the underlying medium. The system of blocks moves owing to prescribed motions of boundary blocks and the underlying medium.

Blocks are perfectly rigid; hence deformation takes place only in fault zones and at block bases in contact with the underlying medium. Relative block displacements take place along fault zones.

Strains in the model are accumulated in fault zones. This reflects strain accumulation due to deformations of plate boundaries. Naturally, considerable simplifications are made in the model, but they are necessary to understand the dependence of earthquake flow on the main tectonic movements in a region and on its structure. This assumption is justified by the fact that the effective elastic moduli in the fault zones are significantly smaller than those within the blocks.

Blocks viscoelastically interact with the underlying medium. The corresponding stresses depend on the value of relative displacement. This dependence is assumed to be linearly elastic. The motion of the medium underlying different blocks may be different. Block motion is defined so that the system is in quasi-static equilibrium.

The interaction of blocks along fault zones is viscoelastic ("normal state") while the ratio of stress to pressure remains below a certain strength level. When this ratio exceeds the critical level in some part of a fault zone, a stress drop ("failure") occurs (in accordance with the dry friction model), possibly causing failure in some parts of other fault zones. These failures produce earthquakes. An earthquake starts a period where the affected parts of fault zones are in a state of creep. This state differs from the normal state by faster growth of inelastic displacements lasting until the ratio of stress to pressure falls below some level. Numerical simulation of this process yields synthetic earthquake catalogs. The model is described in detail in Sect. 3.2. It was used to study the dependence of features of synthetic earthquake flow on structure fragmentation and the boundary movement [KRS97]. Synthetic seismicity was simulated for three groups of structures with increasing structure fragmentation inside each group and for two types of boundary movement. It is characterized by several features, including the frequency–magnitude (FM) relation (the Gutenberg–Richter plot). The results obtained show that the features of synthetic seismicity obtained by numerical simulation depend on block structure geometry and on boundary movement. When the structure fragmentation increases, the slope of the Gutenberg–Richter plot changes monotonically in the same direction for all groups of structures considered provided that the boundary movement is the same. The form of the dependence on geometry is changed drastically, when boundary movement of another type is specified and, what is more, when the dependence of seismic characteristics on structure fragmentation for the boundary movement involving rotation contradicts the views generally accepted. These results are given in Sect. 3.3. General features of synthetic earthquake flow obtained for different types of tectonic motion are also studied in [GR98].

The model of block structure dynamics was studied to seek space–time correlation between synthetic earthquakes (e.g. [GKR+97, RS98, MS99, SV99a]). Numerical experiments with a simple block structure consisting of four blocks show that there is a long-range interaction between synthetic earthquakes. It is detected by statistical analysis of synthetic earthquake catalogs. At the same time, when greater values are assumed for the strength levels of individual faults (which inhibits earthquakes there), earthquake flows on other faults are affected pronouncedly. This means that the long-range interaction found in the observed seismicity can be explained by considering that lithospheric blocks are perfectly rigid in comparison with fault zones, separating them and the underlying medium. Section 3.4 contains the results of this study.

Seismicity in most active regions of the world is caused by the interaction of continents with oceanic plates along subduction zones. Features of earthquake flow differ in different segments of these zones, and the origin of this difference is not yet clear. It is natural to relate these differences, among other factors, to the dip of the subducting slab and to the direction of the relative movement of continental and oceanic plates. The dynamics of a block structure approximating an arc-like subduction zone, which is typical of island arc regions, was modeled to single out the impact of isolated factors on synthetic seismicity [RS99]. This modeling was carried out with different dip angles of the subduction zone, different directions of motion of "a continent" and "an oceanic plate" and distribution of earthquake epicenters was studied together with other characteristics of the synthetic earthquake flow obtained. Basic relations common to different subduction zones were sought in this study. Accordingly, a simplified structure, not imitating a specific subduction zone, has been considered.

The model provides the possibility of considering block structures reflecting the fault geometry observed in active regions. The block structure approximating Sunda Arc (Sunda Isles) was studied to find the dependence of features of synthetic seismicity on the movements specified [RSV98, SRRV99]. Section 3.5 presents the results obtained for the models of arc subduction zones.

The geometry of observed faults and blocks was considered in block models of the Vrancea earthquake-prone region in Romania [PSV97, SVP99, IKS99, SVP00]. Numerical experiments provided the values of the model parameters that yielded the spatial distributions of synthetic epicenters and hypocenters close to those observed in the Vrancea region. FM plots obtained for synthetic and observed catalogs had common features. Assuming this set of parameters as a benchmark, we studied how the features of the synthetic catalog depended on model parameters. For this purpose, the parameters were varied until the following features of the initial synthetic catalog changed considerably: the spatial distribution of epicenters, the slope of the FM plot, the level of seismic activity, the maximum magnitude of events, and the relative activity of the Vrancea subduction zone with respect to the other faults considered. The source mechanism of the synthetic earthquakes was also considered. Strike and dip define the azimuth and the dip angle of the rupture plane, and slip defines the direction of the displacement in this plane. Therefore, in the model, strike and dip are prescribed by the block structure geometry and do not depend on the variation of model parameters. Thus, for synthetic earthquakes, only the dependence of slip on the model parameters was studied, and a comparison was made with observations. The effect of a sinking relic slab beneath Vrancea on intermediate-depth seismicity was also studied. The applications of the block model to the Vrancea region are considered in Sect. 3.6.

Section 3.7 contains the results obtained with the block model of the Western Alps [GS96, GKR$^+$97, VGS00]. This model was based on a morphostructural zoning scheme of the Western Alps. The basic principles of morphostructural zoning are given in Sect. 6.2.1. Several synthetic catalogs of earthquakes for the Western Alps were generated by numerical modeling. The space distribution of epicenters of synthetic earthquakes reflected some features of the observed seismicity distribution. There is a similarity of frequency–magnitude relations between the synthetic and observed seismicity. The concentration of synthetic events is found in places where no large earthquakes were reported in the catalogs, but which were previously identified as highly seismic by pattern recognition algorithms (see Sect. 6.3.2).

3.2 Description of the Model

The definitions used in the block model and its formal mathematical description are given below.

3.2.1 Block Structure Geometry

A block structure is illustrated in Fig. 3.1. A layer of thickness H is confined between two horizontal planes; a block structure covers a limited and simply connected part of this layer. Each lateral boundary of the block structure is part of a plane that intersects the layer. These planes divide the structure into blocks. The parts of these planes located inside the block structure and its lateral faces are called fault planes.

The geometry of a block structure is described by lines where fault planes intersect the upper plane limiting the layer (these lines are called faults) and

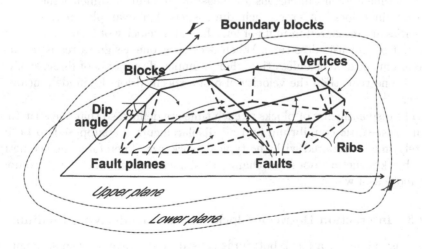

Fig. 3.1. A sketch of a block model

by the dip angle of each fault plane. Fault planes intersect along straight lines and meet the upper and lower planes at points called vertices. A part of such a straight line between two respective vertices is called a rib. Three or more faults cannot have a common point on an upper plane. The direction is specified for each fault, and the dip angle of a fault plane is measured on the left of the respective fault. The part of a fault plane between two ribs corresponding to successive vertices on the fault is called a fault segment. Any fault segment is a trapezoid. The common parts of blocks and the upper and lower planes are polygons; the common part of a block with the lower plane is called a block bottom.

It is assumed that motion is prescribed on the interface between the surrounding medium and the block structure. This interface consists of trapezoids; each is a part of a plane that bounds the structure. This part of the plane is located between two successive ribs (the sides of the trapezoid) and two lines (bases of the trapezoid) where the plane meets the horizontal boundaries of the structure. Motions are prescribed at these trapezoids. One can introduce boundary blocks adjacent to these trapezoids and set the motion of these blocks.

3.2.2   Block Movement

Blocks are assumed to be rigid; hence their relative displacements take place only along the bounding fault planes. Blocks interact with the underlying medium at the lower plane, and any kind of slip is possible.

The movements of block structure boundaries (the boundary blocks) and the medium underlying the blocks are assumed to be external to the structure. The rates of these movements are considered horizontal and known.

Dimensionless time is used in the model; therefore, all quantities that contain time in their dimensions are measured per unit of dimensionless time, so that time does not enter their dimensions. For example, in the model, velocities are measured in units of length, and a velocity of 5 cm means 5 cm per unit of dimensionless time. When necessary, one assigns a realistic value to one unit of dimensionless time. For example, if one unit of dimensionless time is one year, then the velocity of 5 cm, specified for the model, means 5 cm year^{-1}.

The displacements of blocks at each time are defined so that the structure is in quasi–static equilibrium and all displacements are supposed to be infinitely small compared with the typical block size. Therefore, the geometry of a block structure does not change in simulation, and the structure does not move as a whole.

3.2.3   Interaction Between Blocks and the Underlying Medium

The elastic force at a block bottom is caused by the relative displacement of the block and the underlying medium. Considered at a point on the block bottom, the elastic force is assumed to be proportional to the difference between the total relative displacement vector and the vector of slippage (inelastic displacement) at this point.

The elastic force per unit area $\boldsymbol{f}^u = (f_x^u, f_y^u)$ applied to the point with coordinates (X, Y) at some time t is defined by

$$f_x^u = K_u(x - x_u - (Y - Y_c)(\varphi - \varphi_u) - x_a),$$
$$f_y^u = K_u(y - y_u + (X - X_c)(\varphi - \varphi_u) - y_a),$$

$$(3.1)$$

where X_c, Y_c are the coordinates of the geometric center of the block bottom; the motion of the underlying medium adjacent to the block bottom is defined by the translational vector (x_u, y_u) and the angle of rotation φ_u at time t about the geometric center (following the general convention, the positive direction of rotation is anticlockwise); (x, y) and φ are the translational vector of the block bottom and the angle of its rotation about the geometric center at time t; and (x_a, y_a) is the inelastic displacement vector at the point (X, Y) at time t.

The evolution of inelastic displacements at a point (X, Y) is described by the equations

$$\frac{dx_a}{dt} = W_u f_x^u,$$

$$\frac{dy_a}{dt} = W_u f_y^u.$$

$$(3.2)$$

The coefficients K_u and W_u in (3.1) and (3.2) may be different for different blocks.

3.2.4   Interaction Between Blocks Along Fault Planes

Let us consider a point (X, Y) at the fault plane separating blocks i and j; for definiteness, blocks i and j are on the left and right of the fault, respectively. The components Δx, Δy of the relative displacement of the blocks are defined as follows:

$$\Delta x = x_i - x_j - (Y - Y_c^i)\varphi_i + (Y - Y_c^j)\varphi_j,$$
$$\Delta y = y_i - y_j + (X - X_c^i)\varphi_i - (X - X_c^j)\varphi_j,$$

$$(3.3)$$

where $X_c^i, Y_c^i, X_c^j, Y_c^j$ are the coordinates of the geometric centers of the block bottoms, (x_i, y_i), (x_j, y_j) are the translational vectors of the blocks, and φ_i, φ_j are the angles of rotation of the blocks about the geometric centers of their bottoms, all depending on time t.

Relative block displacements, it was assumed, take place only along fault planes; therefore, the displacements along the fault and horizontal planes are related by

$$\Delta_t = e_x \Delta x + e_y \Delta y \ ,$$
$$\Delta_l = \Delta_n / \cos \alpha \ ,$$
$$\text{where} \quad \Delta_n = e_x \Delta y - e_y \Delta x \ . \tag{3.4}$$

Here Δ_t and Δ_l are the displacements along the fault plane parallel (Δ_t) and normal (Δ_l) to the fault line on the upper plane, α is the dip angle of the fault plane, and Δ_n is the horizontal displacement normal to the fault line on the upper plane. It follows from (3.4) that Δ_n is the projection of Δ_t on the horizontal plane (Fig. 3.2a)

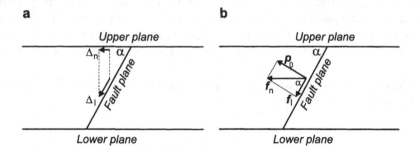

Fig. 3.2. Vertical section of a block structure orthogonal to a fault. Relative displacements of blocks Δ_n and Δ_l (**a**) and forces p_0, f_l, and f_n (**b**)

The elastic force per unit area $\boldsymbol{f} = (f_t, f_l)$ acting along the fault plane at a point (X, Y) is defined by

$$f_t = K(\Delta_t - \delta_t) \ ,$$
$$f_l = K(\Delta_l - \delta_l) \ , \tag{3.5}$$

where δ_t and δ_l are inelastic displacements along the fault plane at the point (X, Y) at time t, parallel (δ_t) and normal (δ_l) to the fault.

The evolution of inelastic displacements at a point (X, Y) is described by the equations

$$\frac{d\delta_t}{dt} = W f_t \ ,$$
$$\frac{d\delta_l}{dt} = W f_l \ . \tag{3.6}$$

The coefficients K and W in (3.5) and (3.6) may be different for different faults. The coefficient K can be considered the shear modulus of the fault zone.

In addition to the elastic force, there is the reaction force normal to the fault plane; the work of this force is zero, because all relative movements are tangent to the fault plane. The elastic energy per unit area at a point (X, Y) is equal to

$$e = [f_t(\Delta_t - \delta_t) + f_l(\Delta_l - \delta_l)]/2 \,. \tag{3.7}$$

From (3.4) and (3.7), the horizontal component of elastic force f_n per unit area normal to the fault is representable as

$$f_n = \frac{\partial e}{\partial \Delta_n} = \frac{f_l}{\cos \alpha} \,. \tag{3.8}$$

It follows from (3.8) that the total force applied at a point on the fault plane is horizontal if there is a reaction force per unit area p_0 (Fig. 3.2b) expressed as

$$p_0 = f_l \tan \alpha \,. \tag{3.9}$$

The reaction force (3.9) is introduced in the model, therefore, there are no vertical components of forces acting on blocks, and there are no vertical displacements of blocks.

Formulas (3.3) are also valid for boundary faults. In this case, one of the blocks separated by the fault is a boundary block. The movement of boundary blocks is prescribed by their translation and rotation about the origin. Therefore, the coordinates of the geometric center of the block bottom in (3.3) are set to zero for any boundary block. For example, if the block numbered j is a boundary block, then $X_c^j = Y_c^j = 0$ in (3.3).

3.2.5 Equations of Equilibrium

The components of the translational vectors of the blocks and the angles of their rotation about the geometric centers of bottoms are found from the condition that the total force and the total moment of forces acting on each block must vanish. This is the condition of quasi-static equilibrium of the system and at the same time the condition of minimum energy. The equilibrium equations include only forces caused by specified movements of the underlying medium and the boundaries of the block structure. In fact, it is assumed that the action of all other forces (gravity, etc.) on the block structure is ruled out and does not cause displacements of blocks.

Formulas (3.1), (3.3–3.5), (3.8), and (3.9) imply that forces applied to blocks depend linearly on the translational vectors and on the angles of rotation. Therefore the system of equations describing the equilibrium is linear and has the following form:

$$\mathcal{A}z = b \,, \tag{3.10}$$

where the components of the unknown vector $z = (z_1, z_2,...,z_{3n})$ are the components of the translational vectors of blocks and the angles of their rotation about the geometrical centers of their bottoms (n is the number of blocks), i.e., $z_{3m-2} = x_m$, $z_{3m-1} = y_m$, $z_{3m} = \varphi_m$ (m is the number of a block, $m = 1, 2,...,n$).

Matrix $\mathcal{A}$ is independent of time, and its elements are defined from formulas (3.1), (3.3–3.5), (3.8), and (3.9). The moment of forces acting on a block is calculated relative to the geometric center of its bottom. Expressions for the elements of matrix $\mathcal{A}$ contain integrals over fault segments and block bottoms. Each integral is replaced by a finite sum, in accordance with the space discretization described in Sect. 3.2.6.

The components of vector b are also defined from formulas (3.1), (3.3–3.5), (3.8), and (3.9). They depend on time explicitly through the movements of the underlying medium and the block structure boundaries and implicitly through inelastic displacements.

3.2.6   Discretization

Time is discretized with a step Δt. The state of the block structure is considered at discrete values of time $t_i = t_0 + i\Delta t$ ($i = 1, 2,...$), where t_0 is the initial time. The transition from the state at t_i to the state at t_{i+1} proceeds as follows: (i) new values of the inelastic displacements x_a, y_a, δ_t and δ_l are calculated from equations (3.2) and (3.6); (ii) translational vectors and rotational angles at t_{i+1} are obtained for boundary blocks and the underlying medium; (iii) the components of b in (3.10) are found, and these equations are used to determine the translational vectors and the angles of rotation for the blocks. The elements of $\mathcal{A}$ in (3.10) are independent of time; hence matrix $\mathcal{A}$ and the associated inverse matrix are calculated only once, at the beginning of modeling.

Formulas (3.1–3.9) describe forces, relative displacements, and inelastic displacements at points of fault segments and block bottoms. Therefore, discretization of these surfaces is required for numerical simulation. The space discretization defined by parameter ε is applied to fault segments and block bottoms. The discretization of a fault segment is performed as follows. Each fault segment is a trapezoid with bases a and b and height $h = H/\sin\alpha$, where H is the thickness of the layer and α is the dip angle of the fault plane. Determine the values $n_1 = \text{ENTIRE}(h/\varepsilon) + 1$ and $n_2 = \text{ENTIRE}[\max(a, b)/\varepsilon] + 1$, and divide the trapezoid into $n_1 n_2$ small trapezoids by two groups of segments inside it; there are $n_1 - 1$ segments parallel to the trapezoid bases and spaced at intervals h/n_1, and $n_2 - 1$ segments connecting the points spaced by intervals of a/n_2 and b/n_2, respectively, on the two bases. Small trapezoids obtained in this way are called cells. The coordinates X, Y in (3.3) and the inelastic displacements δ_t, δ_l in (3.5) are supposed to be the same for all points of a cell and are considered average values over the cell. When introduced in formulas (3.3–3.5), (3.8), and (3.9), they yield the average (over the cell) of the elastic and reaction forces per unit area. The forces acting on a cell are obtained by multiplying the average forces per unit area by the area of the cell.

The bottom of a block is a polygon. Prior to discretization, it is divided into trapezoids (triangles) by segments passing through its vertices and parallel to the Y axis. The discretization of these trapezoids (triangles) is performed in the same way as for fault segments. Small trapezoids (triangles) so obtained are also called cells. Coordinates X, Y and inelastic displacements x_a, y_a in (3.1) are assumed to be the same for all points of a cell.

3.2.7 Earthquake and Creep

Let us introduce the quantity

$$\kappa = \frac{f}{P - p_0} , \tag{3.11}$$

where $f = (f_t, f_l)$ is the vector of the elastic force per unit area given by (3.5); P is assumed the same for all fault planes and can be interpreted as the difference between lithostatic and hydrostatic pressure; p_0 given by (3.9) is the reaction force per unit area.

Three following values of κ are assigned to each fault plane:

$$B > H_f \geq H_s .$$

Let us assume that the initial conditions of the model satisfy the inequality $\kappa < B$ for all cells of fault segments. If, at some time t_i, the value of κ in any cell of a fault segment reaches the level B, a failure ("earthquake") occurs. The failure is considered slippage during which the inelastic displacements δ_t and δ_l in this cell change abruptly to reduce the value of κ to the level H_f. Thus, earthquakes occur in accordance with the dry friction model.

The new values of inelastic displacements in the cell are calculated from

$$\delta_t^e = \delta_t + \gamma f_t ,$$
$$\delta_l^e = \delta_l + \gamma f_l , \tag{3.12}$$

where δ_t, δ_l, f_t, and f_l are the inelastic displacements and the components of the elastic force vector per unit area just before the failure. The coefficient γ is given by

$$\gamma = 1/K - PH_f/[K(|f| + H_f f_l \tan \alpha)] . \tag{3.13}$$

It follows from (3.5), (3.9), (3.11–3.13) that on obtaining the new values of inelastic displacements the value of κ in the cell becomes equal to H_f.

After calculating new values of inelastic displacements for all failed cells, new components of the vector b are found, and the translational vectors and the angles of rotation for all blocks are obtained from the system of equations in (3.10). If κ still exceeds B for some cell(s) of the fault segment, the procedure given above is repeated for this cell (or cells). Otherwise, the state of the block structure at time t_{i+1} is determined as follows: translational vectors, rotational angles (at t_{i+1}) for boundary blocks and for the underlying medium, and the components of b in (3.10) are calculated, and then equations in (3.10) are solved.

The cells of the same fault plane, where failure occurs simultaneously, form a single earthquake. The parameters of the earthquake are defined as follows: (i) the origin time is t_i; (ii) the epicentral coordinates and the source depth are the weighted sums of the coordinates and the depths of cells included in the earthquake (the weight of each cell is given by its area divided by the total area of all cells included in the earthquake); (iii) the magnitude is calculated from

$$M = D \log S + E \,, \qquad (3.14)$$

where D and E are constants and S is the total area of cells (in km^2) included in the earthquake.

It is assumed that the cells in which a failure has occurred are in the creep state immediately after the earthquake. This means that the parameter W_s ($W_s > W$) is used instead of W for these cells in (3.6) to describe the evolution of inelastic displacements; W_s may be different for different fault planes. After each earthquake, a cell is in the creep state as long as $\kappa > H_s$, whereas when $\kappa \leq H_s$, the cell returns to the normal state, and henceforth the parameter W is used in (3.6) for this cell.

3.3 Dependence of a Synthetic Earthquake Flow on Structure Fragmentation and Boundary Movements

The dependence of features of synthetic seismicity obtained in the block model on structure fragmentation and boundary movement was found and studied in [KRS97].

3.3.1 Block Structures and Cases of Boundary Movements Under Consideration

Block Structures. Numerical simulation was carried out for three groups of block structures. The respective schemes of faults on the upper plane are shown in Fig. 3.3. One structure (BS1) belongs to all groups. Its faults on the upper plane form a square of side 320 km divided into four smaller squares. Two other structures of the first (BS12, BS13), the second (BS22, BS23), and the third (BS32, BS33) group are obtained from BS1 by self-similar subdivision [BT94].

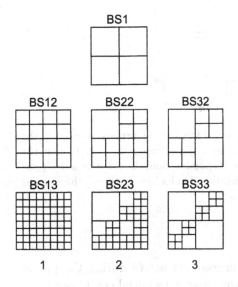

Fig. 3.3. Faults on the upper plane from block structures under consideration: (**1**) the first group (BS1, BS12, BS13); (**2**) the second group (BS1, BS22, BS23); and (**3**) the third group (BS1, BS32, BS33)

The depth of the layer is $H = 20$ km for all of these structures.

The values of the parameters in (3.1) and (3.2) for all blocks of the structures are the following: $K_u = 1$ bar cm^{-1}, and $W_u = 0.05$ cm bar^{-1}.

The values of the parameters in (3.4–3.6), (3.8), and (3.9) for all faults of the structures are the following: $\alpha = 85°$ (a dip angle); $K = 1$ bar cm^{-1}; $W = 0.05$ cm bar^{-1}; and $W_s = 10$ cm bar^{-1}. The thresholds of κ (3.11) are the following for all faults: $B = 0.1$; $H_f = 0.085$; and $H_s = 0.07$.

The parameter P in (3.11) equals 2 kbar.

Movement. The medium underlying all blocks of the structures does not move.

The analysis includes two cases of boundary movement (Fig. 3.4).

The first case represents the progressive movement of the boundaries at the velocity of 10 cm per unit of dimensionless time. The directions of the velocity vectors are shown in Fig. 3.4a. The angle between the velocity vector and the respective side of the square outlining the structure is 10°.

The second case includes the progressive movement and rotation of the boundaries (Fig. 3.4b). Two boundaries move progressively at the velocity of 10 cm per unit of dimensionless time, and the velocity vectors are parallel to the respective sides of the square outlining the structure. The movement of the other two boundaries is rotation about the centers of the respective sides at the angular velocity -0.625×10^{-6} radians per unit of dimensionless time.

Note that in the first case, the distribution of values of $|f|$ and p_0 along a fault segment is expected to be closer to an even function than in the second case.

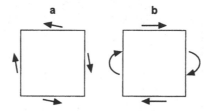

Fig. 3.4. Types of boundary movements considered. The arrows stand for the velocity vectors of the boundaries: (**a**) the first type: the velocity is 10 cm; the angle between the velocity vectors and the respective boundary faults is $10°$; (**b**) the second type: the magnitude of the translational velocities is 10 cm; the angular velocity is 0.625×10^{-6} radians

3.3.2 Results of Modeling

All block structures under consideration were set active during the period of 200 units of dimensionless time, starting from zero initial conditions (zero displacements of the boundary blocks and the underlying medium and zero inelastic displacements for all cells) in both cases of boundary movements.

The values of the discretization parameters in time and space are $\Delta t = 0.001$ and $\varepsilon = 5$ km, respectively.

As a result, synthetic earthquake catalogs were obtained. The magnitude of the earthquake is calculated from (3.14) with the following values of the coefficients:

$$M = 0.98 \log S + 3.93 . \tag{3.15}$$

These values are specified in accordance with [US54]. Note that (3.15) implies a lower limit of magnitudes of synthetic earthquakes. It is determined by the minimum of cells areas obtained through the discretization of fault segments. In accordance with (3.15) and with the value of ε (5 km), this limit must be less than 5.3. Actually, its value depends on the specific structure and varies from 5.12 to 5.18.

The cumulative FM plots for synthetic catalogs obtained for the first case of boundary movements are presented in Fig. 3.5. The shape of the plots agrees with the Gutenberg–Richter law for the observed seismicity, i.e., the logarithm of the number of earthquakes is a linear function of magnitude.

Table 3.1 lists the total number of events (N), the maximum magnitude ($M_{\max}$) given by (3.15), and a slope estimate (b-value) of the frequency–magnitude plot for the catalogs obtained.

The cumulative FM plots for synthetic catalogs obtained in the second case of boundary movements are presented in Fig. 3.6. Table 3.2 lists the values of N, $M_{\max}$, and b for these catalogs.

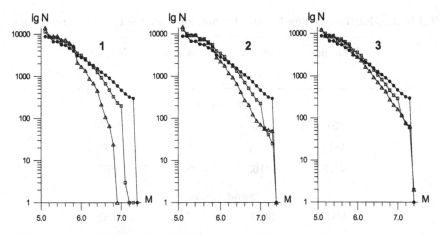

Fig. 3.5. Cumulative frequency–magnitude plots for synthetic catalogs obtained for boundary movement without rotation from three groups of structures: (**1**) first, (**2**) second, and (**3**) third. Curves are marked as follows: dots (obtained from BS1), squares (BS12 and BS22), and triangles (BS13, BS23, and BS33)

3.3.3 Discussion of Results

Figure 3.5 and Table 3.1 show that without rotation, the cumulative FM plots change similarly in each group of self-similar structures when structure fragmentation increases. Earthquake flow features change as follows: the total number of events and the number of small events increase; the number of large events decreases; the slope of the plot increases. The largest difference between the plots for different structures appears in the first group (structures BS1, BS12, BS13).

In experiments with rotations of boundaries (Fig. 3.6 and Table 3.2), FM plots also change similarly in different groups of structures. However, when structure fragmentation increases, several features change in directions opposite to those without rotation: the total number of events and the number of small events decrease; the number of large events increases; the slope of the plot decreases. For this case of boundary movement, the largest difference between the plots for different structures also appears in the first group.

Larger maximum magnitudes for the boundary movement without rotation (compare Tables 3.1 and 3.2) can be explained by the above-mentioned fact that the distributions of $|f|$ and p_0 along a fault segment are closer to an even function. Note that in this case, earthquakes are mainly caused by large values of $|f|$ in (3.11), whereas in the case of boundary movements with rotation, large positive values of p_0 (extension) play the main role.

The slope of the FM plot changes with increasing fragmentation in the first case mainly because the number of larger earthquakes becomes less. In the second case, it is controlled by two factors: first, the number of larger earthquakes increases, and second (for the first and the second groups of structures), the number of small earthquakes decreases.

Table 3.1. Earthquake flow features for boundary movement without rotation

Structure	Number of events	Maximum magnitude	b-value
First Group			
BS1	8801	7.46	0.63
BS12	11417	7.31	0.79
BS13	14163	6.97	1.37
Second Group			
BS1	8801	7.46	0.63
BS22	12692	7.37	0.84
BS23	14670	7.48	1.59
Third Group			
BS1	8801	7.46	0.63
BS32	11986	7.40	0.90
BS33	12366	7.48	1.22

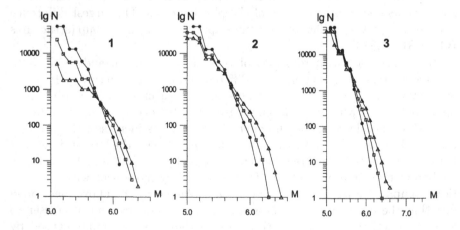

Fig. 3.6. Cumulative frequency–magnitude plots from synthetic catalogs obtained for boundary movement with rotation. Notations are the same as in Fig. 3.5

Table 3.2. Earthquake flow features for boundary movement with rotation

Structure	Number of events	Maximum magnitude	b-value
	First Group		
BS1	59011	6.20	4.12
BS12	23748	6.37	3.39
BS13	5143	6.45	1.90
	Second Group		
BS1	59011	6.20	4.12
BS22	38615	6.31	3.55
BS23	27110	6.53	2.48
	Third Group		
BS1	59011	6.20	4.12
BS32	53532	6.42	3.59
BS33	44645	6.61	2.80

The experience of seismological studies shows that the definition of the magnitude has a significant effect on observed seismic flow features. The magnitude can be defined on the basis of earthquake energy [GR56]

$$M = 0.67 \log E - 7.87 , \qquad (3.16)$$

where E is the energy (in ergs) of an earthquake. Formulas (3.15) and (3.16) agree with the conventional relation between the energy and the fault surface area,

$$E \sim S^{3/2} . \qquad (3.17)$$

In the model under consideration, the sum S of cell areas included in an earthquake is the analogy of the fault surface area. Figure 3.7 shows the relation between S and the energy in the model for synthetic catalogs obtained with structure BS32 in both cases of boundary movement: without rotation (Fig. 3.7a) and with rotation (Fig. 3.7b). Values of E, the difference between the energy of the system before and after an earthquake, are plotted versus S as points corresponding to synthetic earthquakes.

It follows from Fig. 3.7 that the relation between the energy E released through an earthquake and the total area S of the fault plane covered by this earthquake is close to linear in both cases. The most significant deviations from this linear law are observed in the case without rotation (Fig. 3.7a) for large earthquakes. These deviations are mainly due to larger values of E and

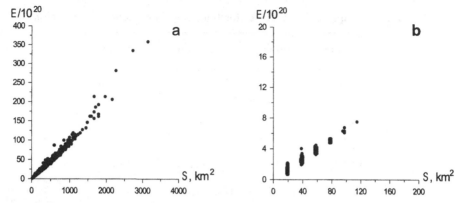

Fig. 3.7. Energy versus the area of events from synthetic catalogs obtained for structure BS32 with boundary movements without (**a**) and with (**b**) rotation

can be explained by the fact that the average energy release per unit area increases for large earthquakes.

The linear dependence between E and S in the model is in conflict with (3.17). It causes nonequivalence of formulas (3.15) and (3.16). Nevertheless, the linear dependence between E and S can be explained by the fact that in the model considered, the energy is distributed along planes and the energy released through an earthquake depends mainly on the total area of the fault plane covered by the earthquake. This is an argument to use (3.15) as a definition of the magnitude in the model, at least when each parameter has the same value for all faults of the block structure.

Note that the relation between the slopes in Figs. 3.5 and 3.6 remains the same when (3.16) is used to define the magnitude instead of (3.15); this is also true for the properties characterizing the dependence of Gutenberg–Richter plots on structure fragmentation. These results show that the features of a synthetic earthquake flow depend on the geometry of the block structure and on its boundary movement. The behavior of feature–geometry dependence is quite different for different cases of boundary movements. Note that for boundary movement with rotation, the dependence of the seismic flow characteristics on the structure fragmentation considered is in contradiction with conventional opinion.

3.4 Space–Time Correlation Between Synthetic Earthquakes

Some features of the observed seismicity connected with the space–time distribution of earthquakes were also detected in the synthetic catalogs. Clustering of events in time and long-range interaction between events are among these features.

3.4.1 Clustering of Synthetic Earthquakes

Investigation of earthquake clustering is exceptionally important for understanding the dynamics of seismicity and specifically for earthquake prediction. That is why this phenomenon was studied by many authors (e.g., [KK78, KKR80, DP84, MD92]). It is vital to clarify whether clustering is caused by specific features of tectonics in a region under consideration or, conversely, whether it is a general phenomenon for a wide variety of neotectonic conditions and reflects general features of systems of interacting blocks that build the seismoactive lithosphere. The clustering of earthquakes in a synthetic catalog obtained from modeling the dynamics of a simple block structure favors the second assumption.

Mechanism of clustering in the model. Consider a block structure consisting of two blocks separated by one fault plane and moving progressively at constant relative velocity directed along the fault and assume that the X axis is the line where the fault plane meets the upper plane; then it follows from (3.4–3.6) and (3.9) that

$$f_x = K(\Delta x - \delta_x) \,,$$
$$f_y = 0, \qquad p_0 = 0 \,,$$
$$\frac{d\delta_x}{dt} = V f_x \,.$$

Here $f = (f_x, f_y)$ is the elastic force per unit area acting along the fault plane, which is the same at all of its points; Δx is the relative displacement of the blocks; δ_x is the inelastic displacement, the same at all points of the fault plane; and p_0 is the density of the reaction force.

Owing to the definition of an earthquake and creep given above, earthquakes in this structure are equally spaced in time, and each earthquake spreads across the whole fault plane because the value of κ defined by (3.11) exceeds the level of B simultaneously across the whole fault plane.

When the movement of blocks is more intricate, the value of κ exceeds the level of B at different times in different parts of the fault plane. As a result, a group (a cluster) of weaker earthquakes occurs in place of a single large one.

Example of clustering of earthquakes in the model. Modeling of the dynamics was done for the block structure consisting of four blocks whose common parts with the upper plane are squares (Fig. 3.8) with a side of 50 km. The thickness of the layer is $H = 20$ km, and the same dip angle of 85° is specified for all fault planes. The parameters in (3.1) and (3.2) have the same values for all blocks: $K_u = 1$ bar cm^{-1}, $W_u = 0.05$ cm bar^{-1}. For all faults, the parameters in (3.5) and (3.6) and the levels for κ (3.11) have the same values: $K = 1$ bar cm^{-1}, $W = 0.05$ cm bar^{-1}, $W_s = 1$ cm bar^{-1}, $B = 0.1$, $H_f = 0.085$, and $H_s = 0.07$. The parameter P in (3.11) is equal to 2 kbar.

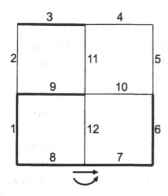

Fig. 3.8. Fault pattern on the upper plane accepted for assessing the clustering of synthetic earthquakes and their long-range interaction. Arrows illustrate movements specified at the boundary consisting of fault segments 7 and 8

Fault segments and block bottoms were discretized with $\varepsilon = 5\,\mathrm{km}$, and $\Delta t = 0.01$ was used for time discretization.

The magnitudes of synthetic earthquakes were calculated from (3.15).

The following boundary movement is assumed: translation of fault segments numbered 7 and 8 (Fig. 3.8) with the velocity components $V_x = 20$ cm and $V_y = -5$ cm and their rotation about the origin with an angular velocity of 10^{-6} radians. Note that this movement is equivalent (as shown in Fig. 3.8) to translation with components $V_x = 20$ cm and $V_y = 0$ and rotation about point $X = 50\,\mathrm{km}$, $Y = 0$ at the same angular velocity. The other parts of the boundary and the underlying medium do not move.

Observe that the dynamics of this block structure differs from that of the two blocks described above. It consists of four blocks and involves rotation.

The structure was set to life with zero initial conditions (zero displacements of boundary blocks and the underlying medium and zero inelastic displacements for all cells). The occurrences of earthquakes (vertical lines) are shown in Fig. 3.9 for individual fault segments and for the whole structure for the time interval of 3 units starting at $t = 480$.

Earthquakes occur on six fault segments. The respective parts of the faults are marked in Fig. 3.8 by thick lines. Segment 9 has only one earthquake during the period under consideration. Earthquakes cluster clearly on fault segments 1, 3, 6, and 7. Segment 8 has the largest number of earthquakes. Here, the clustering appears weaker: the groups of earthquakes are diffuse along the time axis. The pattern for the whole structure looks like that for segment 8, and groups of earthquakes can also be identified.

Clustering of earthquakes for other time intervals does not differ significantly from that presented in Fig. 3.9.

Earthquake clustering found in the model provides a way to use modeling for studying the phenomenon of earthquake clustering in specific seismoactive regions. In particular, one can ascertain the dependence of clustering on the geometry of a block structure and on the values of its parameters.

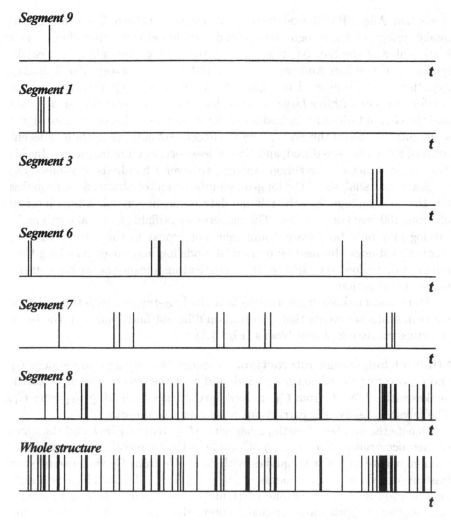

Fig. 3.9. Clustering of synthetic earthquakes. Times of occurrence (vertical lines) are shown in the interval of three dimensionless time units for individual segments numbered as in Fig. 3.8 and for the whole structure

3.4.2   Long-Range Interaction Between Synthetic Earthquakes

Studies of earthquake flows lead to the conclusions that there exists an interaction between earthquakes. Benioff [Ben51], who investigated seismic strain accumulation and release in the period 1904–1950, suggested a hypothesis for interdependence of earthquakes of magnitude $M \geq 8.0$. Duda [Dud65] found that this could be extended to earthquakes of magnitude $M \geq 7.0$. The phenomenon of long-range interaction between earthquakes separated by distances that exceed their aftershock areas considerably was found and studied by Prozorov [Pro91, Pro93, Pro94a, Pro94b] for several active regions.

Press and Allen [PA95] examined a catalog of Southern California earthquakes using pattern recognition methods and found traits that characterize earthquakes of the San Andreas fault system (SA earthquakes) and earthquakes with non-San Andreas attributes (NSA earthquakes). The following hypotheses were suggested to explain these traits: earthquakes in Southern California occur within a larger system that includes at least the Great Basin and the Gulf of California; episodes of activity in these adjacent regions signal subsequent release of the SA type; in the absence of activity in these adjacent regions, SA release is reduced, and NSA release occurs more frequently. In this case, the distance of interaction amounts to several hundreds of kilometers.

Statistical analysis of the long-range interaction of observed earthquakes has the disadvantage that the reliable data cover, in general, a time interval of about 100 years or even less. The patterns identifiable in a real earthquake catalog may only be apparent and may not repeat in the future, whereas synthetic catalogs obtained by numerical modeling may cover very long time intervals that allow us to analyze the statistical significance of the phenomena under consideration.

Statistical methods were used to find the long-range interaction [SV99a] between synthetic events that occurred on different fault planes in the block structure consisting of four blocks (Fig. 3.8).

Study of long-range interaction. Consider two earthquake catalogs C_1 and C_2 covering the same time period, and raise the question whether earthquakes with $M \geq M_2$ from C_2 succeed earthquakes with $M \geq M_1$ from C_1. The following analysis is carried out to answer this question.

Denote the number of earthquakes with $M \geq M_1$ in C_1 by k and the times of their occurrence by $\tau_1, \tau_2, ..., \tau_k$. Consider a time interval $\Delta\tau$ and calculate the number $m(\Delta\tau)$ of earthquakes with $M \geq M_2$ from C_2 that occurred at least in one of the time segments $[\tau_i, \tau_i + \Delta\tau]$, $i = 1, 2, ..., k$. These segments may overlap, and therefore, one earthquake from C_2 may belong to several such segments. Such an earthquake enters the calculation of $m(\Delta\tau)$ only once. Denote by $\eta(\Delta\tau)$ the total duration of time covered by these segments. Obviously, $n(\Delta\tau) = m(\Delta\tau)\Delta\tau/\eta(\Delta\tau)$ is the average number of earthquakes with $M \geq M_2$ from C_2 that occurred in the time period $\Delta\tau$ after earthquakes with $M \geq M_1$ from C_1, and $\lambda(\Delta\tau) = m(\Delta\tau)/\eta(\Delta\tau)$ is the flow rate of these earthquakes during this period. Considering the whole time interval covered by C_2, calculate the unconditional flow rate Λ of earthquakes with $M \geq M_2$ in this catalog and the average number $N(\Delta\tau) = \Lambda\Delta\tau$ of such earthquakes that occurred in the time period $\Delta\tau$.

If $n(\Delta\tau)$ and $\Lambda(\Delta\tau)$ exceed considerably $N(\Delta\tau)$ and Λ, respectively, for some $\Delta\tau$, then one can say that earthquakes with $M \geq M_2$ from C_2 succeed earthquakes with $M \geq M_1$ from C_1, and therefore, there is an interaction between these events. This is the long-range interaction between earthquakes that occurred in two different fault planes with respective catalogs C_1 and C_2.

A random variable $\xi(\Delta\tau)$ is introduced to estimate the statistical significance of $n(\Delta\tau) > N(\Delta\tau)$. It is the number of earthquakes with $M \geq M_2$ from C_2 that occurred in an arbitrary time interval of duration $\Delta\tau$. $N(\Delta\tau)$ is the mean value estimate of $\xi(\Delta\tau)$. Denote by $Q(\Delta\tau)$ the standard deviation estimate of $\xi(\Delta\tau)$. If there is no interaction between the earthquakes with $M \geq M_1$ from C_1 and the earthquakes with $M \geq M_2$ from C_2, then

$$n(\Delta\tau) \approx [\xi_1(\Delta\tau) + \xi_2(\Delta\tau) + ... + \xi_p(\Delta\tau)]/p\,,$$

where $\xi_i(\Delta\tau)$, $i = 1, 2, ..., p$, are random values distributed like $\xi(\Delta\tau)$, $p = $ ENTIRE$(\eta(\Delta\tau)/\Delta\tau)$. The standard deviation of $n(\Delta\tau)$ is estimated by

$$q(\Delta\tau) = Q(\Delta\tau)/\sqrt{p}\,.$$

The value of $q(\Delta\tau)$ measures the deviation of $n(\Delta\tau)$ from $N(\Delta\tau)$. When $n(\Delta\tau) - N(\Delta\tau) > 3q(\Delta\tau)$, the excess of $n(\Delta\tau)$ over $N(\Delta\tau)$ is considered statistically significant.

Assume that the above procedure leads to the conclusion that earthquakes with $M \geq M_2$ that occurred on fault plane F_2 (catalog C_2) succeed earthquakes with $M \geq M_1$ that occurred on fault plane F_1 (catalog C_1). The existence of the long-range interaction between these earthquakes can be verified by the following additional test. Let us increase the value of the threshold B of the fault plane F_1 to inhibit the occurrence of earthquakes there (to lock the fault plane) and repeat the modeling. Denote C_2' as the new catalog of earthquakes that occurred on fault plane F_2. Assume that the procedure described implies that earthquakes with $M \geq M_2$ from C_2' do not succeed earthquakes with $M \geq M_1$ from C_1. This means that earthquakes with $M \geq M_1$ on fault plane F_1 act at origin times of earthquakes with $M \geq M_2$ on fault plane F_2.

Long-range interaction between synthetic earthquakes found in the model. The behavior of the block structure presented in Fig. 3.8 with the values of the model parameters given above and zero initial conditions was computed for 2000 units of dimensionless time.

Consider earthquakes with $M \geq M_2 = 6.6$ that occurred in the union of fault segments 7 and 8 (plane F_2) and earthquakes with $M \geq M_1 = 6.0$ that occurred on fault segment 9 (plane F_1). The analysis shows that earthquakes in F_2 (a total of 202 events) occur more frequently after earthquakes in F_1 (a total of 138 events) than, on average, for the whole time interval of modeling. Figure 3.10a presents the plots of functions $n(\Delta\tau)$ (*solid thick line*), $N(\Delta\tau)$ (*dashed line*), and $N(\Delta\tau) + q(\Delta\tau)$ (*solid thin line*). The relevant plot of the function $\lambda(\Delta\tau)$ is shown in Fig. 3.10b; a dashed horizontal line marks the corresponding value of $\Lambda = 0.101$. It is seen that the difference between $n(\Delta\tau)$ and $N(\Delta\tau)$ appreciably exceeds $q(\Delta\tau)$ for $0.5 < \Lambda\tau < 7.0$. For instance, $n(3.0) = 0.495$, $N(3.0) = 0.303$, and $q(3.0) = 0.047$, and therefore, the difference $n(3.0) - N(3.0) = 0.192$ exceeds $q(3.0)$ by a factor of 4.

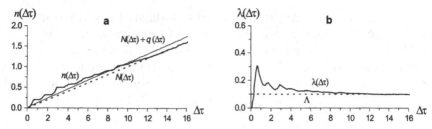

Fig. 3.10. Functions $n(\Delta\tau)$, $N(\Delta\tau)$, $N(\Delta\tau) + q(\Delta\tau)$, and $\lambda(\Delta\tau)$ obtained from different catalogs: (**a**) of segment 9 as C_1 ($M_1 = 6.0$) and (**b**) of segments 7 and 8 as C_2 ($M_2 = 6.6$)

The question arises whether this phenomenon reflects the joint clustering of strong earthquakes from F_1 and F_2. To answer this question, it has been tested whether earthquakes with $M \geq 6.0$ from F_1 succeed earthquakes with $M \geq 6,6$ from F_2, i.e., the times of strong earthquakes from F_1 and F_2 switch places in the analysis given above. Figure 3.11 shows the functions $n(\Delta\tau)$, $N(\Delta\tau)$, $N(\Delta\tau) + q(\Delta\tau)$, and $\lambda(\Delta\tau)$ and the value $\Lambda = 0.069$ obtained in this way. In this case, the function $n(\Delta\tau)$ does not exceed $N(\Delta\tau)$ in the interval $0.5 < \Delta\tau < 7.0$ and is even less than $N(\Delta\tau) - q(\Delta\tau)$, also shown in Fig. 3.11 as a thin line. For instance, $n(3.0) = 0.094$, $N(3.0) = 0.208$, and $q(3.0) = 0.033$; therefore the difference $N(3.0) - n(3.0) = 0.114$ is three times greater than $q(3.0)$. This means that strong earthquakes from F_1 are more seldom after strong earthquakes from F_2 than, on average, for the whole time interval.

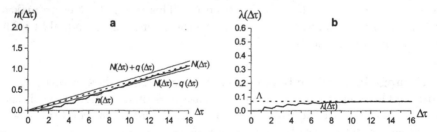

Fig. 3.11. Functions $n(\Delta\tau)$, $N(\Delta\tau)$, $N(\Delta\tau) + q(\Delta\tau)$, and $\lambda(\Delta\tau)$ obtained from different catalogs: (**a**) of segments 7 and 8 as C_1 ($M_1 = 6.6$) and (**b**) of segment 9 as C_2 ($M_2 = 6.0$)

The modeling procedure was repeated with the increased value of $B = 10$ for fault segment 9, i.e., this segment was locked. The catalog of fault segments 7 and 8 obtained in this simulation (150 events) was considered C_2 and was compared with catalog C_1, obtained for segment 9 when segment 9 was not locked. The magnitude thresholds were given as previously, $M_1 = 6.0$ and $M_2 = 6.6$. The functions $n(\Delta\tau)$, $N(\Delta\tau)$, $N(\Delta\tau) + q(\Delta\tau)$, and $\lambda(\Delta\tau)$ and the value of Λ are shown in Fig. 3.12. One can see that $n(\Delta\tau)$ does not exceed $N(\Delta\tau)$.

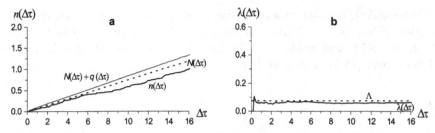

Fig. 3.12. Functions $n(\Delta\tau)$, $N(\Delta\tau)$, $N(\Delta\tau) + q(\Delta\tau)$, and $\lambda(\Delta\tau)$ obtained from different catalogs: (**a**) of segment 9 as C_1 ($M_1 = 6.0$) when segment 9 is unlocked and (**b**) of segments 7 and 8 as C_2 ($M_2 = 6.6$) when segment 9 is locked

Figure 3.13 shows the occurrences of earthquakes with $M \geq 6.0$ from unlocked segment 9 [vertical lines on panel (**a**)] and of earthquakes with $M \geq 6.6$ from segments 7 and 8 [vertical lines on panel (**b**)]. The times of earthquakes with $M \geq 6.6$ from segments 7 and 8, with a locked segment, are presented in panel (**c**). One can see considerable differences between the patterns in panels (**b**) (segment 9 is not locked) and (**c**) (segment 9 is locked).

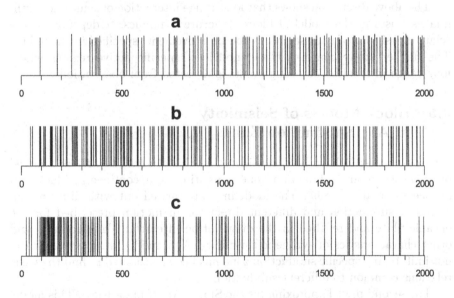

Fig. 3.13. Times of earthquakes that occurred in segment 9 (**a**) and in segments 7 and 8 when segment 9 was unlocked (**b**) and locked (**c**)

There are other examples of long-range interaction between synthetic earthquakes. It was found for synthetic earthquakes that occurred in fault segments 1 (catalog C_1) and 6 (catalog C_2) with $M_1 = 5.8$ (112 events) and $M_2 = 5.5$ (144 events). Figure 3.14 shows the functions $n(\Delta\tau)$, $N(\Delta\tau)$,

$N(\Delta\tau) + q(\Delta\tau)$, and $\lambda(\Delta\tau)$ and the value of $\Lambda = 0.073$. It is seen that the function $n(\Delta\tau)$ exceeds $N(\Delta\tau)$ for $\Delta\tau \geq 3.5$. For instance, $n(6.5) = 0.627$, $N(6.5) = 0.471$, and $q(6.5) = 0.068$, i.e., the difference $n(6.5) - N(6.5) = 0.156$ exceeds $q(6.5)$ more than twice.

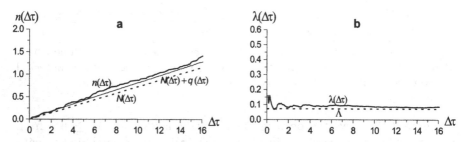

Fig. 3.14. Functions $n(\Delta\tau)$, $N(\Delta\tau)$, $N(\Delta\tau) + q(\Delta\tau)$, and $\lambda(\Delta\tau)$ obtained from different catalogs: (**a**) of segment 1 as C_1 ($M_1 = 5.8$) and (**b**) of segment 6 as C_2 ($M_2 = 5.5$)

Long-range interaction is not evident for other pairs of segments.

The above discussion shows that long-range interaction of synthetic earthquakes exists in the model of block structure dynamics. It depends on the relative positions of fault segments and on movements specified in the model. The further study concerns block models of specific active regions where long-range interaction of strong events is detected from observations.

3.5   Block Models of Seismicity in Arc Subduction Zones

Two block models have been considered in this study. The first is a model of an abstract arc subduction zone consisting of a continental block and an oceanic plate [RS99]. The modeling was carried out with different dip angles of subduction and different directions of motion prescribed for the oceanic plate and resulted in the distribution of earthquake epicenters and other characteristics of synthetic seismicity. A simplified structure that did not imitate any specific subduction zone has been considered to find the basic relations common to different subduction zones.

The second model approximates the Sunda Arc (Sunda Isles). This model elucidates the dependence of features of synthetic seismicity on the movements specified [RSV98, SRRV99]. In this case, the block model of a real-world region is considered as in [GS96, SSR99, PSV97, SVP99, SVP00, VGS00].

3.5.1   A Model of an Abstract Arc Subduction Zone

Description of the model. The block structure consists of one arcuate block A (Fig. 3.15). The structure is bounded by two horizontal planes

$H = 100\,\mathrm{km}$ apart, a depth comparable to the thickness of the lithosphere. Planes intersecting the layer between these horizontal planes form the lateral boundaries of block A numbered 1 through 8 (Fig. 3.15). Block A stands for an island arc and the adjoining edge of the continent or the back-arc basin. The island arc subduction zone where the continent interacts with the oceanic plate is represented in the model by the system of joined fault planes numbered 1–5 (Fig. 3.15). Their successive traces intersect at an angle γ of about 15° (Fig. 3.15b). The total bend of the arc is about 60°.

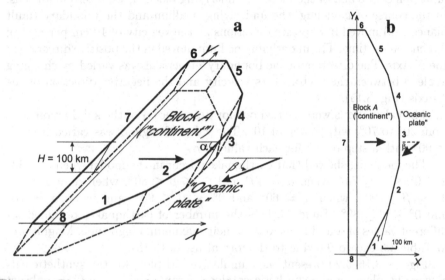

Fig. 3.15. Block structure for the model of an abstract arc subduction zone (a) and its faults on the upper plane (b). Fault planes are identified by numbers 1 through 8, arrows show velocity directions for the medium underlying block A and for the boundary formed by fault planes 1–5

Fault planes 1 through 5 and 7 have the same dip angle α (Fig. 3.15). The angle α, the slope of the subduction zone, was varied from 30° to 70°. The dip angle of fault planes 6 and 8 is 85°.

Velocities were specified for the underlying medium and block boundaries 1–5. The velocity of the boundary models the movement of an oceanic plate bordering a continent.

The same values of the constants in (3.1), (3.2), (3.5), and (3.6), $K_u = K = 1\,\mathrm{bar\,cm^{-1}}$, $W_u = W = 0.05\,\mathrm{cm\,bar^{-1}}$, were specified for the underlying medium and for fault planes 1–5. $K = 0$ was assigned to fault planes 6–8, which are introduced solely to limit the structure. It follows from (3.5) that any displacements do not produce forces at these fault planes.

The values $\varepsilon = 10\,\mathrm{km}$ and $\Delta t = 0.001$ were used for discretization.

The value of P in (3.11) is 2 kbar. The levels for κ (3.11) had the same values for all faults: $B = 0.1$, $H_f = 0.085$, and $H_s = 0.07$. The value $W_s = 10 \, \text{cm} \, \text{bar}^{-1}$ was specified for all faults.

The magnitude of synthetic earthquakes was calculated from (3.15).

Results of simulation. The dynamics of the block structure was modeled across 400 units of dimensionless time, starting from zero. At zero time, all inelastic displacements, the displacements of the block boundary formed by fault planes 1–5 and of the medium underlying block A, were zero in all cells. In the course of modeling, the underlying medium and the boundary (fault planes 1–5) moved in opposite directions at the velocity of 10 cm per unit of dimensionless time. The underlying medium moved in the positive direction of the X axis. The direction of the boundary movement was varied by changing angle β between the vector of its velocity and the negative direction of the X-axis (Fig. 3.15).

The calculations were carried out for dip angles α of the subduction zone from 30° to 70°, with a step of 10°. The value of angle β was varied from 0° to 90° with a step of 10° for each value of α.

The modeling showed that earthquakes occur in the model when $\alpha = 30°$ and $0° \le \beta \le 70°$, when $\alpha = 40°$ and $0° \le \beta \le 80°$, when $\alpha = 50°$ and $0° \le \beta \le 90°$, when $\alpha = 60°$ and $0° \le \beta \le 90°$, and when $\alpha = 70°$ and $0° \le \beta \le 80°$. Table 3.3 lists the number of earthquakes obtained for different values of α and β, as well as their maximum magnitude. Magnitudes in Table 3.3 exceed 9 owing to the formal use of (3.15).

Figures 3.16–3.20 present the cumulative FM plots for the synthetic catalogs. The plots for the synthetic catalogs obtained with $\beta = 0°$ are absent because they contain too many events of the same magnitude (Table 3.4). When $\alpha \le 60°$, it is the maximum magnitude given in Table 3.3 for the relevant catalog.

When $\alpha \le 60°$ and β is small, the largest earthquakes of the synthetic catalogs cover the whole area of fault planes 1–5; this follows from comparing Table 3.3 with Table 3.5, where the magnitudes of earthquakes completely covering several faults (from 1 to 5), are calculated by (3.15) for the values of α considered.

The spatial distribution of large synthetic earthquakes is given in Figs. 3.21–3.25. Projections on the Y axis of areas covered by large earthquakes (with $M \ge 7.5$ when $\alpha \le 60°$ and with $M \ge 7.0$ when $\alpha = 70°$) of the synthetic catalogs are shown in these figures by vertical segments parallel to ordinate axes, and abscissas of these segments correspond to ordinal numbers of earthquakes. One can see from Figs. 3.21–3.25 that the interval of numerical modeling (400 units of dimensionless time) is enough to detect general features of the space–time distribution of the earthquake epicenters in the synthetic catalogs obtained.

Table 3.3. The number N and the maximum magnitude M_{max} of synthetic earthquakes for different values of α and β

	α									
β	30°		40°		50°		60°		70°	
	N	M_{max}	N	M_{max}	N	M_{max}	N	M_{max}	N	M_{max}
0°	10	9.33	19	9.23	28	9.15	26	9.10	21	8.82
10°	68	9.22	58	9.12	64	9.04	66	8.99	23	8.96
20°	784	9.12	318	9.12	324	9.04	213	8.99	200	8.80
30°	1228	9.07	1447	9.20	4236	8.99	1182	9.09	858	8.79
40°	1383	8.91	2937	8.99	1357	8.92	1655	9.01	728	8.82
50°	1379	8.64	1596	8.91	3072	9.15	2152	8.95	1042	8.52
60°	1299	8.38	1945	8.86	307	8.77	332	8.72	1786	8.39
70°	1531	8.10	709	8.59	1178	8.83	627	8.83	1838	8.14
80°			681	8.54	181	8.58	530	8.63	424	7.64
90°					39	8.51	854	8.29		

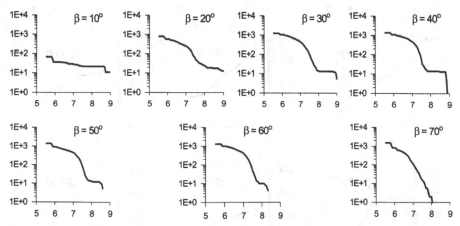

Fig. 3.16. Cumulative FM plots for the synthetic catalog obtained with $\alpha = 30°$. Numbers of earthquakes and magnitudes are plotted on the vertical and horizontal axes, respectively

The results of the modeling described above suggest several features of correlation between synthetic seismicity and the prescribed tectonics of the subduction zone.

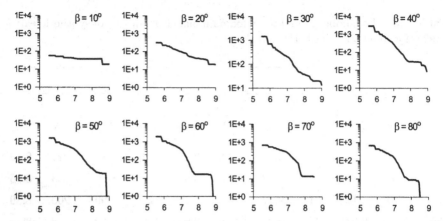

Fig. 3.17. Cumulative FM plots for the synthetic catalog obtained with $\alpha = 40°$. Numbers of earthquakes and magnitudes are plotted on the vertical and horizontal axes, respectively

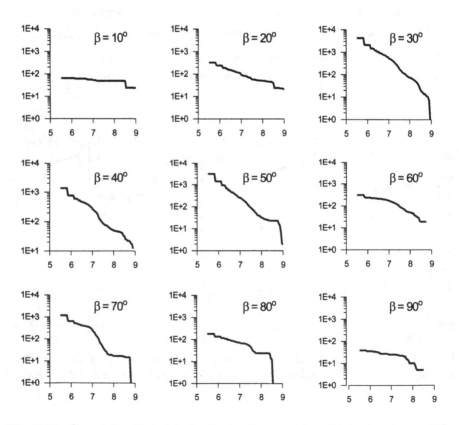

Fig. 3.18. Cumulative FM plots for the synthetic catalog obtained with $\alpha = 50°$. Numbers of earthquakes and magnitudes are plotted on the vertical and horizontal axes, respectively

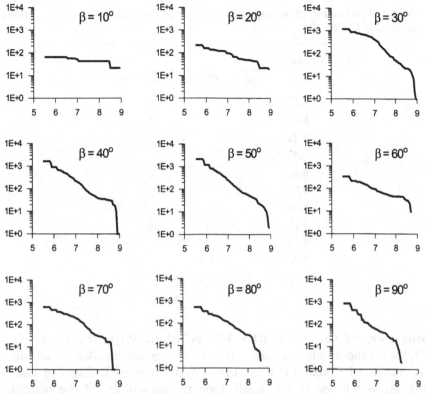

Fig. 3.19. Cumulative FM plots for the synthetic catalog obtained with $\alpha = 60°$. Numbers of earthquakes and magnitudes are plotted on the vertical and horizontal axes, respectively

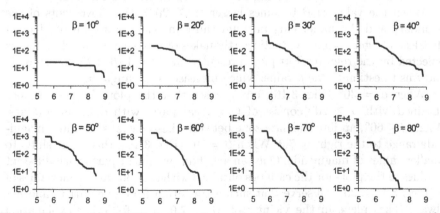

Fig. 3.20. Cumulative FM plots for the synthetic catalog obtained with $\alpha = 70°$. Numbers of earthquakes and magnitudes are plotted on the vertical and horizontal axes, respectively

Table 3.4. The number of earthquakes versus magnitude in the synthetic catalogs obtained with $\beta = 0°$

Magnitude	α				
	30°	40°	50°	60°	70°
8.37				1	
8.38				2	
8.42			1		14
8.43			2		
8.46				2	
8.51			2		
8.82					7
9.10				21	
9.15			23		
9.23		19			
9.33	10				

Parameters of the cumulative FM relation. When β is small (0° or 10°), the synthetic catalogs consist mainly of large earthquakes for all values of α considered (Table 3.4, Figs. 3.16–3.20 for $\beta = 10°$). This fact can be explained as follows. If β is small, then the movements of the boundary and the underlying medium do not cause substantial rotation of block A. Therefore magnitudes f of the elastic stress in all cells of faults 1–5 are similar, and as a rule, κ (3.11) exceeds the threshold B simultaneously in a large number of the cells giving rise to a large synthetic earthquake.

When the value of β becomes larger ($\beta \geq 20°$), the movements of the boundary and the underlying medium cause increased rotation of block A that leads to generating synthetic earthquakes of various magnitudes. This is reflected by the relevant FM plots presented in Figs. 3.16–3.20. These plots allow us to estimate the b-value, which increases with angle β.

When $\alpha = 30°$ (Fig. 3.16), one can see that the plots for the catalogs obtained with $\beta \geq 20°$ consist of two linear parts with different b-values. When $\beta \leq 60°$, an intermediate zone between these parts is seen in a magnitude range to the right of 7.0. When $\beta = 70°$ (Fig. 3.16), this range shifts to smaller values of magnitude. One can also find two linear parts with different b-values in the plots for the catalogs obtained with $\alpha > 30°$ and some values of β (Figs. 3.17–3.20). Observe that such a transition from one part to another always takes place in the vicinity of $M = 7.0$. The following explanation of this fact is suggested. In accordance with the structure depth (100 km) and the limitation (10 km) on linear cell sizes, the whole area of one vertical column of cells varies from 2000 km^2 ($\alpha = 30°$) to 1064 km^2 ($\alpha = 70°$), and it follows from (3.15) that the magnitude of an earthquake covering one vertical column of cells has to fall between 6.90 and 7.17. Therefore, the change in the b-value can be caused by the passage from earthquakes formed by cells belonging to one vertical column to earthquakes formed by cells belonging to several vertical columns.

Many cumulative FM plots presented in Figs. 3.16–3.20 have slopes reduced in the vicinity of the largest magnitude. This reflects an abnormal multiplicity of largest events that can be interpreted as "characteristic earthquakes."

Areas covered by large earthquakes. Figures 3.21–3.25 and Table 3.3 show that the area covered by the largest earthquakes from the synthetic catalogs obtained with the values of α considered decreases, as a rule, when the value of β increases. It follows from comparing Table 3.3 with Table 3.5 that the area covered by a the largest earthquake alters from the whole area of all fault planes 1–5 (when $\alpha \leq 60°$ and $\beta = 10°$) or the area of fault planes 2–5 (when $\alpha = 70°$ and $\beta = 10°$) to about the area of one fault plane (when $40° \leq \alpha \leq 60°$) or much less (when $\alpha = 30°$ or $\alpha = 70°$).

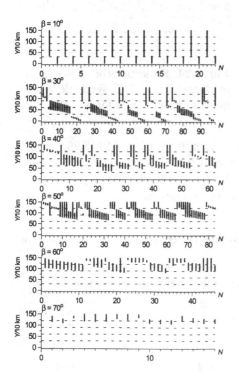

Fig. 3.21. Projections of areas covered by large ($M \geq 7.5$) synthetic earthquakes on the Y axis for $\alpha = 30°$

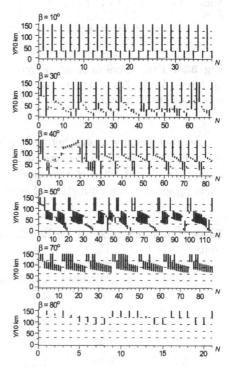

Fig. 3.22. Projections of areas covered by large ($M \geq 7.5$) synthetic earthquakes on the Y axis for $\alpha = 40°$

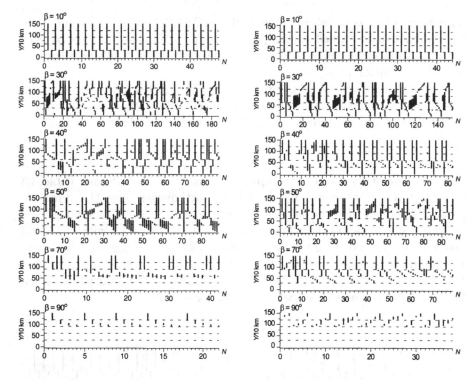

Fig. 3.23. Projections of areas covered by large ($M \geq 7.5$) synthetic earthquakes on the Y axis for $\alpha = 50°$

Fig. 3.24. Projections of areas covered by large ($M \geq 7.5$) synthetic earthquakes on the Y axis for $\alpha = 60°$

Another feature of synthetic seismicity seen from Figs. 3.21–3.25, is that larger earthquakes tend to occur at larger values of Y. This tendency becomes more pronounced when β increases, i.e., larger earthquakes take place in arc parts opposite to the direction of oceanic plate motion.

Table 3.5. Magnitudes of earthquakes completely covering a whole number of faults

Angle	Number of faults completely covered by an earthquake				
α	1	2	3	4	5
30°	8.69	8.95	9.10	9.22	9.33
40°	8.56	8.83	9.00	9.12	9.23
50°	8.48	8.76	8.92	9.04	9.15
60°	8.43	8.71	8.87	8.99	9.10
70°	8.40	8.69	8.83	8.95	9.06

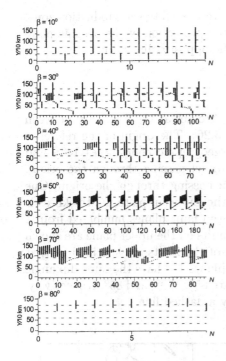

Fig. 3.25. Projections of areas covered by large ($M \geq 7.5$) synthetic earthquakes on the Y axis for $\alpha = 70°$

One can also see the migration of earthquakes in Figs. 3.21–3.25. When angle α, the slope of the subduction zone, is 30° or 40° (Figs. 3.21 and 3.25), smaller events following largest earthquakes migrate in the negative direction of Y (from fault 5 to fault 1). When $\alpha = 40°$ and $\beta = 40°$ (Fig. 3.22), there is also migration in the opposite direction in the initial part of the synthetic catalog. When $\alpha = 50°$, both kinds of migration take place in synthetic catalogs obtained with $30° \leq \beta \leq 50°$ (Fig. 3.23). For larger values of β (Fig. 3.23), there is only the first kind of migration. Both kinds of migration can also be found in synthetic catalogs obtained with $\alpha = 60°$ and $\beta = 30°, 50°$, or 90° (Fig. 3.24) and in synthetic catalogs obtained with $\alpha = 70°$ and $\beta = 30°$ or 40° (Fig. 3.25). If $\alpha = 70°$ and $\beta = 50°$ or 70° (Fig. 3.25), there is only the second kind of migration (from fault 1 to fault 5).

It can be inferred how the synthetic seismicity obtained in the abstract model of a subduction zone depends on its slope α and on the angle β between velocities of plates.

1. Seismic activity becomes higher when the slope α increases from 30° to 40°–50°, decreases slightly with a further increase in α, and drops when α exceeds a certain critical value at about 70°.

2. Seismic activity as a function of the direction difference β has a peak at about 40°.

3. The b-value for synthetic catalogs increases with β.

4. Seismicity migrates along the island arc related to the subduction zone. In most cases, migration is directed like the projection of the oceanic plate velocity on the arc; in some cases it has the opposite direction.

3.5.2   Model of the Sunda Arc

Description of the model. The Sunda Arc lies at the boundary between the Eurasian and Australian plates (Fig. 3.26). The origin of the reference coordinate system is the point with the geographic coordinates 14°S and 90°E. The X axis is the east-oriented parallel passing through the origin. The Y axis is the north-oriented meridian passing through the origin. One block is considered to represent a part of the Eurasian plate. Its thickness is $H = 130$ km. This block is bounded by seven fault planes numbered from 1 through 7 (Fig. 3.26). The corresponding faults on the upper plane are shown in Fig. 3.26 by solid lines. The directions of the faults are specified so that the block is on the left of each fault. Fault planes 1–4 have the same dip angle of 21°. The dip angles of fault planes 5–7 are 85°, 159°, and 85°, respectively. The block bottom is shown in Fig. 3.26 by a dashed line.

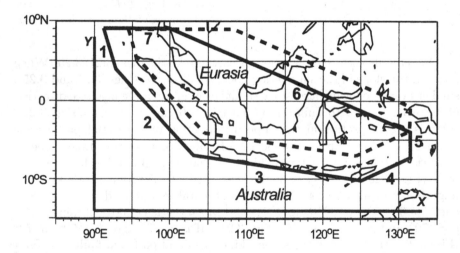

Fig. 3.26. Model of the Sunda Arc: Configuration of faults on the upper plane (*solid lines*) and on the lower plane (*dashed lines*); numbers 1 through 7 identify the faults

Fault planes 1–4 form the boundary between the Eurasian and Australian plates. The same values of the constants in (3.5), and (3.6) are specified for these fault planes: $K = 1\,\mathrm{bar\,cm^{-1}}$, $W = 0.025\,\mathrm{cm\,bar^{-1}}$. For fault planes 5–7, $K = 0$, and (3.5) implies that any displacements do not produce forces at these fault planes. For the underlying medium, $K_u = 1\,\mathrm{bar\,cm^{-1}}$, and $W_u = 0.025\,\mathrm{cm\,bar^{-1}}$.

The values $\varepsilon = 15\,\text{km}$ and $\Delta t = 0.001$ were used for discretization.

The value of P in (3.11) is 2 kbar. The levels for κ (3.11) have the same values for all faults: $B = 0.1$, $H_f = 0.085$, and $H_s = 0.07$. $W_s = 5\,\text{cm}\,\text{bar}^{-1}$. The magnitude of synthetic earthquakes is calculated from (3.15).

The HS2-NUVEL1 model [GG90] is used to specify the movements. The velocities of several points within Eurasia and Australia in the vicinity of their interface are taken from HS2-NUVEL1. The following velocities of blocks are obtained by the least-squares method. The block (Eurasia) moves at the translational velocity $V_x = -0.604\,\text{cm}$, $V_y = 0.396\,\text{cm}$ and the angular velocity 0.06×10^{-8} rad. The boundary block adjacent to fault planes 1–4 (Australia) moves at the translational velocity $V_x = 1.931\,\text{cm}$, $V_y = 6.560\,\text{cm}$ and the angular velocity 0.42×10^{-8} rad. In both cases, the coordinate origin is the center of rotation, and its positive direction is counterclockwise. Therefore, the movement of Australia relative to Eurasia is translation at the velocity $V_x = 2.535\,\text{cm}$, $V_y = 6.164\,\text{cm}$, and rotation about the origin at the angular velocity 0.36×10^{-8} rad. One unit of dimensionless time stands for one year.

Results of modeling. The dynamics of the block structure was modeled across 200 units of dimensionless time, starting from zero. At zero time, zero displacements of the block boundary formed by fault planes 1–4 and of the underlying medium were assumed, as well as zero inelastic displacements in all cells. In the case considered basic, the movement of the boundary formed by fault planes 1–4 consisted of translation $V_x = 2.535\,\text{cm}$, $V_y = 6.164\,\text{cm}$, and rotation about the origin at the angular velocity 0.36×10^{-8} rad; respective values for the underlying medium were $V_x = -1.025$ cm, $V_y = -3.340\,\text{cm}$, 0.13×10^{-8} rad, the rotation about the geometric center of the block bottom. This movement of the underlying medium was specified to make the block immobile and therefore, to make the relative movement of the boundary and the block identical to the movement of the boundary.

The seismicity observed in the region is compared with the stable part of the synthetic earthquake catalog from 100 to 200 units of dimensionless time. In accordance with the movements specified, this period corresponds to 100 years.

Figure 3.27 presents the cumulative FM plots for the synthetic catalog and the observed seismicity. One can see that the slopes of the curves (b-values) are close, but the synthetic curve is shifted to larger magnitudes. The value of the shift is about 1. The synthetic curve with magnitudes reduced by 1 is also shown in Fig. 3.27. It is rather close to the curve for the observed seismicity.

The epicenter maps of earthquakes with $M \geq 6.0$ observed in the Sunda Arc and the synthetic epicenters with $M \geq 7.0$ are shown in Figs. 3.28 and 3.29. In both cases, the largest earthquakes occur in the eastern part of the arc.

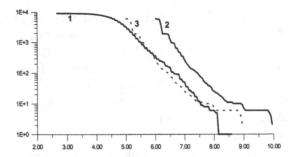

Fig. 3.27. Cumulative FM plots for the observed seismicity of the Sunda Arc (curve 1), for the synthetic catalog (curve 2), and for the synthetic catalog with magnitudes reduced by 1 (curve 3)

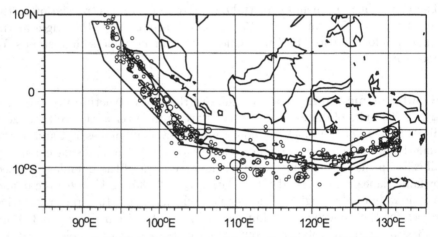

Fig. 3.28. Seismicity observed in the Sunda Arc, $M \geq 6.0$

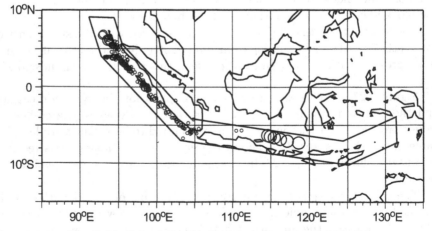

Fig. 3.29. Seismicity obtained from the model, $M \geq 7.0$

Rundquist et al. [RVR98,RSV98] found the migration of earthquakes with $M \geq 6.0$ observed in the Sunda Arc. The migration is directed from the eastern to the western part of the arc. The same method [VS79] applied to the synthetic catalog indicated the migration of synthetic earthquakes with $M \geq 7.0$ in the same direction.

Several experiments were done to study the dependence of the cumulative FM plot for the synthetic catalog on the movements specified in the model. In the first group of experiments, the angular velocity of the boundary formed by fault planes 1–4 was varied, and its translational velocity was the same as that in the basic case. In the second group of experiments, the translational velocity of the boundary varied, and its angular velocity was the same as that in the basic case. In each experiment, the movement of the medium underlying the block was specified to make the block immovable.

Changing the angular velocity of the boundary. The angular velocity took on the following values (in rad): 0.09×10^{-8}, 0.18×10^{-8}, 0.54×10^{-8}, and 0.72×10^{-8}. Figure 3.30 shows cumulative FM plots obtained for the synthetic catalogs. The curve obtained in the basic case is also shown for comparison. One can see that when the angular velocity is significantly less than that in the basic case, the b-value of the FM plot is also less. Larger b-values are obtained for larger angular velocities. This is in line with the results obtained by Keilis-Borok et al. [KRS97]; see also Sect. 3.3.

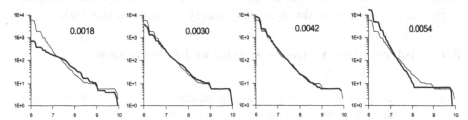

Fig. 3.30. Dependence of the cumulative FM plots from synthetic earthquake catalogs on the Australia angular velocity relative to Eurasia. The plot for the basic case (*thin lines*) is compared with plots for different values of the angular velocity (*thick lines*). The respective values of the angular velocity are shown in 10^{-6} rad yr^{-1}

Changing the translational velocity of the boundary. In these experiments, the direction of the translational velocity was varied, and its magnitude remained the same as that in the basic case. The angle between the east-oriented parallel and the translational vector is about 68° in the basic case. Modeling was carried out with the following values of this angle: 28°, 48°, 88°, and 108°. Figure 3.31 presents cumulative FM plots obtained for the respective synthetic earthquake catalogs. The plot for the basic case is also shown for comparison. One can see how the b-value of the FM plot depends on the direction of the boundary translational velocity.

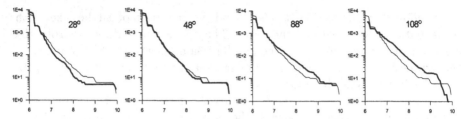

Fig. 3.31. Dependence of the cumulative FM plots from synthetic earthquake catalogs on the direction of the Australia translational movement relative to Eurasia. The plot for the basic case (*thin lines*) is compared with plots obtained for different values of the angle between the direction of the movement and the east-oriented parallel (*thick lines*)

The above discussion shows that the model of the Sunda Arc with movements specified in HS2-NUVEL1 yields synthetic seismicity having certain common features with observations; they include the locations of larger events, the direction of migration of earthquakes, and the b-value of the FM plot.

3.6   Models of Block-and-Fault Dynamics of the Vrancea Region (the Southeastern Carpathians)

Vrancea was selected as an object for modeling because it is a relatively small seismoactive region with a high level of seismic activity. The strongest European earthquakes of the twentieth century occurred in this region.

3.6.1   Introduction to the Seismicity and Geodynamics of the Region

The earthquake-prone Vrancea region is situated at a bend of the Eastern Carpathians and is bounded on the north and northeast by the East European platform, on the east and south by the Moesian platform, and on the west by the Transylvanian and Pannonian basins (Fig. 3.32). The epicenters of mantle earthquakes in the Vrancea region are concentrated within a very small area (about 40×80 km, Fig. 3.33a), and the distribution of the epicenters is much denser than that of intermediate-depth events in other intracontinental regions. The projection of the foci on the NW–SE vertical plane across the bend of the Eastern Carpathians (Fig. 3.33b) shows a seismogenic body in the form of a square box about 100 km long, about 40 km wide, and extending to a depth of about 180 km. Beyond this depth, the seismicity ends suddenly: a seismic event represents an exception beneath 180 km [Tri90, TDRL91, OB97].

As early as 1949, Gutenberg and Richter [GR54] drew attention to the remarkable source of shocks in the depth range of 100 km to 150 km in the Vrancea region. According to historical data, there have been 16 large

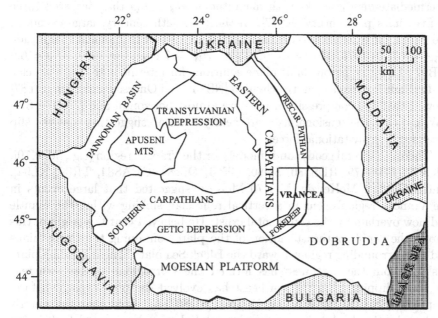

Fig. 3.32. Tectonic sketch of the Carpathian area, modified after [IPN00]

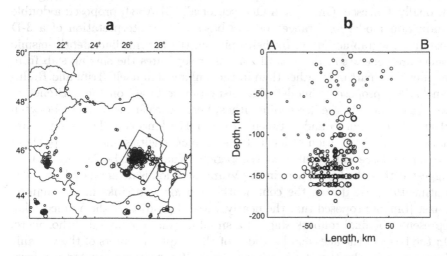

Fig. 3.33. Map of the observed seismicity in Vrancea: (a) Epicenters of Romanian earthquakes with magnitude greater than 4 that occurred since 1900; (b) Hypocenters of the same Romanian earthquakes projected onto the vertical plane AB along the NW-SE direction. Several catalogs have been combined to prepare the figure [VNK+96]

intermediate-depth shocks with magnitudes $M_S > 6.5$ that occurred three to five times per century [KS77]. In the twentieth century, large events in the depth range of 70 to 170 km occurred in 1940 with moment magnitude $M_W = 7.7$, in 1977, $M_W = 7.4$, in 1986, $M_W = 7.1$, and in 1990, $M_W = 6.9$ [OB97]. Using numerous fault-plane solutions for intermediate-depth shocks, Radu [Rad67], Nikolaev and Shchyukin [NS75], and Oncescu and Trifu [OT87] showed that the compressional axes are almost horizontal and directed SE–NW and that the tensional axes are nearly vertical, suggesting that the slip is caused by gravitational forces.

There are several geodynamic models for the Vrancea region (e.g., [McK70], [McK72, FBB+79, RDSS80, SD80, CE84, Onc84, OBAS84, TR89, KL94], [Lin96, GF98]). McKenzie [McK70, McK72] suggested that large events in the Vrancea region occur in a vertical relic slab sinking within the mantle and now overlain by the continental crust. He believed that the origin of this slab is the rapid southeast motion of the plate containing the Carpathians and the surrounding regions toward the Black Sea plate. The overriding plate pushing from the northwest has formed the Carpathian orogen, whereas the plate dipping from the southeast has evolved the pre-Carpathian fore-deep [RDSS80]. Shchyukin and Dobrev [SD80] suggested that the mantle earthquakes in the Vrancea region are related to a deep-seated fault going steeply down. The Vrancea region was also considered [FBB+79] a place where an oceanic slab detached from the continental crust is sinking gravitationally. Oncescu [Onc84] and Oncescu et al. [OBAS84] proposed a double subduction model for Vrancea on the basis of the interpretation of a 3-D seismic tomographic image. In their opinion, the intermediate-depth seismic events are generated in a vertical zone that separates the sinking slab from the immobile part of it rather than in the sinking slab itself. Trifu and Radulian [TR89] proposed a model of a seismic cycle based on the existence of two active zones in the descending lithosphere beneath the Vrancea between 80 and 110 km deep and between 120 and 170 km deep. These zones are marked by a distribution of local stress inhomogeneities and are capable of generating large earthquakes in the region. Khain and Lobkovsky [KL94] suggested that the lithosphere in the Vrancea region is delaminated from the continental crust during the continental collision and sinks in the mantle. Linzer [Lin96] proposed that the nearly vertical position of the Vrancea slab represents the final rollback stage of a small fragment of oceanic lithosphere. On the basis of the ages and locations of the eruption centers of the volcanic chain and also the thrust directions, Linzer [Lin96] reconstructed a migration path of the retreating slab between the Moesian and East-European platforms. Most recently Girbacea and Frisch [GF98] suggested a model of subduction beneath the suture followed by delamination. The model can explain the location of earthquake hypocenters and of calc-alkaline volcanics, surface structure, and an accretionary wedge.

According to these models, the cold (hence denser and more rigid than the surrounding mantle) relic slab beneath the Vrancea region sinks due to gravity. The active subduction ceased about 10 Ma ago; thereafter, only some slight horizontal shortening was observed in the sedimentary cover [WLN98]. Hydrostatic buoyancy forces help the slab to subduct, but viscous and frictional forces resist the descent. At intermediate depths, these forces produce an internal stress with one principal axis directed downward. Earthquakes occur in response to this stress. These forces are not the only source of stress that leads to seismic activity in Vrancea; the process of slab descent may cause seismogenic stress by mineralogical phase changes and dehydration of rocks, which possibly leads to fluid-assisted faulting [IPN00].

3.6.2   Block Structure of the Vrancea Region: Model *A*

In accordance with [Ari74], the main structural elements of the Vrancea region are (i) the East-European plate; (ii) the Moesian plate, (iii) the Black Sea; and (iv) the Intra-Alpine (Pannonian-Carpathian) subplates (Fig. 3.34). The fault separating the East-European plate from the Intra-Alpine and Black Sea subplates and the fault separating the Intra-Alpine and Black Sea subplates have dip angles significantly different from 90° [Moc93]. The main directions of the movement of various plates are shown in Fig. 3.34.

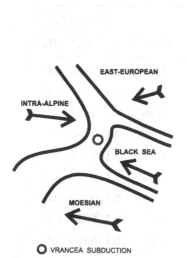

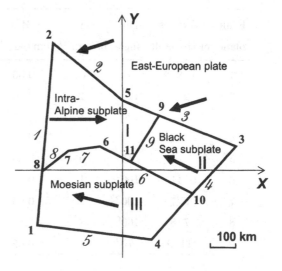

Fig. 3.34. Gross kinematic model proposed for the double subduction process in the Vrancea region (modified after [Moc93])

Fig. 3.35. Block structure used in numerical modeling. Vertices marked by numbers (**1–11**), faults (*1–9*), and blocks (**I–III**) are shown in the figure. The movements of blocks and the underlying medium are indicated by arrows outside and inside the structure, respectively

This information is sufficient to set a block structure serving as a coarse approximation of the Vrancea region and to specify movements necessary to model the dynamics of this block structure.

Figure 3.35 shows the pattern of faults on the upper plane of the block structure used to model the Vrancea region. The point with the geographic coordinates 44.2°N and 26.1°E is chosen as the origin of the reference coordinate system. The X axis is the east-oriented parallel passing through the origin, and the Y axis is the north-oriented meridian passing through the origin.

The layer is $H = 200$ km thick which agrees with the depth of the deepest earthquakes in the Vrancea region. Vertices of the block structure numbered 1–7 have the following coordinates (in km): $(-330, -210)$; $(-270, 480)$; $(450, 90)$; $(110, -270)$; $(0, 270)$; $(-90, 90)$; $(-210, 75)$. Vertices 8–11 have the following relative positions on the faults to which they belong: 0.3, 0.33, 0.5, 0.667. The relative position of each vertex is the ratio of its distance from the initial point of the fault to its length. Vertices 1, 5, 3, and 10 are considered initial points of the respective faults. The structure contains nine fault planes. The values of the parameters for these fault planes are given in Table 3.6, and the values of the parameters for the three blocks that form the structure are given in Table 3.7.

Table 3.6. Parameters of fault planes

Fault plane	Vertices of the fault	Dip angle	K $\mathrm{bar\,cm}^{-1}$	W $\mathrm{cm\,bar}^{-1}$	W_s $\mathrm{cm\,bar}^{-1}$	B	H_f	H_s
1	1, 8, 2	45°	0	0.00	0.00	0.1	0.085	0.07
2	2, 5	120°	1	0.50	1.00	0.1	0.085	0.07
3	5, 9, 3	120°	1	0.05	0.10	0.1	0.085	0.07
4	3, 10, 4	45°	0	0.00	0.00	0.1	0.085	0.07
5	4, 1	45°	0	0.00	0.00	0.1	0.085	0.07
6	10, 11, 6	100°	1	0.05	0.10	0.1	0.085	0.07
7	6, 7	100°	1	0.05	0.10	0.1	0.085	0.07
8	7, 8	100°	1	0.05	0.10	0.1	0.085	0.07
9	11, 9	70°	1	0.02	0.04	0.1	0.085	0.07

The movement of the underlying medium is progressive. The components of its velocity (V_x, V_y) are specified for blocks in accordance with the directions of the main movements in the Vrancea region shown in Fig. 3.34 and are given in Table 3.7. Dimensionless time is used in modeling, and the values of W and W_s in Table 3.6 as well as the values of W_u and the velocities (V_x, V_y) are measured per unit.

Table 3.7. Parameters of blocks

Block	Block vertices	K_u, bar cm^{-1} [see (3.1)]	W_u, cm bar^{-1} [see (3.2)]	V_x, cm	V_y, cm
1	2, 8, 7, 6, 11, 9, 5	1	0.05	25	0
2	3, 9, 11, 10	1	0.05	−15	7
3	4, 10, 11, 6, 7, 8, 1	1	0.05	−20	5

The boundary consisting of fault planes 2 and 3 moves progressively at the same velocity: $V_x = -16$ cm, $V_y = -5$ cm. The boundary fault planes 1, 4, and 5 do not mark any real geologic features of the Vrancea region and are introduced only to limit the block structure; they are assigned $K = 0$ (Table 3.6). Therefore owing to (3.5) and (3.8), all forces at these fault planes are equal to zero. The value of P in (3.11) is 2 kbar. The magnitude of synthetic earthquakes is calculated from (3.15). The values of the parameters for the discretization in time and space are $\Delta t = 0.001$ and $\varepsilon = 7.5$ km, respectively.

3.6.3 Comparing Vrancea Seismicity with the Results from Model A

The model A synthetic earthquake catalog was computed for the period of 200 dimensionless time units starting from zero initial conditions (zero displacements of boundary blocks and the underlying medium and zero inelastic displacements for all cells). The synthetic catalog contained 9439 events with magnitudes varying from 5.0 to 7.6. The minimum value of the magnitude corresponds, in accordance with (3.15), to the minimum area of one cell. The maximum value of the magnitude in the synthetic catalog is 7.6.

Vrancea is a relatively small region with a high level of seismic activity, mainly occurring at an intermediate depth. In the twentieth century, four catastrophic earthquakes of magnitudes 7 or more (defined as maximum magnitudes given in catalogs) have occurred there (Table 3.8). Note that the maximum magnitude (7.6) in the synthetic catalog is close to that ($M = 7.4$) observed in reality (Table 3.8).

Figure 3.36 presents the observed seismicity for the period 1900–1995, and Fig. 3.37 shows the distribution of epicenters from the synthetic catalog. The majority of synthetic events occur on fault plane 9 (cluster A in Fig. 3.37), which corresponds to the subduction zone of Vrancea where most of the observed seismicity is concentrated (cluster A in Fig. 3.36). All large synthetic earthquakes concentrate here, and the same phenomenon is seen in the distribution of observed events.

Some events occur on fault plane 6, and they appear as a cluster of epicenters (cluster B in Fig. 3.37) located southwest of the main seismic area and separated from it by a nonseismic zone. An analogous cluster of epicenters

Table 3.8. Large earthquakes of Vrancea, 1900–2000

Date	Time	Hypocenter			Magnitude
		Latitude	Longitude	Depth (km)	
1940/11/10	01 h 39 m	45.80°N	26.70°E	133	7.4
1977/03/04	19 h 21 m	45.78°N	26.80°E	110	7.2
1986/08/30	21 h 28 m	45.51°N	26.47°E	150	7.0
1990/05/30	10 h 40 m	45.83°N	26.74°E	110	7.0

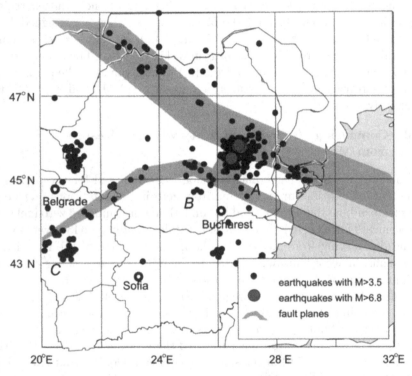

Fig. 3.36. Map of observed seismicity in Vrancea in the period 1900–1995. Gray areas are the projections of fault planes with $K \neq 0$ on the upper plane

can be seen on the map of observed seismicity (cluster B in Fig. 3.36). The third cluster of events (cluster C in Fig. 3.37) groups on fault plane 8 and corresponds to cluster C of the observed seismicity in Fig. 3.36.

There are several additional clusters of epicenters on the map of the observed seismicity (Fig. 3.36) that are absent in the synthetic catalog. This is not surprising because only a few main seismic faults of the Vrancea region are included in the model. A block structure reflecting a real fault system

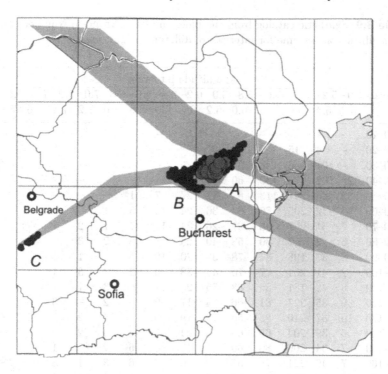

Fig. 3.37. Map of synthetic seismicity obtained from modeling the Vrancea block structure dynamics. Gray areas are the projections of fault planes with $K \neq 0$ on the upper plane

at hand in greater detail is needed to obtain a more realistic distribution of synthetic epicenters. Nevertheless, even a very simple structure consisting of only three blocks reproduces the main features of distribution of observed seismicity in space.

Table 3.9 presents the number of synthetic earthquakes versus magnitude and time. Modeling starts from zero conditions, and some time is needed for quasi-stabilization of stresses. It is possible to estimate the stabilization time from the time–magnitude histogram (Table 3.9). Starting from 60 units of time, the distribution of the number of events versus magnitude and time looks stable. The stable part of the synthetic catalog, from 60 to 200 units of dimensionless time only, enters the analysis below.

The Gutenberg–Richter frequency–magnitude law for observed seismicity states that the logarithm of the number of earthquakes depends linearly on the magnitude. Figure 3.38 depicts FM plots for the observed seismicity in Vrancea and for the synthetic catalog. The curve constructed from the synthetic catalog (*dashed line*) is close to linear and has approximately the same slope as the curve constructed from the observed seismicity (*solid line*).

Table 3.9. Synthetic catalog from the block model of Vrancea: Number of events versus dimensionless time for intervals of different magnitude

Time intervals	Magnitude intervals														Total
	5.0–5.2	5.2–5.4	5.4–5.6	5.6–5.8	5.8–6.0	6.0–6.2	6.2–6.4	6.4–6.6	6.6–6.8	6.8–7.0	7.0–7.2	7.2–7.4	7.4–7.6	7.6–7.8	
0–10	–	–	–	–	–	–	–	–	–	–	–	–	–	–	0
10–20	–	4	15	5	–	–	–	–	–	–	–	–	–	–	24
20–30	–	101	556	148	37	4	–	–	–	–	–	–	–	–	846
30–40	28	102	378	130	60	23	4	–	–	–	–	–	–	–	725
40–50	39	66	247	113	83	47	27	7	1	–	–	–	–	–	630
50–60	6	42	205	76	62	36	24	14	5	1	–	–	–	–	471
60–70	7	51	227	98	67	26	24	11	7	1	2	1	–	–	522
70–80	5	42	207	80	58	40	23	8	8	2	3	–	–	–	476
80–90	8	37	198	87	78	37	20	10	5	–	1	1	–	–	482
90–100	2	41	201	77	70	43	38	9	2	3	3	1	–	–	478
100–110	8	42	181	86	68	35	20	11	6	3	1	–	–	–	461
110–120	9	35	227	81	66	34	17	9	6	2	–	3	–	–	489
120–130	10	37	210	88	56	29	19	9	–	4	2	–	1	–	465
130–140	6	39	204	65	69	33	24	5	4	3	2	2	–	–	456
140–150	7	34	206	80	60	37	15	4	6	2	1	1	–	–	453
150–160	7	49	221	76	67	32	20	7	6	3	1	3	–	–	492
160–170	15	37	227	92	64	23	21	9	5	5	1	1	–	–	500
170–180	12	55	219	62	66	25	19	14	–	1	2	1	–	–	476
180–190	11	42	192	88	68	23	21	7	9	–	–	1	–	1	463
190–200	7	55	220	98	81	31	21	7	7	3	–	–	–	–	530
Total	187	911	4341	1630	1180	558	345	141	77	33	19	15	1	1	9439

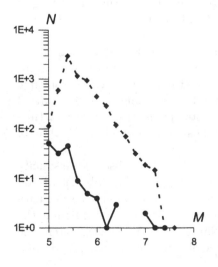

Fig. 3.38. FM plots for the real (*solid line*) and synthetic (*dashed line*) catalogs

There is a gap in the observed seismicity in the magnitude interval $6.5 < M < 7.0$ that is not visible in the synthetic catalog. This gap may appear because the time window for which relevant observations are available is too short.

It is possible to associate one unit of dimensionless time with a real time interval from frequency–magnitude plots and the duration of the real catalog (95 years); 140 units of dimensionless time correspond to about 7000 years, or equivalently 1 unit to about 50 years. This estimate implies that the velocity of the tectonic movement is about 5 mm per year.

The above estimate makes the duration of the stable part in the synthetic catalog 70 times longer than the total period of the observed catalog. According to this scale, the synthetic catalog was divided into 70 parts, and these parts were compared with each other and with the real catalog. Table 3.10 shows how the number of events depends on the magnitude for these 70 parts.

The real catalog contains 71 events with $M \geq 5.4$, the maximum magnitude is 7.4, and there are four large events (Table 3.8). The number of events with $M \geq 5.4$ in each of the 70 parts of the synthetic catalog varies from 53 to 94; the average number of such events is 68, and the maximum magnitude varies from 6.0 to 7.6 (Table 3.10). If a synthetic event with $M \geq 6.8$ is considered large, then the number of large earthquakes varies from zero to four. There are no large events in 29 parts, one large event in 20 parts, two large events in 16 parts, three large events in 4 parts, and four large events in 1 part.

Four large events (as in the observed present-day seismicity) occurred in only one part of the synthetic catalog. This may indicate that nowadays we live in an "active period" of the Vrancea region.

It is interesting to compare FM plots for periods with and without large earthquakes. The trend and the intensity of earthquake flows for the real (*solid line*) and the synthetic (*dashed line*) catalogs for the period (part 49 in Table 3.10) without large earthquakes (Fig. 3.39) are quite similar. There is a gap in the number of earthquakes across the magnitude range from 6.4 to 6.8 in the plot for the period (part 16 in Table 3.10) with four large earthquakes (Fig. 3.40), a pattern similar to that observed. Such a gap is a typical phenomenon for parts with several large shocks (Table 3.10).

The temporal distribution of large ($M > 6.8$) synthetic earthquakes in the period from 60 to 200 units of dimensionless time (7000 years) is presented in Fig. 3.41. One can see strong irregularity in the flow of these events.

For example, groups of large earthquakes occur periodically in the interval from 70 to 120 units of dimensionless time, with a return period of about 6–7 units (300–350 years). The unique part (part 16 in Table 3.10) of the synthetic catalog with four large shocks belongs to this time interval.

The periodic occurrence of a single large earthquake with a return period of about 2 units, or 100 years, is typical of the interval from 120 to 140

Table 3.10. Number of events versus magnitude for 70 parts of the synthetic catalog obtained from the block model of Vrancea

Part #	Magnitude intervals													Total	M_{max}
	5.4–5.6	5.6–5.8	5.8–6.0	6.0–6.2	6.2–6.4	6.4–6.6	6.6–6.8	6.8–7.0	7.0–7.2	7.2–7.4	7.4–7.6	7.6–7.8	$M \geq$ 7.8		
1	22	18	17	5	5	4	–	–	–	–	–	–	0	71	6.4
2	25	18	9	2	2	2	–	–	–	1	–	–	1	59	7.2
3	26	30	18	9	6	3	1	–	1	–	–	–	1	94	7.1
4	25	12	8	8	5	–	4	–	–	–	–	–	0	62	6.6
5	31	20	15	2	6	2	2	1	1	–	–	–	2	80	7.0
6	24	14	5	10	2	2	3	1	–	–	–	–	1	61	6.8
7	19	20	12	12	6	2	–	–	1	–	–	–	1	72	7.0
8	22	8	13	2	8	2	1	–	1	–	–	–	1	57	7.0
9	25	19	10	8	6	–	3	–	–	–	–	–	0	71	6.6
10	23	19	18	8	1	2	1	1	1	–	–	–	2	74	7.1
11	23	15	14	9	3	3	–	–	–	–	–	–	0	67	6.4
12	21	25	16	2	3	3	1	–	–	–	–	–	0	71	6.6
13	22	19	11	10	4	–	1	–	1	1	–	–	2	69	7.2
14	22	13	24	9	6	1	1	–	–	–	–	–	0	76	6.6
15	20	15	13	7	4	3	2	–	–	–	–	–	0	64	6.6
16	22	15	18	5	9	2	–	1	2	1	–	–	4	75	7.2
17	25	12	15	6	2	2	1	–	–	–	–	–	0	63	6.6
18	23	18	9	10	7	1	1	–	–	–	–	–	0	69	6.6
19	20	19	12	10	5	2	–	2	1	–	–	–	3	71	7.1
20	23	13	16	12	3	2	–	–	–	–	–	–	0	69	6.4
21	21	12	12	10	1	2	1	–	–	–	–	–	0	59	6.6
22	20	24	13	7	5	–	1	2	–	–	–	–	2	72	6.9
23	18	12	9	3	4	1	4	1	1	–	–	–	2	53	7.0
24	22	21	14	7	6	4	–	–	–	–	–	–	0	74	6.4
25	21	17	20	8	4	4	–	–	–	–	–	–	0	74	6.4
26	30	19	14	8	2	3	2	–	–	–	–	–	0	78	6.6
27	30	10	17	5	4	2	2	–	–	2	–	–	2	72	7.3
28	16	15	10	9	4	2	–	1	–	1	–	–	2	58	7.2
29	23	17	14	6	5	1	1	1	–	–	–	–	1	68	6.8
30	30	20	11	6	2	1	1	–	–	–	–	–	0	71	6.6
31	21	19	7	6	2	2	–	1	1	–	–	–	2	59	7.0
32	17	19	8	5	4	2	–	1	–	–	1	–	2	57	7.4
33	19	14	11	4	3	1	–	–	1	–	–	–	1	53	7.0
34	24	15	17	6	7	1	–	1	–	–	–	–	1	71	6.9
35	20	21	13	8	3	3	–	1	–	–	–	–	1	69	6.8
36	23	17	19	6	3	2	2	–	–	–	–	–	0	72	6.6
37	28	17	4	7	7	1	1	–	1	2	–	–	3	68	7.3
38	31	7	18	4	7	1	–	1	–	–	–	–	1	69	6.8
39	17	11	12	8	3	1	1	1	–	–	–	–	1	54	6.9
40	26	13	16	8	4	–	–	1	1	–	–	–	2	69	7.1

Table 3.10. (Continued)

Part #	Magnitude intervals													Total	M_{max}
	5.4–5.6	5.6–5.8	5.8–6.0	6.0–6.2	6.2–6.4	6.4–6.6	6.6–6.8	6.8–7.0	7.0–7.2	7.2–7.4	7.4–7.6	7.6–7.8	$M \geq 7.8$		
41	22	13	12	11	1	1	–	1	1	–	–	–	2	62	7.0
42	18	18	10	8	3	1	1	–	–	1	–	–	1	60	7.3
43	24	15	13	6	2	1	2	1	–	–	–	–	1	64	6.8
44	31	20	11	8	4	–	2	–	–	–	–	–	0	76	6.6
45	22	14	14	4	5	1	1	–	–	–	–	–	0	61	6.6
46	25	17	14	8	1	1	3	1	–	1	–	–	2	71	7.2
47	24	16	17	6	4	2	2	–	1	–	–	–	1	72	7.0
48	25	14	12	6	4	1	–	2	–	1	–	–	3	65	7.2
49	26	19	11	9	2	2	–	–	–	–	–	–	0	69	6.4
50	26	10	13	3	9	1	1	–	–	1	–	–	1	64	7.2
51	20	18	19	6	3	3	–	–	–	–	–	–	0	69	6.4
52	31	20	5	3	2	1	3	1	–	1	–	–	2	67	7.2
53	28	14	9	7	3	1	–	–	–	–	–	–	0	62	6.4
54	27	18	15	4	6	2	–	2	1	–	–	–	3	75	7.0
55	30	22	16	3	7	2	2	2	–	–	–	–	2	84	6.9
56	27	5	11	2	2	5	–	–	2	–	–	–	2	54	7.0
57	26	12	14	9	–	–	–	–	–	–	–	–	0	61	6.0
58	19	17	17	5	6	3	–	–	–	1	–	–	1	68	7.3
59	25	9	12	7	6	3	–	1	–	–	–	–	1	63	6.8
60	25	19	12	2	5	3	–	–	–	–	–	–	0	66	6.4
61	23	22	14	7	4	3	4	–	–	–	–	–	0	77	6/6
62	19	14	11	5	5	2	1	–	–	1	–	1	2	59	7.6
63	19	22	14	3	4	–	–	–	–	–	–	–	0	62	6.2
64	23	16	15	6	5	1	2	–	–	–	–	–	0	68	6.6
65	25	14	14	2	3	1	2	–	–	–	–	–	0	61	6.6
66	24	23	11	5	6	1	2	1	–	–	–	–	1	73	6.8
67	22	20	15	4	1	1	2	1	–	–	–	–	1	66	6.8
68	25	15	16	8	4	2	1	–	–	–	–	–	0	71	6.6
69	23	21	20	8	7	2	1	–	–	–	–	–	0	82	6.6
70	19	16	15	5	2	1	1	1	–	–	–	–	1	60	6.8

units. There is no periodicity in the occurrence of large earthquakes in the remaining parts of the synthetic catalog.

These results show that it is necessary to be careful when using seismic cycle for predicting of the occurrence of a future large earthquake, because the available observations cover only a very short time interval, in comparison with the timescale of tectonic processes. This is particularly relevant for the regularity of maxima in the seismic activity of the Vrancea region discussed in [NVE+95].

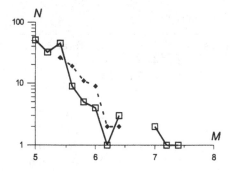

Fig. 3.39. FM plots for the real catalog (*solid line*) and for part 49 (Table 3.10) of the synthetic catalog without large earthquakes (*dashed line*)

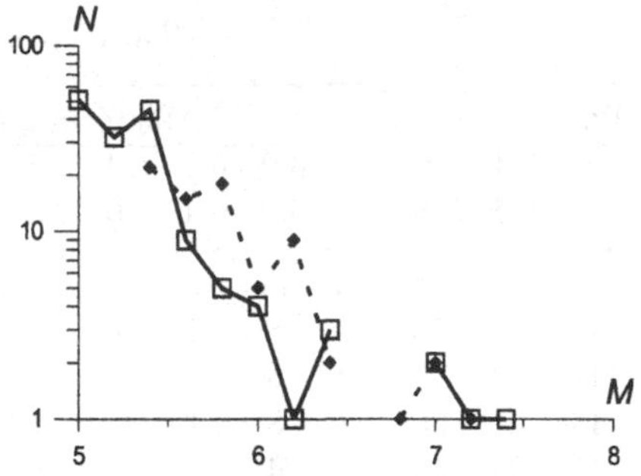

Fig. 3.40. FM plots for the real catalog (*solid line*) and for part 16 (Table 3.10) of the synthetic catalog with four large earthquakes (*dashed line*)

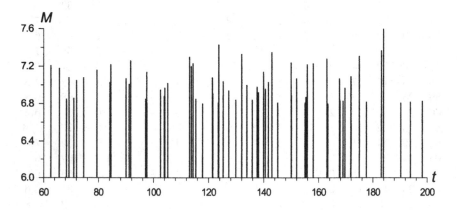

Fig. 3.41. Temporal distribution of large earthquakes in the synthetic catalog for the interval from 60 to 200 units of dimensionless time (about 7000 years)

These results show that the simulation of the block structure dynamics in the Vrancea region results in a synthetic catalog that has features similar to those observed. The values of the model parameters yielding an agreement between synthetic and observed catalogs can be useful for estimating the velocities of tectonic movements and the values of physical parameters inherent in dynamic processes in fault zones. Assume that a segment approximating the real seismic flow at sufficient accuracy is identified in the synthetic catalog. Then the part of the synthetic catalog immediately following this segment could be used to predict the future behavior of seismicity in the region at hand.

3.6.4   Numerical Tests on Model *A* Parameters

Various numerical tests were carried out to study the dependence of synthetic earthquake catalogs on the values of model *A* parameters [SVP99]. The set of parameters given in Sect. 3.6.3 was assumed as a benchmark, because it results in a synthetic catalog close to that observed in the Vrancea region. Some features of synthetic catalogs deviate considerably from those of the benchmark in varying boundary block velocities and the rate of inelastic displacements.

The velocities of the boundary blocks and of the underlying medium represent the rates and directions of tectonic motions in the region. The inelastic coefficient W of the displacement rate represents the mobility of the fault; the smaller the coefficient, the less mobile or, equivalently, more locked the fault. Therefore, the value of W controls the part of the energy released by earthquakes and by viscous displacements in the fault. When W is large enough, earthquakes do not occur at all, and we have only creep.

The following features are considered: (1) the spatial distribution of epicenters; (2) the slope of the FM plot; (3) the level of seismic activity (the number of events and their total source area); (4) the maximum magnitude of the events that occurred; and (5) the relative activity of fault plane 9, which represents the Vrancea subduction zone. The last feature is characterized by two values: the first is the ratio (in percent) of the number of events on fault plane 9 to the number of events in the whole structure; the second is the ratio (in percent) of the total source area of events on fault plane 9 to the total source area of events in the whole structure.

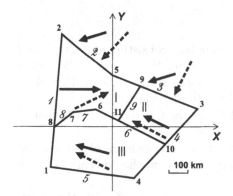

Fig. 3.42. A sketch of the block structure for the test of velocity variation. *Solid arrows* stand for velocities from the basic case and *dashed arrows* show the velocities specified in the test

Figures 3.42 and 3.43 present an example of variations in the space distribution of synthetic earthquake epicenters in changing the velocities of the boundaries and the underlying medium. The velocities specified in this test are shown in Fig. 3.42 by *dashed arrows*, and *solid arrows* stand for the velocities specified in the basic case (Sect. 3.6.3). In this example, the part of the boundary consisting of fault plane 2 moves progressively at the velocity

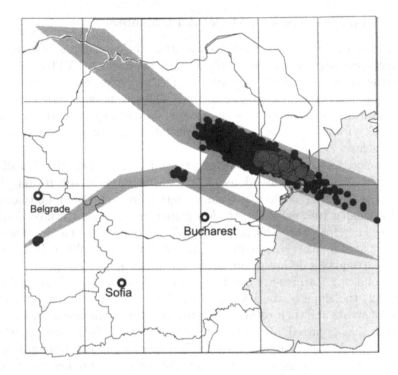

Fig. 3.43. Map of synthetic epicenters obtained when the velocities of the structure boundary and the underlying medium were specified as shown in Fig. 3.42 by dashed arrows. Gray areas are the projections of fault planes with $K \neq 0$ on the upper plane

$V_x = V_y = -16$ cm, and the part of the boundary consisting of fault plane 3 moves progressively at the velocity $V_x = -10$ cm, $V_y = -5$ cm. The underlying medium moves progressively at the velocities: $V_x = 25$ cm, $V_y = 0$ cm (under block 1); $V_x = -15$ cm, $V_y = 7$ cm (under block 2); $V_x = -20$ cm, $V_y = 5$ cm (under block 3). Figure 3.43 shows the space distribution of synthetic epicenters obtained in this example.

Spatial distributions of synthetic earthquake epicenters obtained from similar tests indicate their substantial dependence on the movements specified.

The results of the tests also indicate the possibility of using the procedure of block structure dynamics modeling to reconstruct the ranges of some parameters used to describe real regional tectonics.

The seismic activity of a fault depends on the relative velocities of tectonic motions along it, and these motions are interconnected in a fault system. Therefore, the spatial distribution of seismicity can be used not only as a characteristic for comparing the activity within different faults, but also for reconstructing block motions. It is even possible to formulate the inverse problem: to reconstruct the motions of boundary blocks and the underlying medium, which produce driving forces in the model, based on the observed epicenter distribution and other features of seismicity.

3.6.5 Source Mechanisms of Synthetic Seismicity

In this section, we consider the source mechanism of synthetic earthquakes obtained from model A. The mechanism is described by three angles: strike, dip, and slip. Strike and dip define the azimuth and the dip angle of the rupture plane, and slip defines the direction of the displacement in this plane. Therefore, strike and dip in model A are prescribed by the block structure geometry and do not depend on the variation of model parameters. For fault plane 9, that represents the Vrancea subduction zone whose strike is 212° and dip is 70°. Thus, for synthetic earthquakes, only the dependence of slip on the variation of model parameters is studied, and a comparison is made with observations.

Consider the vector change δu of the inelastic displacement in a cell where the failure occurs and set $\delta u = \gamma f$ where γ is determined by (3.13) and $f = (f_t, f_1)$ is the vector of the elastic force per unit area given by (3.5). The synthetic earthquake mechanism is the vector ΔU, defined as the weighted sum, with weights proportional to the areas of failed cells, of vectors δu for cells forming the earthquake. It follows from the definition of f that ΔU is located in the fault plane where the earthquake occurs. The slip angle for synthetic earthquakes is determined by the direction of the vector ΔU. The largest Vrancea earthquakes (see Table 3.11) have similar source mechanisms [CMTC94], and the average value of dip (67°±5°) agrees well with that (70°) of fault plane 9, whereas the average strike (237° ± 3°) differs slightly from that in the model (212°).

Table 3.11. Source mechanisms of the largest Vrancea earthquakes as given in the Harvard catalog [CMTC94]

Date	Epicenter		Mechanism		
	Latitude	Longitude	Strike	Dip	Slip
1977/03/04	45.78°N	26.80°E	235°	62°	92°
1986/08/30	45.51°N	26.47°E	240°	72°	97°
1990/05/30	45.83°N	26.74°E	236°	63°	101°

The slip in the basic case (Sect. 3.6.3) for the 20 largest synthetic earthquakes that occurred on fault plane 9 averages 99° ± 1°, close to the average observed value (97° ± 5°).

The slip of synthetic earthquakes depends on model parameters. In studying this dependence, the following model parameters were varied, as in

Sect. 3.6.4: the directions of movements of boundary fault planes 2 and 3, the directions of movement of the medium underlying the blocks, and the coefficients of the inelastic displacement rate W on fault planes 2 and 9. As for the basic case, the source mechanisms for the 20 largest synthetic earthquakes (the magnitude cutoff is given in Table 3.12), which occur on fault plane 9, are determined in this parametric test.

Table 3.12. Dependence of slip angle of synthetic earthquakes on variations of model parameters

Varied parameter	Magnitude cutoff	Slip, degrees
Basic case	7.0	99 ± 1
Direction of movement		
Boundary block 1, $-110°$	6.9	102 ± 3
Boundary block 1, $+50°$	6.9	99 ± 1
Boundary block 2, $-140°$	6.0	-70 ± 74
Boundary block 2, $+70°$	5.6	94 ± 6
Medium underlying block 1, $-50°$	5.8	94 ± 1
Medium underlying block 1, $+50°$	6.3	102 ± 4
Medium underlying block 2, $-50°$	6.5	101 ± 3
Medium underlying block 2, $+50°$	6.6	101 ± 2
Medium underlying block 3, $-50°$	5.0	96 ± 0.1
Medium underlying block 3, $+50°$	7.0	96 ± 6
Coefficient W		
Fault plane 2, 0.05	5.8	89 ± 6
Fault plane 2, 0.7	7.0	99 ± 1
Fault plane 9, 0.01	6.8	99 ± 1
Fault plane 9, 0.05	5.9	98 ± 2

Table 3.12 lists the slip angles of synthetic earthquakes and the respective standard deviations obtained for parameter values bounding the ranges of their variation. The second column in Table 3.12 contains angles between two vectors: the velocity used in the basic case and the velocity used in the test (the positive direction is counterclockwise).

The largest earthquakes on fault plane 9 have similar slip (the standard deviation is small), and its value is close to that obtained with the basic case; the exception is represented by the variation in the direction of the boundary block 2 velocity. This exception, wide variations of synthetic mechanisms

Table 3.13. Dependence of the synthetic slip angle on the rotation of the boundary block 2 velocity vector

Velocity vector rotation	Magnitude cutoff	Slip, degrees
0° (basic case)	7.0	99 ± 1
−40°	7.0	98 ± 2
−60°	6.6	102 ± 3
−80°	6.2	81 ± 58
−100°	6.1	−30 ± 90
−140°	6.0	−70 ± 74

for the direction of boundary block 2 velocity, can be explained by a very significant change (140°) in this parameter.

The values of slip and their standard deviations for different directions of the boundary block 2 velocity are given in Table 3.13. The slip remains nearly constant when the velocity vector rotates through less than −60°. The deviation in slip values increases for rotations through wider angles, and earthquakes with different mechanisms occur, though all events are of the overthrust type, as in the basic case. When the velocity vector rotates through −140°, most earthquakes have an extension mechanism.

Figure 3.44 shows how the number of synthetic earthquakes is distributed across the slip angle for different directions of the boundary block 2 velocity. Nearly all earthquakes in the basic case are the overthrust type. When

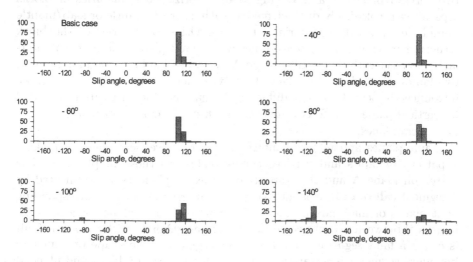

Fig. 3.44. Distribution of the number of synthetic earthquakes versus slip angle for different directions of the movement of the second boundary block

the velocity vector rotates clockwise, extension earthquakes appear, and two maxima emerge in the bottom bar chart in Fig. 3.44; the larger corresponds to extension, and the smaller to overthrust earthquakes. Thus, the modeling of block structure dynamics shows that earthquakes with markedly different mechanisms may occur on the same fault.

Synthetic source mechanisms obtained with the basic case of the Vrancea block model (Sect. 3.6.3) indicate that earthquakes on fault plane 9 have nearly constant slip angles close to the average values obtained from seismicity observed in the Vrancea subduction zone. Nevertheless, earthquakes with different types of the source mechanism (different values of slip) can occur within the same fault plane when certain specific velocities are specified in the model.

Numerical tests described in Sect. 3.6.4 show that the direction of the velocity vectors assigned to the western part of the East European plate (boundary block 1) and to the Black Sea subplate (boundary block 2) have little effect on the main features of the synthetic catalog. On the contrary, changes in the directions of motion of the Intra-Alpine and Moesian subplates have a profound effect on all features of synthetic seismicity.

The slip angle is reasonably stable under variations of model parameters, even when they vary within a wide range.

3.6.6   Block Structure of the Vrancea Region: Model B

The effects of intermediate depth seismicity in the region are modeled by a block structure containing a sinking slab (see Fig. 3.33b). The structure of the blocks is a limited and simply connected part of a layer bounded by two vertical planes **A** and **B** (Fig. 3.45). Horizontal boundaries of blocks separating rheologically distinct media, such as crust/mantle or slab/mantle boundaries, are called material interfaces. Other boundaries of the block structure are called fault planes. Lines where fault planes and material interfaces intersect with the vertical plane **A** are called faults. A common point of two faults is called a vertex. Note that these definitions are similar to those introduced in Sect. 3.2.1. The difference is that here the structure is bounded by vertical planes and forces, together with displacements of blocks that lie in vertical planes.

Fault planes and material interfaces intersect vertical plane **B** making a pattern of faults and vertices corresponding to those in plane **A**. The vertex on plane **A** and the respective vertex on plane **B** are connected by a segment called a rib; it is located at the intersection of the corresponding fault planes or material interfaces. The part of a fault plane (or a material interface) between two ribs corresponding to neighboring vertices on the fault is called a fault segment (or a boundary segment). The shape of fault and boundary segments is a trapezoid. The common part of a block and plane **B** is a polygon; it is called a block flank.

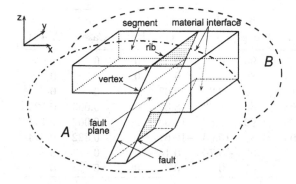

Fig. 3.45. Block structure geometry and definitions used in model B

The configuration of blocks and faults on plane **A** used in the model is presented in Fig. 3.46. The model consists of two rigid blocks: the continental crust (block **I**) and the sinking slab overlain by the continental crust (block **II**). The Z axis is pointing upward from the roof of the model ($z = 0$); the X axis points to the right ($0 \le x \le 350$ km). The vertices of the block structure numbered 1–9 have the following coordinates (in km): (350, 0); (0, 0); (0, −33.3); (227.5, −33.3); (240, 0); (181, −160); (206, −160); (240, −66.6); (350, −66.6). The block structure contains eight faults. The values of the model parameters are listed in Table 3.14.

Fault plane **1** imitates Earth's surface; fault planes **2** and **8** are "passive," no shocks are observed there. Fault planes **1**, **2**, and **8** are immobile; $K = 0$ on them; therefore, all forces are nil on these fault planes. Fault planes **3**, **5**, **6**, **7**, and the part of fault plane **4** confined between vertices *4* and *6* move

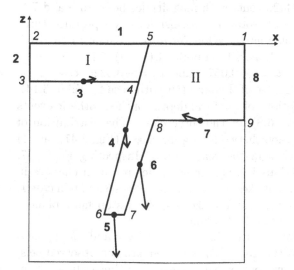

Fig. 3.46. Block structure used for modeling. Vertices (numbers *1–9*), faults (**1–8**), and blocks (**I** and **II**) are indicated in the figure. The arrows stand for the velocities of boundaries

Table 3.14. Model parameters of fault planes

Fault plane	Vertices	K, $\mathrm{bar\,cm}^{-1}$	W, $\mathrm{cm\,bar}^{-1}$	α, deg	V_x, cm	V_z, cm	ω Case 1	Case 2	Case 3
1	1, 5, 2	0	0	90	0	0	0	0	0
2	2, 3	0	0	85	0	0	0	0	0
4	3, 4	1	0.1	90	0.2	0.07	0	0	0
4^a	5, 4, 6	1	0.05	85	4.73	−8.82	0	−0.063	−0.126
5	6, 7	1	0.05	90	1.82	−17.93	0	−0.067	−0.114
6	7, 8	1	0.05	85	15.63	−19.61	0	−0.322	−0.644
7	8, 9	1	0.1	90	−0.74	0.13	0	0	0
8	9, 1	0	0	85	0	0	0	0	0

^{a} V_x, V_z and ω are prescribed for the part of fault plane 4 between vertices *4* and *6*

at prescribed velocities (V_x, V_z) specified in Table 3.14. These velocities are found from the numerical model of the sinking slab developed by Ismail-Zadeh et al. [IPN00]. All deformations occur on these fault planes. We assume that the horizontal thickness of blocks is 60 km, which is close to that observed, and $P = 2\,\mathrm{kbar}$. The magnitude of earthquakes is calculated from the relationship $M = D \log S + E$, where the constants take the values $D = 0.98$ and $E = 3.93$ [US54] and S is the sum of the cell areas (in km^2) included in the earthquake.

3.6.7 Synthetic Features of Model B and Vrancea Seismicity

Model B was set active across 300 units of dimensionless time. Modeling started from zero initial conditions, and some time was needed for quasi-stabilization of stress. We consider only the stable part of the resulting synthetic catalog. It contains 96442 events with magnitudes between 5 and 7.1.

To examine the effect of slab rotation on seismicity, we produce three synthetic catalogs for different angular velocities ω of boundaries (in 10^{-6} radians) about the origin. Case 1: $\omega_i = 0$ on faults 4 ($i = 4$), 5 ($i = 5$), and 6 ($i = 6$); case 2: $\omega_4 = -0.0631$, $\omega_5 = -0.0571$, and $\omega_6 = -0.3225$; case 3: the velocities are increased by a factor of 2 compared with case 2 (Table 3.14). Figure 3.47 shows the distributions of the focal depths of the synthetic events (with magnitudes greater than 6.8) for the three cases. The distribution of earthquake hypocenters in the synthetic catalog for case 1 (Fig. 3.47, case 1) is in good agreement with that in the real catalog. Inspecting Fig. 3.47, we conclude that small variations in angular velocities result in changes in seismicity. Large synthetic earthquakes occur on fault planes 5 and 6 in case 1. The slab rotation tends to reduce earthquake magnitudes on fault plane 6 and to concentrate larger events on fault plane 5.

Figure 3.48 shows FM plots for the observed seismicity and for the synthetic catalogs. According to the Gutenberg–Richter law for observed seismicity, the logarithm of the number of earthquakes is a linear function of

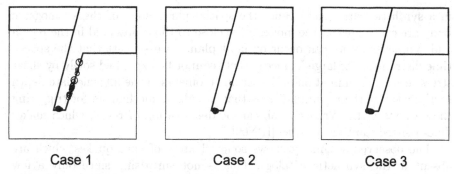

Fig. 3.47. Maps of synthetic seismicity in the modeled region: three cases

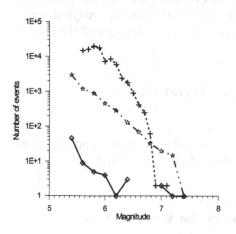

Fig. 3.48. FM plots for the real catalog (*solid line marked by squares*), for synthetic catalogs obtained with boundary movements without rotation (*dashed line marked by crosses*), and from model A (*dot-dashed line marked by stars*)

magnitude. The synthetic curves in case 1 (*dashed line*) and that obtained from model A (*dot-dashed line*) are nearly linear, and they have approximately the same slopes as the observed curve (*solid line*) in the range of magnitudes from 5.5 to 6.5. The observed seismicity displays a gap in the magnitude interval $6.5 < M < 7.0$; it is not visible in the synthetic catalog. The gap seems to be caused by the short time interval for which relevant observations are available.

The results of the analysis show that the synthetic catalogs obtained by modeling block structure dynamics have features similar to those of the real earthquake catalog. Several numerical tests for various model parameters show that the spatial distribution of synthetic events is significantly sensitive to the directions of block movements. Changes in synthetic seismicity due to small variations in slab rotation are in overall agreement with the hypothesis of Press and Allen [PA95] that small changes in the direction of plate motion control the pattern of seismic release.

The maximum value of the magnitude in the synthetic catalog is 7.1, whereas there were events with larger magnitudes in the Vrancea region. It should be noted that the magnitude depends on the number of cells included in a synthetic earthquake. Since the spatial parameters of the seismogenic body are considered in the model to be close to those observed in the region and synthetic events can occur on fault planes, we suggest that the space–time distribution of large Vrancea events cannot be explained solely by shear stress release. Examination of the seismic moments of the intermediate-depth earthquakes in the region indicates that a realistic mechanism for triggering these events in the Vrancea slab can be dehydration of rocks, which makes fluid-assisted faulting possible [IPN00].

The observed seismicity shows some clusters of earthquakes which are absent in the synthetic catalogs. This is not surprising since only a few main seismic faults of the Vrancea region are included in the model. A more realistic distribution of synthetic events calls for more detailed 3-D structures of blocks containing additional fault planes. Nevertheless, the analyzed structure containing only two blocks reproduces the main features of the spatial distribution of the real seismicity.

3.7 Modeling Block Structure Dynamics of the Western Alps

The results below were obtained for the block structure approximating a morphostructural zoning scheme of the Western Alps [GS96, GKR+97, VGS00].

3.7.1 Block Structure Approximating a Morphostructural Scheme of the Western Alps

It is natural to use the morphostructural scheme of the region as a basis for developing a model of its block structure dynamics because both the model and the morphostructural zoning exploit the concept of the hierarchical block structure of the lithosphere [AGG+77]. The basic principles of morphostructural zoning are given in Sect. 6.2.1. The scheme of morphostructural zoning for the Western Alps was initially constructed for recognition of earthquake-prone areas ([CGG+85], see also Sect. 6.3.2.

The source morphostructural scheme (Fig. 6.8) was simplified by removing low rank lineaments from the map. Only the lineaments of the first and second ranks were left and produced a rough morphostructural scheme given in Fig. 3.49. The block structure for numerical modeling was designed on the basis of this rough scheme. It contained 28 faults and 8 blocks. The block boundaries (faults) were linearized, compared with the rough morphostructural scheme, but the model still agreed with the structure of the region. Figure 3.50 depicts the pattern of faults of the block structure on the upper plane.

The thickness of the layer is $H = 30\,\mathrm{km}$ agreeing with data on the depth of the Moho discontinuity that varies from 25 to 40 km in the Western Alps [Men79]. Figure 3.50 shows the dip angles of the fault planes. The parameters

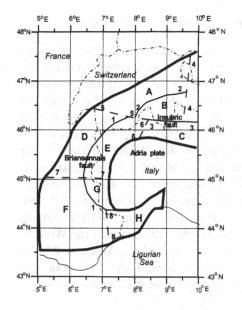
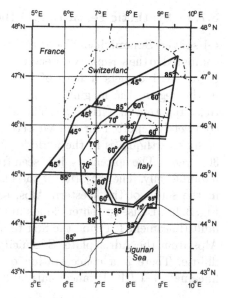

Fig. 3.49. A simplified map of the morphostructural lineaments in the Western Alps: lineaments of the first rank (*thick lines*) and the second rank (*thin lines*), longitudinal (*solid lines*) and transverse (*dashed lines*)

Fig. 3.50. The block structure studied by numerical modeling. Numerals are the dip angles of respective fault planes

in (3.5) and (3.6) and the levels of κ (3.11) are the same for all faults: $K = 1\,\mathrm{bar\,cm^{-1}}$, $W = 0.05\,\mathrm{cm\,bar^{-1}}$, $W_s = 10\,\mathrm{cm\,bar^{-1}}$, $B = 0.1$, $H_f = 0.085$, and $H_s = 0.07$. The parameters in (3.1) and (3.2) are also the same for all blocks: $K_u = 1\,\mathrm{bar\,cm^{-1}}$; $W_u = 0.05\,\mathrm{cm\,bar^{-1}}$.

The values $\varepsilon = 5\,\mathrm{km}$ and $\Delta t = 0.001$ are accepted for discretization.

The value of P in (3.11) is $2\,\mathrm{kbar}$.

The Western Alps are a continental collision zone within the Mediterranean orogenic mobile belt. The regional stress field is controlled by the interaction between the Adria and Western European plates. According to plate tectonics, the Adria plate moves northwest, causing compression in the Western Alps [McK70]. The part of the boundary shown as a double line in Fig. 3.50 (this part corresponds to the boundary between the Alps and the Apennines) moves progressively while other parts of the boundary, as well as the underlying medium, do not move. The velocity of the boundary movement is $10\,\mathrm{cm}$. The angle between the velocity vector and the north-looking meridian varies between 30 and $90°$.

3.7.2   Synthetic Features and the Seismicity of the Region

Observed Seismicity. The synthetic seismicity obtained in the model was compared to the seismicity observed in the Western Alps, as given in [GHDB94]. Figure 3.51 depicts a map of earthquakes with $M \geq 4.0$ that occurred in the region; the magnitude of an earthquake is taken as the highest of those reported in [GHDB94]. The $M \geq 4.0$ earthquakes have certainly been completely reported for the Western Alps since 1968; Fig. 3.52 gives a map of such earthquakes in the period 1 January 1968 to 30 June 1998, i.e., for 30.5 years. The epicenters, it is seen from Figs. 3.51 and 3.52, are distributed nonuniformly. The majority of the instrumental epicenters are concentrated in the south of the Western Alps, south of transverse lineament 7-7 (see Fig. 3.49). Dense epicenter clouds in this area extend along the Brianconnais fault (lineament 1-1) and the first rank lineament that separates the Western Alps from the Adria plate. Seismicity in the north of the Western Alps is diffuse. There is a concentration of epicenters around the intersection of lineament 1-1 (the Brianconnais fault) and transverse lineament 5-5.

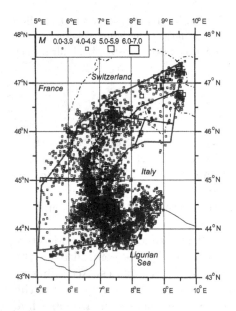

Fig. 3.51. Seismicity of the Western Alps

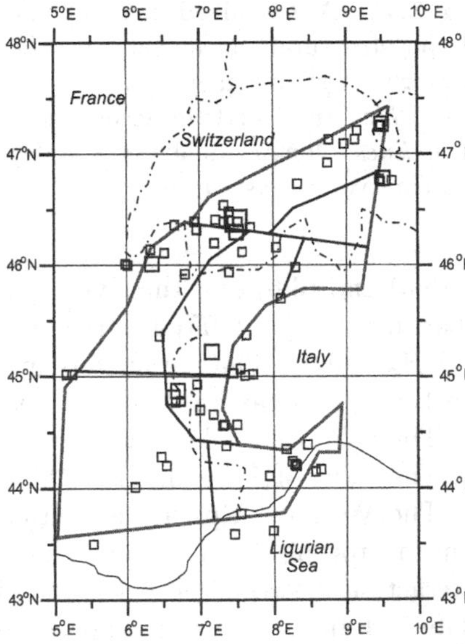

Fig. 3.52. Epicenters of earthquakes that occurred in the Western Alps from 1968–1998

Figure 3.53 shows a cumulative FM plot for the events whose epicenters are shown in Fig. 3.52.

The largest magnitude recorded during the period of instrumental observation is 6.1 for the 25 January 1946 earthquake in the Western Alps

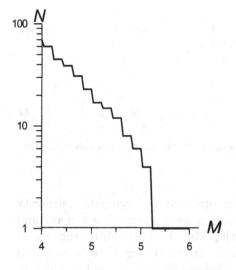

Fig. 3.53. Cumulative FM plot of observed seismicity for the Western Alps (1968–1998)

($\varphi = 46°N$, $\lambda = 7.5°E$). According to catalogs containing historical earthquakes [Cap81,Vog79], more than 20 earthquakes with an intensity of shaking of VIII–X have occurred in the region. This intensity may well have been produced by earthquakes of magnitude 7.0 or even higher.

Comparing synthetic and observed seismicity. Synthetic earthquake catalogs were obtained through block structure dynamics modeling for the period of 200 units of dimensionless time starting from zero initial conditions. The magnitudes of the synthetic earthquakes are calculated from (3.15).

Consider the boundary (marked by a double line in Fig. 3.50) consisting of fault planes that separate the Western Alps from the Adria plate. Several cases were computed for different values of the angle γ between the velocity of this boundary and the northward meridian [VGS00]. This angle took on the following values: 30, 45, 60, and 90°.

To eliminate the initial unstable period in the synthetic catalogs, only their parts in the interval of 100 to 200 units of dimensionless time were examined for comparison with the observed seismicity. Figure 3.54 presents cumulative FM plots for the four cases under consideration. The lowest magnitude of the synthetic earthquakes is determined by the magnitude of an earthquake that consists of a single cell; this must be of order 5.3 in accordance with (3.15) and the limitation on linear cell size (5 km). The highest magnitudes generated were 7.04 to 7.2. The total number of earthquakes decreases when the value of γ increases from 30 to 90°. The relative number of large earthquakes ($M \geq 6.5$) in the synthetic catalogs appreciably diminishes from case 1 ($\gamma = 30°$) to case 3 ($\gamma = 60°$) and slightly increases in case 4 ($\gamma = 90°$) compared with case 3. The FM plots show a significant decrease in the slope of the maximum magnitudes in case 4. This may be treated as the presence of "characteristic" earthquakes.

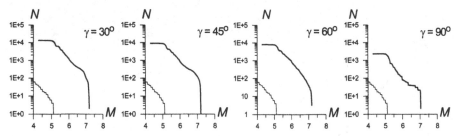

Fig. 3.54. Cumulative FM plots based on synthetic catalogs for four cases; the thin lines show the FM plot from Fig. 3.53

The comparison of FM relations for observed and synthetic seismicity shows that synthetic catalogs contain many more events, their maximum magnitude is greater than that of earthquakes recorded during the period of instrumental observation. The latter circumstance may have been due to two facts: first, (3.15) overestimates the magnitude and, second, the period covered by instrumental observation is short and does not contain the largest earthquakes possible in the region. The difference in the number of events is due to the fact that the time period covered by a synthetic catalog can be several hundreds, or even thousands, of times as long as the period of instrumental observation. A more accurate interpretation of the time period concerned is difficult because it depends on the definition of magnitude in the model [relation (3.15)].

Shapes and b-value were chosen as the criteria for comparing synthetic and observed FM plots. The greatest similarity is obtained in case 3, i.e., when $\gamma = 60°$. This direction of motion for the boundary is in agreement with data available for the Adria plate [McK70, AJ87].

Spatial distributions of synthetic seismicity for the four cases are shown in Fig. 3.55. Obviously, complete agreement between synthetic and observed seismicity has not been achieved in any. In particular, the seismicity in the north of the Western Alps and the earthquakes in the Brianconnais fault zone (lineament 1–1 in Fig. 3.49) have not been reproduced.

The emerging differences between synthetic and observed seismicity can be explained, among other things, by a more complex geodynamics of the region compared with that assumed for the model. Faults of lower ranks than those considered in the model take part in earthquake generation. The motion prescribed in the model for the part of the outer boundary of the block structure cannot fully represent the real complexity of block motions in the region.

The majority of events generated in all cases are confined to lineaments of the first rank that separate the Western Alps from the Adria plate and the Ligurian Sea basin. Cases 1–3 (Fig. 3.55, $\gamma = 30, 45, 60°$) with northwestern motion produced synthetic seismicity that is in agreement with the observations for the lineament of the first rank that bounds megablock E.

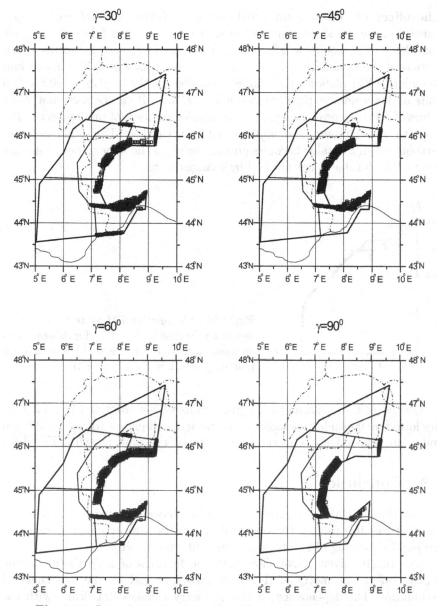

Fig. 3.55. Spatial distributions of synthetic epicenters in four cases

It should be noted that many synthetic earthquakes, also the largest ones, occur on the eastern extension of that lineament, which is the boundary of megablock C where no large earthquakes have yet been recorded. We note that the study of the recognition of earthquake-prone areas ([CGG+85]; see also Sect. 6.3.2) identified the same place as likely to generate earthquakes of $M \geq 5.0$.

The effect of cell size on synthetic seismicity. The value of the parameter ε, the cell size for computation, was set equal to 5 km. To find out how the results would be affected by a change in ε, case 2 (northwestern motion of the boundary, $\gamma = 45°$) was computed for $\varepsilon = 2.5\,\text{km}$ and 1 km. Cumulative FM plots for the synthetic catalogs in the interval of 100 to 200 units of dimensionless time are shown in Fig. 3.56. It can be seen that lower values of ε did not produce significantly smaller numbers of earthquakes in the magnitude range common to all three synthetic catalogs. The total number of earthquakes in a catalog becomes greater owing to the appearance of smaller events. The b-value is not affected by changes in ε.

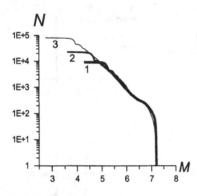

Fig. 3.56. Cumulative FM plots for synthetic seismicity in case 2 ($\gamma = 45°$) for different discretization: $\varepsilon = 5$ km (curve 1), $\varepsilon = 2.5$ km (curve 2), and $\varepsilon = 1$ km (curve 3)

One notes that relation (3.15) gives values of the order of 4.7 and 3.9 for the lowest magnitude threshold of synthetic earthquakes when $\varepsilon = 2.5$ km and 1 km, respectively. This can be seen in the FM plots shown in Fig. 3.56.

3.8   Conclusion

The synthetic catalogs obtained by modeling block structure dynamics possess features intrinsic to the observed seismicity: linearity of FM relation, temporal clustering of events, long-range interaction, etc.

The results given above show that the features of a synthetic catalog depend on the geometry of the block structure and on the movement of boundaries. The dependence on the geometry changes drastically with the types of boundary movement. In particular, when the boundary movement includes rotation, the slope of the FM plot depends on structure fragmentation in a way contradictory to the currently accepted view. The features of a synthetic earthquake catalog depend on prescribed movements of block structure boundaries and of the underlying medium. This dependence reflects the fact that the seismic activity of a fault is controlled by the rate of relative tectonic movements along the fault and these movements are caused by the movements of blocks that form the structure. Therefore, the spatial distribution of seismicity can be used to compare activity on different faults and also to reconstruct block motion. It is even possible to formulate the inverse problem: to reconstruct block motions using the observed distribution of epicenters and other features of seismicity. Such reconstruction becomes possible by using the model of block structure dynamics. Numerical experiments carried out with varying values of model parameters show that the spatial distribution of epicenters is sensitive to the direction of block motion. The results of experiments allow us to hope that it is possible to use the procedure of modeling block structure dynamics to reconstruct block motions from the observed distribution of epicenters and other features of seismicity.

The values of model parameters yielding some similarity between synthetic and real catalogs can be useful for estimating the rates of tectonic movements and the physical parameters involved in dynamic processes within fault zones. The modeling of block structure dynamics could be a useful tool in studying relations between the geometry of faults and block movements and earthquake flow, including premonitory seismicity patterns, testing existing earthquake prediction algorithms, and developing new ones.

4 Earthquake Prediction

V. Kossobokov and P. Shebalin

4.1 Introduction

The scientific research aimed at predicting earthquakes started in the second half of nineteenth century, when seismology reached the level of a recognized scientific discipline. The desire to find tools that would permit forecasting the phenomenon under study is so natural that, as early as 1880, John Milne, a famous British engineer-seismologist and the inventor of the seismometer, defined earthquake prediction as one of the pivotal problems of seismology and discussed possible precursors of large earthquakes. In more than 100 years of earthquake prediction research, there were alternating periods of high enthusiasm and critical attitudes.

Earthquake prediction remains a challenging and controversial problem [GJK97, Wys97a]. Although many observations reveal unusual changes of geophysical fields in the approach of a large earthquake, most of them report a unique case history and lack a systematic description [Wys91]. The latter makes an earthquake prediction method hardly reproducible and, therefore, testable by an independent investigator even when such a method was proposed a long time ago and its ex post facto applications are the subject of numerous publications in prestigious scientific journals.

Sir Charles Richter, whose critical attitude toward predicting earthquakes is often cited in discussions on the subject, has written a one third page note [Ric64] commenting on the publication by V.I. Keilis-Borok and L.N. Malinovskaya [KM64] that described observation of a general increase in seismic activity prior to large earthquakes. He noted "a creditable effort to convert this rather indefinite and elusive phenomenon into a precisely definable one", marked as important a confirmation of "the necessity of considering a very extensive region including the center of the approaching event", and outlined "difficulty and some arbitrariness, as the authors duly point out, in selecting the area which is to be included in each individual study". However at that time the information database for earthquake prediction research remained sparse and fragmentary and, therefore, did not allow meaningful testing of hypotheses about phenomena claimed precursory to large earthquakes in any systematic way.

The situation changed drastically with the installation of standard seismographs organized into the World Wide Standard Seismograph Network in the mid-1960s. The growing data flow (Fig. 4.1) on all (!) earthquakes above a sufficiently low magnitude level encouraged scientists who study the interdependence of earthquake sequences and precursory phenomena. At about the same time, national earthquake prediction projects were launched in the USSR, the US, China, and other countries. The success of Chinese seismologists in predicting, in actual practice, the devastating 1975 Haicheng earthquake [ZPD+84] stimulated further design of methods for diagnosing large approaching quake. Most of the methods (regretfully nearly all) suggested at that time hardly found due confirmation in the following years. The catastrophic 1976 Tangshan earthquake caused hundreds of thousands casualties in China and turned out to be a cold shower of disillusion not only for Chinese seismologists. The necessity of strict formulations and stringent methods of testing complex hypotheses proposed by seismology that would distinguish a lucky guess from a reliable prediction became evident as never before.

In the 1970s the progress in formalizing the description and pattern recognition of earthquake-prone areas (see Chap. 6), which delivers a termless zero-approximation prediction of large earthquakes, resulted in better

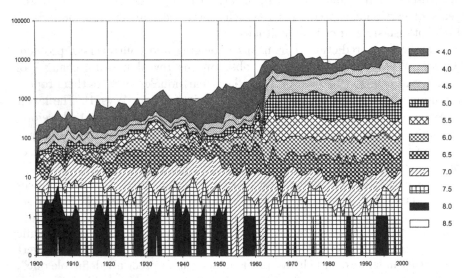

Fig. 4.1. Annual number of earthquakes in the NEIC Global Hypocenters Data Base System by time and magnitude, 1900–2000. Each band corresponds to a half-magnitude range of M_{max}. These bands stacked from higher ranges provide the number of earthquakes above a certain threshold. One can observe (1) sharp changes at times when Californian (1932) and the World Wide Standard Seismograph Networks (1963) were installed, (2) the stability of the number of events with magnitude 7.0 and above from the beginning of the century, and (3) the uniform width of the bands since 1963 in agreement with the Gutenberg–Richter relationship down to magnitude 5.0. This frequency–magnitude graph demonstrates the stability of the global seismic observations from the mid-1960s to the present

understanding of the certain universality of the seismic process when the tectonic environment changes. A distinctive similarity of criteria in the zero approximation provided an encouraging foundation for a systematic search of universal patterns in the low-magnitude seismic sequence in the approach of a large quake.

Upon publication of "Long-term earthquake prediction: Methodological recommendations" in 1986 [Sad86], along with twenty years of accumulated global seismic data of high quality and completeness, a novel understanding of seismic processes basically emerged (see Chap. 1). Seismic processes consti- tute an essential part of Earth's lithospheric dynamics, where the lithosphere is viewed as a self-organized hierarchy of blocks of different sizes; the relative displacements of blocks show up in the seismic activity of their boundaries, i.e., complex, in many aspects, self-similar systems of faults [Kei90a]. The behavior of such a dynamic system, in general, is controlled by nonlinear laws and is chaotic to a great extent. In principle, it is impossible to predict the behavior of such a system in detail. The prediction of extreme events in dynamics of such a system is possible, at least for the relatively near future and after a heavy round of averaging the complex set of observable variables. The accuracy of prediction can be improved hierarchically by bringing into analysis new observable variables from lower levels of averaging under a re- fined resolution of the system. It is rather evident that efficient prediction in such a complex system can hardly be based on a single phenomenon. Although some seismological precursors, like "burst of aftershocks" described below and in [KKR80], were statistically proved in forward application, their spacetime uncertainty is far from adequate for efficient practical usage. Let us stress once again that most of the seismological precursors (see the list of proposed seismological precursors in [Wys91]) are not defined with the certainty required for scientific testing; thus they fall outside the rank of scientific hypotheses and create numerous and, to our great disappointment, fruitless debates on the predictability of earthquakes *per se* [Nat99].

The algorithms suggested in 1986 [Sad86] for intermediate-term prediction of large earthquakes were derived in the application of pattern recognition methods to a set of observable integral variables measured in a seismic area. The results of their retrospective testing in other regions worldwide encour- aged an experimental real-time test of the algorithms. The existing on-line data permitted us to launch such an experiment at about that time (with a few reservations for delays at different stages for compiling global and regional earthquake catalogs). Regretfully, the discontinuity of seismic data in the Caucasus and former Asian Republics that happened on the collapse of the Soviet Union did not allow us to perform the experiment to its full extent. These regions constituted an essential part of the test related to possible rescaling of the algorithms to predict earthquakes of magnitudes lower than in the originally determined range (in some regions, magnitude 5 earthquakes were considered strong). Nevertheless, recent results of systematic prediction of magnitude 8.0 events or more as well as of magnitude 7.5 or more world-wide [HKD92,KHD97,KRKH99] permit us to conclude the statistical validity and reliability of the methodology proposed.

Let us illustrate the potential of the hierarchical approach to earthquake prediction with a typical example from recent history (Fig. 4.2). On 16 October 1999, an earthquake of magnitude $M_s = 7.4$ struck the Hector Mine area in Southern California. The seismic patterns preceding this earthquake allowed predicting it with medium-range, intermediate-term accuracy [RKH99], i.e., 100 km and 1.5 years by using the prediction scheme of the ongoing real-time testing of the M8-MSc algorithms [KRKH99] extended to smaller magnitudes.

The earthquake location in space and time satisfies the following Predictions:

Zero approximation: The epicenter is located within 37.5 km from one of 73 intersections of morphostructural lineaments in California and Nevada determined by Gelfand et al. [GGK+76] as earthquake-prone for magnitudes 6.5 or above. Since 1976, fourteen such earthquakes occurred, all in a narrow vicinity of these intersections.

First approximation: Algorithm M8 [KK90] aimed at $M6.5+$ events in circles of investigation centered at the 73 D-intersections and run on 1 July 1999 determines alarms in two of them. In both circles, the alarms started in July 1998 and have not yet expired.

Second approximation: The MSc algorithm [KKS90], when applied to the area of alarm in the two circles of investigation, narrows down the prediction to a location between 34.68° N–33.82° N and 117.23° W–116.17° W. As at the beginning of the alarm, i.e., July 1998, the area is larger by 48 km to the north and 24 km to the east. As time goes on, the area shrinks gradually to the 96 by 96-km square (as on 1 July 1999), where the epicenter of the Hector Mine earthquake and most of its aftershocks occurred on 16 October.

On the grounds of intermediate-term predictions, a swarm of low-magnitude earthquakes, of which the strongest did happen 6 hours in advance of the Hector Mine main shock in its future aftershock zone, would have certainly been recognized as an argument for a short-term alert, as happened prior to the 1975 Haicheng earthquake in China.

The high statistical significance of intermediate-term earthquake prediction methods achieved in the ongoing experimental testing worldwide confirms the following paradigms:

● Precursory seismic patterns exist.

● The dimension of an area where precursory seismic patterns appear is considerably (about ten times) larger than that of the source zone of the incipient strong earthquake.

● Many precursory seismic patterns are similar in regions of fundamentally different seismic and tectonic environments.

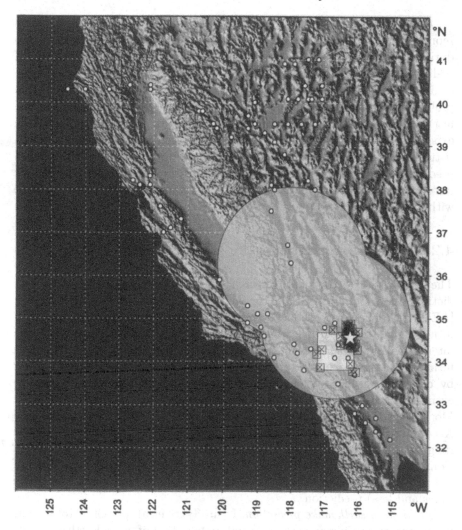

Fig. 4.2. The M8-MSc predictions on 1 July 1999 and the 16 October 1999 Hector Mine earthquake [RKH99]. *Zero approximation:* A termless prediction of earthquake-prone areas for magnitude 6.5 or more determined by pattern recognition methods [GGK+76] are marked by *open circles. First approximation:* An intermediate-term prediction by algorithm M8 [KK90], on 1 July 1999 determines alarms in two highlighted circles. *Second approximation:* A significant reduction of uncertainty by the MSc algorithm [KKS90] down to the 96 × 96-km additionally highlighted square, where the epicenter of the main shock (*a star*) and most of its aftershocks (*crossed squares*) occurred

- Some precursory seismic patterns are universal. Their analogs take place in advance of extreme catastrophic events in other complex nonlinear systems.

Although of limited accuracy, the intermediate-term predictions made by reproducible algorithms may help to reduce a substantial part of the losses during a catastrophic earthquake by prompt escalation and de-escalation of realistic low-cost intermediate-term civil defense measures [KKMV70].

In this chapter, we first describe our vision of the earthquake prediction problem (Sect. 4.2) and a general approach to a sequential hierarchical solution of the question when and where the largest earthquakes should be expected (Sect. 4.3). Section 4.4 presents reproducible algorithms whose statistical significance and performance were tested and confirmed in forward prediction experiments. In Sect. 4.5, the algorithms developed recently are introduced as the subject of future testing in advance prediction. We conclude with a general discussion.

4.2   What Is an Earthquake Prediction?

The United States National Research Council, Panel on Earthquake Prediction of the Committee on Seismology, suggested the following definition [AEH+76, p.7]: "An earthquake prediction must specify the expected magnitude range, the geographical area within which it will occur, and the time interval within which it will happen with sufficient precision so that the ultimate success or failure of the prediction can readily be judged. Only by careful recording and analysis of failures as well as successes can the eventual success of the total effort be evaluated and future directions charted. Moreover, scientists should also assign a confidence level to each prediction." Accordingly, one can identify an earthquake prediction of a certain magnitude range by the duration of the time interval and/or by territorial specificity. Commonly, temporal classification loosely distinguishes long-term (tens of years), intermediate-term (years), short-term (tens of days), and immediate (days and less) prediction.

Rethinking earthquake prediction, Lynn Sykes et al. [SSS99] wrote: "The public perception in many countries and, in fact, that of many Earth scientists is that earthquake prediction means short-term prediction, a warning of hours to days. They typically equate a successful prediction with one that is 100% reliable. This is in the classical tradition of the oracle. Expectations and preparations to make a short-term prediction of a great earthquake in the Tokai region of Japan have this flavor. We ask instead are there any time, spatial and physical characteristics inherent in the earthquake process that might lead to other modes of prediction and what steps might be taken in response to such predictions to reduce losses?"

Following common perception, many investigators usually overlook spatial modes of predictions and concentrate their efforts on predicting the "exact" fault segment to rupture (e.g., Parkfield earthquake prediction experiment [BL85]), which is by far more difficult and might be an unsolvable problem. Being related to the rupture size L of the incipient earthquake, such modes can be summarized in a classification that distinguishes wider prediction ranges in addition to the "exact" location of a source zone (Table 4.2). The accuracy of prediction might improve when independent observations are brought into the analysis.

Table 4.1. Classification of predictions

Temporal, in years		Spatial, in source zone size	
Long-term	10	Long-range	Up to 100 L
Intermediate-term	1	Middle-range	5–10 L
Short-term	0.01–0.1	Narrow	2–3 L
Immediate	0.001	Exact	1 L

From the viewpoint of such a classification, the earthquake prediction problem might be approached by a hierarchical, step-by-step prediction technique, which accounts for multiscale escalation of seismic activity to the main rupture [Kei90a,KMU99]. It starts with recognition of earthquake-prone zones (see Chap. 6) for earthquakes of a number of magnitude ranges, then follows with determination of long- and intermediate-term areas and times of increased probability, and, finally, may come out with an exact short-term or immediate alert.

Finally, it is worth noting that earthquake prediction algorithms suggested in 1986 [Sad86] and their offspring described below in this chapter are in complete agreement with the 1976 general definition and essentially provide predictions of at least intermediate-term, middle-range accuracy. In contrast, probability mappings like those by [JK99, KJ00], which might be useful for many other practical applications, are not earthquake predictions in this sense, unless one specifies exactly the probability cutoff and the expected magnitude range for a given mapping.

4.3   Reproducible Prediction Algorithms

In a broader sense, the problem of reproducible methods for analyses in seismology was posed and outlined a long time ago [Kei64]. Although thousands of observations have been claimed to precede large earthquakes, seismology does not have many quantitative definitions of "precursors". More than a decade ago, 31 candidates were submitted to the Subcommission on Earthquake Prediction of the International Association of Seismology and Physics of the Earth's Interior [Wys91]. None of them fully satisfied the guidelines, mainly due to the inability of the authors to provide a precise definition of the observed phenomenon. We regret that the situation did not change in the second round of evaluation [Wys97b].

Some years earlier, the methodology described in detail in the following sections of this chapter was presented to the U.S. National Earthquake Prediction Evaluation Council (NEPEC) [Upd89]. NEPEC recommended that the U.S. Geological Survey undertake an extensive evaluation of this approach to the problem. The methodology is based on general concepts of pattern recognition that automatically imply strict definitions and reproducible prediction results. The quantifiable uncertainty that remains is due to essential uncertainties in the data, which are unavoidable.

Citing Christopher Scholz [Sch97], "Predicting earthquakes is as easy as one-two-three. Step 1: Deploy your precursor detection instruments at the site of the coming earthquake. Step 2: Detect and recognize the precursors. Step 3: Get all your colleagues to agree and then publicly predict the earthquake through approved channels."

The sections below show that some "precursor detection instruments" are already deployed worldwide and their records are available for general use, some precursors are already detected and, finally, some earthquakes were already "publicly predicted", although our colleagues continue to debate whether or not it is possible [GJK97, Wys97a, Nat99]. People who earn their living by giving advice to the public on important economic problems do not have the luxury of waiting for decades before making decisions based on the available evidence. The forecasts and predictions made by seismologists already have a large economic impact, and as the methods of predicting and forecasting improve along with seismological data, these economic effects will increase.

4.3.1   Data for Precursor Detection

Catalogs of earthquakes remain the most objective record of seismic activity on Earth. Many discoveries in geophysics, including the revolutionary plate tectonics hypothesis would hardly be possible without seismic evidence. Figure 4.1 shows the annual number of earthquakes from the NEIC Global Hypocenters Data Base CD-ROM and its updates through 2000. This frequency–magnitude graph demonstrates temporal variations in global seismicity during the twentieth century. One can easily observe several "historic changes" of which the most dramatic reflects deployment of the World Wide Standard Seismograph Network in 1963. From a statistical viewpoint, from about that time, the catalog is surprisingly consistent in reporting earthquakes of magnitude 5.0 and above: the magnitude bands have almost the same width in logarithmic scale in agreement with the Gutenberg–Richter relationship, i.e., $\log N \sim (8 - M)$. One can see also that the list of earthquakes above 7.0 in the NEIC GHDB is probably complete from the beginning of the century. Such remarkable stability in the annual number of earthquakes suggests stationary global underlying processes on a timescale of decades and encourages research aimed, specifically, at intermediate-term prediction of earthquakes.

It is common knowledge that catalogs have some errors [HC94]. It is desirable to have independent records of seismicity to identify and eliminate errors. In some cases, where a number of catalogs have common overlap, the pattern recognition technique may help in detecting possible errors and duplicate entries [She87]. Such a technique permits developing automated procedures that could reduce the percentage of errors (see, e.g., [She92]). The analysis of the frequency–magnitude graph of the catalog for consistency, as well as special searches for duplicates and possible errors, are the essential preliminary part in every application of the methodology described below.

In general, errors in the data can be neutralized in two ways: first, by postponing the analysis until the data are refined and second, by robust analysis of existing data within the limits of their applicability. We follow the second way and use routine catalogs of earthquakes to describe the dynamics of seismic regions, derive *precursory seismic patterns* in the approach of large earthquakes and make predictions based on these determinations.

4.3.2   General Scheme of Data Analysis

We first define a *strong earthquake* as one we aim to predict, by the condition that its magnitude $M \geq M_0$. Naturally, the magnitude scale we use should reflect the size of earthquake sources. Accordingly, M_s is usually taken for larger magnitudes, and m_b is used for smaller ones, for which M_s determinations are not available. For many catalogs, one can use the maximum reported magnitude (e.g., we do so when using the National Earthquake Information Center/U.S. Geological Survey Global Hypocenters' Data Base [GHDB94]).

In most cases, the choice of M_0 is predetermined by the condition that the average recurrence time of strong earthquakes must be sufficiently long in the territory considered. To establish a value of M_0 for a seismic territory, we consider values of M_0 with an increment of 0.5, unless the actual distribution of earthquake size suggests a natural cutoff magnitude that determines earthquakes, sometimes called characteristic [SC84]. The analysis may distinguish a number of intervals $M_0 \leq M < M_0 + 0.5$ and deliver a hierarchy of predictions.

Areas of investigation, whose size depends on M_0, overlie the territory of a seismic region under study. For natural reasons, their size is proportional to the rupture size $L = L(M_0)$ of the incipient earthquake. Each area of investigation at a given time is an object for pattern recognition based on recent seismic dynamics. There is difficulty in selecting such areas, a task to be included in each individual study, even if their shape and position are objectively motivated. One can expect to avoid this difficulty by using areas of regular shape, e.g., circular, sampled regularly through the whole seismic region under study. For natural reasons, among which heterogeneity of epicenter distribution might be most important, such scanning also requires certain testing and justification.

To characterize seismic dynamics in each area of investigation, we determine robust trailing averages, so-called *traits*, that, in line with the theory of nonlinear dynamic systems, are expected to experience significant variations in the approach to a singularity (critical point, catastrophe). Qualitatively, traits usually depict an extreme level and variation of activity, temporal clustering, and spatial concentration of lower magnitude quakes.

The theory predicts a significant difference of an informative trait distribution in periods associated with extreme events and outside them. Such differences provide the basis for a search for *precursory seismic patterns*, that eventually determine *criteria* that distinguish so-called *"time of increased probability"* (TIP) from periods when strong earthquakes should not be expected. A diversity of premonitory seismic patterns was found by detecting such a difference.

Note that one or several informative traits can be used for prediction. Moreover, a combination of traits can be informative for prediction if some or even all of them show unsatisfactory performance when applied separately.

Once detected, a precursory seismic pattern is the subject of testing in experimentally predicting strong earthquakes. For this purpose, we design a prediction algorithm that implements criteria and recognizes, at a given time, areas of investigation in TIP, where strong earthquakes should be expected within a certain alarm time. We count one "success" if a strong earthquake happens inside the space–time limits of an alarm, and one "failure to predict" if a strong earthquake happens outside the limits. The accumulated statistics of successes and failures, on one hand, and the alarm area fraction, on the other hand, provide information for statistical conclusions about the efficiency and reliability of the algorithm and, in its turn, of the originating precursory seismic pattern.

Extensive retrospective testing essentially simulates forward prediction from the data available at the stage of developing the algorithm. We found that this testing is useful for establishing the limits of stability and applicability of the algorithm. However, the ultimate test of any prediction method is advance prediction of phenomena. Inevitably, each advance prediction experiment requires many years of tedious bookkeeping due to the infrequent occurrence of significant earthquakes. The procedures of such bookkeeping should be sufficiently transparent, so that other interested parties can repeat and/or revise *a posteriori* results of the prediction experiment [HKD92]. At the same time, these procedures must be free from human intervention and must be a black-box version of the algorithm in which potentially adjustable parameters of the algorithm are fixed. The source of the input data must be defined in advance, and the algorithm must resolve ambiguities in these data automatically, which is far from being a simple task. All of this may explain why we are not aware of many studies on testing predictions. But we do repeat that there is no other way to achieve unbiased statistical justification of a predictive method and a precursory pattern except for actual prediction of earthquakes. The more predictions, fulfilled or not fulfilled, we make, the more confidence in accepting or rejecting the underlying hypotheses we obtain.

Obviously, this scheme is open for including other traits and other data that are not necessarily seismological.

4.3.3 Major Common Characteristics of Premonitory Seismic Patterns

Robustness. We have to search for patterns common in a wide variety of regions and magnitude ranges as well as within sufficiently long time periods; otherwise any conclusive testing of earthquake prediction algorithms becomes practically impossible in our lifetime. Accordingly, premonitory seismic patterns are given robust definitions in which the diversity of circumstances and minutest details are averaged, and additional predictive power is retained.

Timescale. The trailing averages, used to determine a dynamic state of an investigated area, are calculated over time intervals as long as time spans of isolated alarms. These intervals are independent of M_0, and according to the Gutenberg–Richter relation, earthquakes with smaller magnitudes occur more frequently: the average time between earthquakes of magnitude M is proportional to 10^{BM}. This is not a contradiction, since the Gutenberg–Richter law refers to a given region, the same for all magnitudes, whereas premonitory patterns are defined for an area with a linear dimension $L(M)$ proportional to 10^{aM}. The average time between earthquakes in such an area is proportional to $10^{(B-a\nu)M}$, where ν is the fractal dimension of earthquake epicenters in the locality. The existing estimates of parameters B (about 1), a (between 0.4 and 1), and ν (between 1.2 and 2) do not contradict the hypothesis that $B - a\nu$ is close to zero as if earthquakes with different magnitudes have about the same recurrence time in their own sites. An accurate estimation of $B - a\nu$ favors this hypothesis [KM94]. At the same time, we must note that such time scaling is not the only possible one, so that for some premonitory seismic patterns, the timescale may change with M_0.

Normalization. To ensure application of a prediction algorithm without changing adjustable parameters, it is necessary to use normalized sequences of earthquakes in regions of different seismic activity. In the studies presented here the earthquake sequence in an area of investigation is normalized by the minimum cutoff magnitude M_{min} defined by one of two conditions: $M_{min} = M_0 - a$ or $\tilde{N}(M_{min}) = b$, where $\tilde{N}(M_{min})$ is the average annual number of earthquakes with magnitude $M \geq M_{min}$ and parameters a and b are constants common for all areas. Note that if the second condition is applied, the long-term frequency of earthquakes considered is the same in all areas (even if they differ in seismic activity), whereas M_{min} may be different.

4.3.4 Statistical Significance and Efficiency of Predictions

A statistical conclusion on the efficiency and reliability of an earthquake prediction algorithm can be drawn in the following way.

Let T and S be the total time and territory considered; A_t is the territory covered by alarms at time t; $\zeta \times \mu$ is a measure on $T \times S$ (we consider here a direct product measure $\zeta \times \mu$ reserving the general case of a time - space dependent measure for future, more sophisticated null hypotheses); N counts the total number of strong earthquakes with $M \geq M_0$ within $T \times S$ and k counts how many of them are predicted. The time–space occupied by alarms, $A = \bigcup_T A_t$, as a percentage of the total space–time considered is

$$ p = \int\limits_A \mathrm{d}(\zeta \times \mu) / \int\limits_{T \times S} \mathrm{d}(\zeta \times \mu) \ . $$

The statistical significance level of the prediction results equals

$$ 1 - \mathrm{B}(k - 1,\, N,\, p) \ , $$

where B is the cumulative binomial distribution function.

Let us choose the measure $\zeta \times \mu$. We assume a uniform measure ζ for time; it corresponds to the Poisson recurrence of earthquakes. For space, we assume a measure μ proportional to the spatial density of epicenters. Specifically, the measure μ of an area is proportional to the number of earthquake epicenters from a sample catalog, for example, for earthquakes above a certain magnitude cutoff M_c. This is a natural empirical spatial measure of seismicity useful also in estimating the seismic hazard [KT96, KKTM00]. When using the literal measure of territory in km^2, one may overestimate the statistical significance of the results obtained by equalizing areas of high and low seismic activity or, at the extreme, areas where earthquakes happen and do not happen.

The actual empirical distribution of earthquake locations is the best present-day estimate of where earthquakes can occur. The recipe of using measure μ and calculating p is the following. Choose a sample catalog. Count how many events from the catalog are within the territory considered; this will be your denominator. At a given time, count how many events from the catalog are within the area of alarm; this will be your numerator. Integrate the ratio over the time of prediction experiment.

This simple recipe has a nice analogy that justifies using statistical tools available since Blaise Pascal (1623–1662).

Seismic Roulette. Consider a roulette wheel with as many sectors as the number of events in your sample catalog, one sector for each event. Make your bet according to a prediction: determine which events are within the area of alarm and put one chip in each of the corresponding sectors. Turn the wheel. If the seismic roulette is not perfect, one can win systematically.

Besides the statistical significance, the most important characteristic of a prediction algorithm is its efficiency and corresponding usefulness of the predictions. G.M. Molchan [Mol97] suggested a link between prediction and its potential and optimal use by introducing a theory of prediction strategies (see Chap. 5 for details). In a nutshell, the lower envelope Γ of the error point set $\{n, \tau\}$, where n is the fraction of failures to predict and τ is the rate of alarm, characterizes the efficiency of the prediction algorithm [Mol97]. It demonstrates how far from a random guess are predictions that result from the algorithm. A trade-off between n and τ depends on the choice of adjustable parameters and the loss function γ, which may be very different when different preparedness problems and corresponding measures are considered in response to the prediction. The point where γ and Γ touch one another determines both the minimum achievable loss and the optimal set of adjustable parameters in the prediction algorithm.

The error envelope Γ is a desirable signature of the prediction algorithm, but by no means a compulsory one. Even without it, one can benefit from a prediction algorithm: given a single error point A, attributed to a prediction algorithm, one can always construct Γ connecting the three points $O(0, 1)$, $A(n, \tau)$, and $P(1, 0)$, and use it to define the optimal strategy. On the other hand, constructing the lower envelope Γ assumes extended variation of the adjustable parameters and therefore, delivers a demonstration of the stability level for the algorithm in retrospective analysis. There are other ways to justify the stability, for example, by applying it to simulated catalogs of earthquakes [Pro78, HC94, SCL97, SK99, SZK00].

It is also worth remembering that suggested estimations applied to statistics achieved in a retrospective prediction experiment could be biased due to possible discard, deliberate or unintentional, of some free parameters from the null hypothesis.

4.4 Validated Precursory Seismic Patterns

The following seismic patterns were given unambiguous definitions and their predictive value was validated by prospective worldwide tests. It is not yet clear whether some single simple premonitory pattern may compete in performance with prediction algorithms that combine several traits describing the dynamics of a seismic region in the approach of a large earthquake. Nevertheless, we begin with two simple ones, pattern Σ and "burst of aftershocks", and then turn to intermediate-term prediction algorithms, which were the subject of extensive testing for a decade or more in a real-time prediction mode. Algorithms M8 and MSc receive a fair amount of attention, because testing them is unprecedented in rigor and coverage and the relevant predictions are communicated, with due discretion, to several dozens of leading scientists and administrators in many countries. Among earthquakes already predicted are the last seven great ones of magnitude 8 or more.

4.4.1   Pattern Σ

We already mentioned *pattern Σ* [KM64] in the introduction to this chapter. It was defined as the trailing total sum of the source areas of earthquakes of medium size. The sum, coarsely estimated by the function $\Sigma(t) = \sum_{t-s}^{t} 10^{dM}$ with parameter d chosen so that each summand corresponds to the area of an earthquake of medium size, $M_{\text{cutoff}} \leq M < M_0 - q$, that occurred in the time interval from $t - s$ to t; this sum is roughly proportional to $E^{2/3}$, where E is the energy of the earthquake and M_{cutoff} and q are fixed constants. An alarm is declared for a certain period τ, when this sum rises closely to the source area $\widetilde{\Sigma}$ of a *single* earthquake of magnitude M_0.

Although this is the first reported unambiguously defined premonitory pattern featuring worldwide similarity and long-range correlations, testing it in advance prediction was not set up and carried on. The difficulty of widespread systematic testing is the same as for other similar characteristics [Var89, BV93], namely, a certain ambiguity in selecting the area of study and prediction. It is enough to mention that Bowman et al. [BOS+98], who claimed precursor validity at a 95% confidence level for the increase in the cumulative Benioff strain prior to major earthquakes in California, erroneously oversimplified the null hypothesis in their retrospective test overlooking evident clustering of small earthquakes and a free choice of both magnitude cutoff and the time span of analysis. If these simplifications are corrected, the common level of confidence at 95% is not reached. Clearly, it would be premature to attribute some higher order regularities, like log-periodicity [SS95], on top of a power-law rise, although they are suggested by several models [ALM82, GN94, SS95, AMCN95]. The abrupt rise of seismic activity prior to and its drop after strong earthquakes is described in a detailed case history of 11 magnitude 6.8 events in California [KLKM96].

4.4.2   Burst of Aftershocks or Pattern B

One of most evident characteristics of a seismic sequence is clustering of earthquakes in space and time. The most clear and therefore well-recognized phenomenon is the direct cascade of aftershocks. It was found that a large, abnormal number of aftershocks after a moderate-size main shock indicates the approach of a major event [KKR80]. Such a burst of activity was named "burst of aftershocks", or *pattern B*, defined as a main shock in a medium magnitude range, $M_{\text{cutoff}} \leq M < M_0 - q$, with a large number of aftershocks. The latter requires a definition of aftershocks with cutoff magnitude M_{aft} and time span e after the main shock to determine whether an event must be included in the aftershock count; also, a constant C sets the level of large. An alarm is declared for a certain period τ when the count of some main shock rises above C.

The aftershocks of a given main shock are defined as all smaller earthquakes that follow it within a certain time $T(M)$ and distance $R(M)$, where

$T(M)$ and $R(M)$ are arbitrary, usually step functions of the magnitude of the main shock M [GK74, KKR80]. Such a crude definition is appropriate for many analyses of seismic dynamics (e.g., intermediate-term prediction algorithms described below) stripping the earthquake sequence in the area under study from evident clustering of events, which, presumably, allows an easier search for precursory activation of the seismic region. More sophisticated definitions and methods of aftershock identification are available [MD92, Rea85, DF91] and, potentially, their use may improve the overall performance of earthquake prediction methods.

Pattern B was tested extensively in retrospective applications in many regions worldwide [Kei82]. The drawback of ambiguity in selecting the area of study and large territorial and spatial uncertainty remains, although certain low-cost mitigation measures of civil defense type in response to a pattern B alarm might already be justifiable. The performance of pattern B can be improved by counting weighted sums of aftershocks [AK89] that estimate either the energy or rupture area of the aftershock sequence or by setting an alarm when a sufficient number of main shocks reveal pattern B [MDRD90]. So far, the latter is the only simple premonitory seismic pattern, whose statistical significance in a strict sense is established. With this exception, the rates of successes and failures for other simple seismic patterns, like swarms [CGK+77, KLJM82] or a generalized pattern B [CCG+83], have not been adequately explored systematically.

Advances in studying pattern B eventually encouraged a wider and more systematic search for premonitory seismic patterns of a complex nature. It became clear enough that the accuracy of prediction achieved by a simple precursory seismic pattern can be improved by bringing into the analysis other observable patterns and that efficient prediction of strong earthquakes can be based on a complex combination of phenomena.

4.4.3 Algorithm M8

This intermediate-term earthquake prediction method was designed by the retrospective analysis of dynamics in the seismicity preceding the greatest, magnitude 8.0 or more, earthquakes worldwide, hence its name. Its prototype [KK84] and the original version [KK86] were tested retrospectively in the vicinities of 143 points, of which 132 are recorded epicenters of earthquakes of magnitude 8.0 or greater from 1857 to 1983. In 1986, algorithm M8 was tested in retrospective application to predict earthquakes of smaller magnitudes, down to 6.5 [Kos86] by using independent regional seismic databases (note that the USGS/NOAA global database available to us at that time covered only the period through 1983). By 1990, the list of the territories, where the original and other versions of algorithm M8 have been applied, extended to 19 regions listed in Table 4.2 [KK90].

Table 4.2. Summary of TIPs diagnosed by algorithm M8 (after [KK90])

Region	M_0	Time period	N_T/N^a	V_{TIPs}^b	N_{suc}/N_{all}^c
Learning					
1. World	8.0	1967–1982	5/7	5	7/16
Testing of the Original Version					
2. Central America	8.0	1977–1986	1/1	16	1/2
3. Kuril Islands and Kamchatka	7.5	1975–1987	2/2	17	2/3
4. Japan and Taiwan	7.5	1975–1987	5/6	20	6/8
5. South America	7.5	1975–1987	3/3	18	3/8
6. Western U.S.	7.5	1975–1987	-/-	5	0/1
7. Southern California	7.5	1947–1987	1/1	12	1/1
8. Western U.S.	7.0	1975–1987	2/2	24	2/2
9. Baikal and Stanovoy Range	6.7	1975–1986	-/-	0	-/-
10. Caucasus	6.5	1975–1987	2/3	12	2/2
11. East Central Asia	6.5	1975–1987	4/5	24	5/6
12. Eastern Tien Shan	6.5	1963–1987	4/4	27	5/5
13. Western Turkmenia	6.5	1979–1986	-/-	0	-/-
14. Apennines	6.5	1970–1986	1/1	10	1/1
15. Koyna reservoir	4.9	1975–1986	1/1	42	1/1
Testing of Modified Versions					
16. Greece	7.0	1973–1987	3/3	18	4/5
17. Himalayas	7.0	1970–1987	2/2	8	3/4
18. Vrancea	6.5	1975–1986	2/2	58	2/2
19. Vancouver Island	6.0	1957–1985	4/4	20	5/7
Regions 1–19 together			39/44 (89%)	18	49/72
Regions 2–15 together			25/28 (89%)	16	28/38

[a] N and N_T are the number of all earthquakes and their number within TIPs
[b] V_{TIPs} is the space–time fraction of TIPs
[c] N_{all} and N_{suc} are the number of all and successful TIPs, respectively

Algorithm M8 is based on a simple physical scheme of prediction briefly described below. The values of the constant parameters that enter the algorithm are listed at the end of the description.

Prediction is aimed at earthquakes of magnitude M_0 and higher. If the data permit, we consider different values of M_0 with a step of 0.5. Overlapping circles of diameters $D(M_0)$ scan the territory of the seismic region under study. The sequence of earthquakes with aftershocks removed is considered within each circle. Denote this sequence by $\{t_i, m_i, h_i, b_i(e)\}$, $i = 1, 2, ...,$ where t_i is the origin time, $t_i \leq t_{i+1}$; m_i is the magnitude; h_i is focal depth; and $b_i(e)$ is the number of aftershocks of magnitude M_{aft} or greater during the first e days. The sequence is normalized by the lower magnitude cutoff $\underline{M} = M_{\min}(\widetilde{N})$, where $\widetilde{N}$ is the standard value of the average annual number of earthquakes in the sequence.

Several running averages are computed for this sequence in the trailing time window $(t - s, t)$ and magnitude range $M_0 > m_i \geq \underline{M}$. They depict different measures of the intensity in earthquake flow, its deviations from the long-term trend, and clustering of earthquakes. The averages include

- $N(t) = N(t \mid \underline{M}, s)$, the number of main shocks of magnitude $\underline{M}$ or larger in $(t - s, t)$;
- $L(t) = L(t \mid \underline{M}, s, t_0)$, the deviation of $N(t)$ from the longer term trend, $L(t) = N(t) - N_{\text{cum}}(t-s)(t-t_0)/(t-t_0-s)$, where $N_{\text{cum}}(t) = N(t \mid \underline{M}, t - t_0)$ is the cumulative number of main shocks with $M \geq \underline{M}$ from the beginning of the sequence t_0 to t;
- $Z(t) = Z(t \mid \underline{M}, \overline{M}, s, \alpha, \beta)$, linear concentration of main shocks $\{i\}$ from the magnitude range $(\underline{M}, \overline{M}) = (M_{\min}(\widetilde{N}), M_0 - g)$ and interval $(t-s, t)$; the linear concentration is estimated as the ratio of the average source diameter l to the average distance r between sources; and
- $B(t) = B(t \mid \underline{M}, \overline{M}, s', M_{\text{aft}}, e) = \max_{\{i\}}\{b_i|\}$, the maximum number of aftershocks (i.e., a measure of earthquake clustering). The sequence $\{i\}$ is considered in the trailing time window $(t - s', t)$ and in the magnitude range $(\underline{M}, \overline{M}) = (M_0 - p, M_0 - q)$.

Each of the functions N, L, and Z is calculated twice with $\underline{M} = M_{\min}(\widetilde{N})$, for $\widetilde{N} = 20$ and $\widetilde{N} = 10$. As a result, the earthquake sequence is given a robust averaged description by seven functions: N, L, Z (twice each), and B.

"Very large" values are identified for each function from the condition that they exceed Q percentiles (i.e., they exceed Q percent of the encountered values).

An alarm or a TIP, time of increased probability, is declared for 5 years when at least six out of seven functions, including B, become "very large" within a narrow time window $(t - u, t)$. To stabilize prediction, this criterion is checked at two consecutive moments, t and $t + 0.5$ years. In the course of a forward application, the alarm can extend beyond or terminate in less than 5 years when updating causes changes in the magnitude cutoffs and/or the percentiles.

The following standard values of parameters indicated above are prefixed in the algorithm M8: $D(M_0) = [exp(M_0 - 5.6) + 1]°$ in degrees of meridian (this is 384, 560, 854 and 1333 km for $M_0 = 6.5$, 7.0, 7.5, and 8, respectively), $s = 6$ years, $s' = 1$ year, $g = 0.5$, $p = 2$, $q = 0.2$, $u = 3$ years, and $Q = 75\%$ for B and 90% for the other six functions. Usually, the average diameter l of the source is estimated by $(n)^{-1}\sum_{\{i\}} 10^{\beta(M_i - \alpha)}$, where n is the number of main shocks in $\{i\}$, $\beta = 0.46$ to represent the linear dimension of source, and $\alpha = 0$ (which does not restrict generality), and the average distance r between sources is set proportional to $(n)^{-1/3}$. The performance of the algorithm can be improved by estimating the linear concentration of main shocks more accurately [RK96].

Running averages are defined in a robust way, so that a reasonable variation of parameters does not affect predictions. At the same time, the discrete character of seismic data and strict usage of the prefixed thresholds result in a certain discreteness of the alarms.

It is worth mentioning that, qualitatively, algorithm M8 uses a rather traditional description of a dynamic system adding dimensionless concentration (Z) and a characteristic measure of clustering (B) to the phase space of rate (N) and rate differential (L). The algorithm recognizes a *criterion* defined by extreme values of the phase space coordinates, as the vicinity of a system singularity. When a trajectory enters the criterion, the probability of an extreme event increases to a level sufficient to predict it effectively. The choice of the M8 criterion determines a specific intermediate-term rise, an inverse cascade [GKZN00], of seismic activity at the middle-range distance.

4.4.4   Algorithm MSc or "The Mendocino Scenario"

This second approximation prediction method [KKS90] was designed by retrospective analysis of the detailed regional seismic catalog prior to the Eureka earthquake (1980, $M = 7.2$) near Cape Mendocino in California, hence its name abbreviated to MSc. Given a TIP diagnosed for a certain territory U at time T, the algorithm is designed to find a smaller area V within U, where the predicted earthquake can be expected. To execute the algorithm, one needs a reasonably complete catalog of earthquakes with magnitudes $M \geq M_0 - 4$, which is lower than the minimum threshold usually used by M8. When this condition does not hold, we assume that the dynamics of earthquakes available in the database inherits behavior from lower levels of the seismic hierarchy. The detection of the MSc criteria in such a case is more difficult and might result in additional failures to predict.

The essence of MSc can be summarized as follows. Territory U is coarse-grained into small squares of size $s \times s$. Let (i, j) be the coordinates of the centers of the squares. Within each square (i, j) the number of earthquakes $n_{ij}(k)$, aftershocks included, is calculated for consecutive, short, time windows u months long, starting from time $t_0 = T - 6$ years onward, to include earthquakes that contributed to the TIP's diagnosis; k is the sequence number of a time window. In this way, the time–space considered is divided into small boxes (i, j, k) of size $(s \times s \times u)$. "Quiet" boxes are singled out for each small square (i, j); they are defined by the condition that $n_{ij}(k)$ is below the Q percentile of n_{ij}. The clusters of q or more quiet boxes connected in space or in time are identified. Area V is the territorial projection of these clusters.

The standard values of parameters adjusted for the 1980 Eureka earthquake are as follows: $u = 2$ months, $Q = 10\%$, $q = 4$, and $s = 3D/16$, where D is the diameter of the circles used in algorithm M8.

Qualitatively, the MSc algorithm outlines such an area of the territory of alarm where the activity, from the beginning of the seismic inverse cascade recognized by algorithm M8 in declaration of the alarm, is continuously high but infrequently interrupted for a short time. Such interruption must have a sufficient temporal and/or spatial span. The phenomenon, which is used in the MSc algorithm, might reflect the second (possibly, shorter term and, definitely, narrow-range) stage of the premonitory rise of seismic activity near the incipient source of a main shock. The anomalous quiescence used in the definition of the *precursory intermittent pattern* in the dynamics of the seismic region should not be mixed with a prolonged state of *"seismic quiescence"*, advocated by [WH88].

4.4.5   Global Testing of Algorithms M8 and MSc

After successfully predicting the Loma Prieta 1989 earthquake, J.H. Healy, V.G. Kossobokov, and J.W. Dewey [HKD92, Har98] designed a rigid test to evaluate algorithm M8. By that time, all of the components necessary for a reproducible real-time prediction – i.e., the database and unambiguous definition of the algorithms – were specified in publications [GHDB89, KK90, KKS90], and since 1991, the algorithm has been applied each half-year in a real time prediction mode to monitor the seismic dynamics of the entire Circumpacific. More extended testing, for all seismically active territories on Earth, where seismic data were enough to run the standard version of algorithm M8 [HKD92, Kos97] in two magnitude ranges defined by $M_0 = 8.0$ and 7.5, was carried on in parallel [KRKH99].

Earthquake catalog. The National Earthquake Information Center/U.S. Geological Survey Global Hypocenters Database [GHDB89] and its routine updates timely describe the dynamics of seismic regions on Earth with the remarkable stability (see Fig. 4.1). This database encompasses records from several global and regional catalogs. It is routinely updated with the Preliminary Determinations of Epicenters (PDE). The PDE-Weekly becomes available with a delay of about 1 month, and the final version of the catalog, PDE-Monthly, is published with a longer delay. In the reorganization, the NEIC has reconsidered the data processing routine so that PDE-Weekly became virtually the final version of the catalog.

The NEIC GHDB is complete for magnitude 4.5–5 events since 1963, when the World Wide Standard Seismograph Network was established to record all significant earthquakes and explosions of even lower magnitudes. The GHDB has allowed a systematic prospective application of algorithm M8 on the global scale since 1985; we need at least 12 years of data for the stable evaluation of the seven functions and an additional 10 years to form a reasonable evaluation of their percentiles. For the catalog preprocessing, which includes elimination of possible duplicates and identification of aftershocks, we used the algorithm [She92], which automates the technique first described in [KKR80, ZS87]. The aftershock identification rules are the same as formulated in [KKR80]. Also note that M8 prediction results are practically insensitive to small variations of these rules, whereas the MSc algorithm does not require the identification of main shocks and aftershocks.

Since publication of [HKD92], the global catalog and predictions by M8 and MSc are updated every 6 months, usually with a 3-month delay, last time through the end of 1998. Since July 1999 it became possible to update and distribute global predictions with a delay of about 2 weeks. The PDE-Weekly compiled within 4 weeks of the occurrence of an earthquake has reached the level of completeness of the monthly listings (PDE) and opened the possibility of eliminating the delay by combining the PDE, PDE-W, and Quick Epicenter Determination (QED) data. The QED catalog is complete in reporting large earthquakes, although it lacks magnitude 4–5 events, which practically cannot bias determination of the M8-MSc up-to-date predictions.

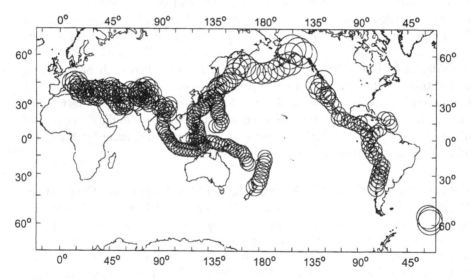

Fig. 4.3. Global testing of algorithms M8 and MSc, $M_0 = 8.0$: Circles of investigation

Prediction of largest earthquakes, $M \geq 8$. We applied algorithm M8 and then MSc in the areas of current alarm in 262 overlapping circles of investigation shown in Fig. 4.3. Specifically, 170 circles were selected from the larger number that scans nearly uniformly the Circumpacific and its surroundings, whereas the other 92 circles are taken from the Alpine-Himalayans and Burma (25 in the Mediterranean, 25 in Asia Minor and Iran, 28 in Pamirs-Hindukush, and 14 in Burma). Thus, we may conclude that the completeness of the NEIC GHDB is sufficient for application of the original version of M8 in 80–90% of the major seismic belts. A sample prediction, as of the date of writing this text, is given in Fig. 4.4 (a complete set of predictions from 1985–2000 can be viewed at http://mitp.ru/predictions.html). Earthquakes of magnitude 8.0 or more are expected in 16 circles, each of radius 667 km, that form seven compact areas of alarm. In the second approximation, the MSc algorithm outlines nine smaller areas inside alarms, where great earthquakes are most likely. It is worth mentioning that, due to the very low recurrence rate of great earthquakes, most of the areas, perhaps all of them, will not be confirmed until the next update. However, the alarms last for many years (about five on the average in accordance with the formal definition) and correspond to areas and times where the increased probability of occurrence is already confirmed in the test. The probability gain depends on locality and varies from 2–3 in regions of extremely high activity, like Tonga-Kermadek, to 20–100 in regions where the recurrence of great earthquakes is much lower than the average, like Sumatera.

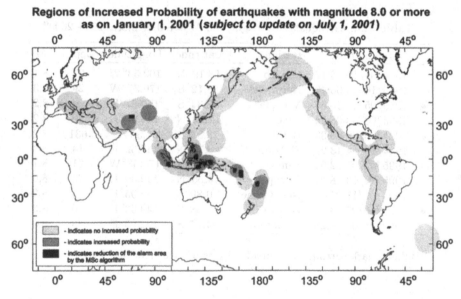

Fig. 4.4. Global testing of algorithms M8 and MSc, $M_0 = 8.0$: Areas of alarm as on 1 January 2001. Circular areas of alarm in the first approximation, i.e., from algorithm M8, are shaded light gray; rectangular areas determined in the second approximation by algorithm MSc are shaded dark gray

The performance of both algorithms is illustrated in Fig. 4.5 for one segment of the Circumpacific, namely, from Kamchatka to the Marianas Trench. For the whole territory where prediction is made, the M8 alarms cover, on average, one-third of its length at any given time, whereas MSc reduces this number to 10%. All eight earthquakes of magnitude 8.0 or greater, which

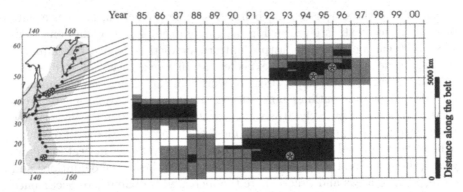

Fig. 4.5. Global testing of algorithms M8 and MSc, $M_0 = 8.0$: Space–time distribution of alarms from Kamchatka to the Marianas. Circles of investigation (*light*) and their centers (*heavy dots*) are shown on the left; the space–time distribution of alarms from 1985–2000 is on the right (*dark* for M8 and *darker* for MSc). The space coordinate is the distance along the belt. Great earthquakes are marked by *stars*

Table 4.3. Earthquakes of magnitude $M = 8.0$ or more, 1985–2000[a]

	Date and time	Region	Latitude	Longitude	Depth	M
	1985/09/19 13:17	Mexico	18.19°N	102.53°W	27	8.1
	1986/10/20 06:46	Kermadek	28.12°S	176.37°W	29	8.3
•	1989/05/23 10:54	Macquarie	52.34°S	160.57°E	10	8.2
	1993/08/08 08:34	Guam	12.98°N	144.80°E	59	8.2
•	1994/06/09 00:33	Bolivia	13.84°S	67.55°W	631	8.2
	1994/10/04 13:22	Shikotan	43.77°N	147.32°E	14	8.3
	1995/04/07 22:06	Samoa	15.20°S	173.53°W	21	8.1
	1995/12/03 18:01	Iturup	44.66°N	149.30°E	33	8.0
	1996/02/17 05:59	New Guinea	00.89°S	136.95°E	33	8.2
•	1998/03/25 03:12	Balleny	62.88°S	149.53°E	10	8.3
	2000/06/04 16:28	Sumatera	04.77°S	102.05°E	33	8.0

[a]Bullets mark earthquakes outside circles of investigation

occurred in the area in the time span 1985–2000 (Table 4.3), are predicted by M8 and only one of them, the 1996 New Guinea, is missed in the second approximation given by MSc. Table 4.4 summarizes the success-to-failure score; two time intervals are distinguished there, 1985–1997 and 1992–1997. Seismic data for any of the two intervals were not available to the authors when they designed the M8 and MSc algorithms. Since 1985, the database has become sufficient for the forward prediction considered, and in 1992, upon publication of [HKD92], the rigid framework of the test was formally established and the test has begun in a real-time mode.

Table 4.4. Performance of earthquake prediction algorithms M8 and M8-MSc:
Magnitude $M = 8.0$ or more

Test period	Strong earthquakes Predicted by		Total	Percentage of alarms Circumpacific		Worldwide		Confidence level, %[a]	
	M8	MSc		M8	MSc	M8	MSc	M8	MSc
1985–2000	8	7	8	37.2	20.0	34.9	18.0	99.96	99.99
1992–2000	6	5	6	34.1	18.4	30.2	15.3	99.84	99.90

[a]The significance level estimates use the most conservative measure μ of the alarm
volume accounting for the empirical distribution of epicenters in the Circumpacific

Let us describe in detail each case history of all 11 great earthquakes;
the description is instructive because even the three cases that fall out of
consideration in the test reveal a clear pattern of a precursory rise of activity
in its generalized definition expressed by algorithm M8.

Each of Figs. 4.6–4.16 displays a prediction map along with epicenters of
the great main shock and its first aftershocks (on the left); on the right, it
depicts the space–time diagram of seismic activity in a circle of radius 667 km
where the alarm was in progress when the great earthquake happened and
below it presents the functions of algorithm M8. On each diagram, arrows
indicate the great shock, and small circles stand for smaller magnitude earth-

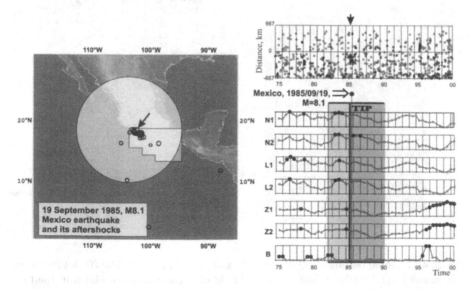

Fig. 4.6. Global testing of algorithms M8 and MSc, $M_0 = 8.0$: The 19 September
1985 Mexican earthquake. Observe the highlighted circular areas of alarm in the
first approximation determined from algorithm M8 and the highlighted rectangular
areas of alarm in the second approximation determined from algorithm MSc

quakes used by the algorithm for determining the TIP. The distance along
the seismic belt measured in kilometers from the center of the circle is plotted
on the vertical axis. Time is plotted along the horizontal axis.

1985/09/19. Space–time diagram for the 19 September 1985 Mexican earthquake (Fig. 4.6). One can observe a localized long-term quiescence in the 200-
km vicinity of the coming epicenter. It started in the middle of 1979 and then
was shortly interrupted by a magnitude 7.4 earthquake in 1981 and another
one at the end of 1983. From the beginning of 1984, the activity rose dramatically in the area of the future epicenter. It was so noticeable that it resulted
in the rise of M8 functions to the level required to declare a TIP, although
they are evaluated on a larger territory. McNally et al. [MGS86] asserted
that such a combination of quiescence and activation was observed before six
earthquakes of magnitude 7 or more along the Central American trench. The
corresponding TIP was determined in July 1985. The MSc algorithm lacks
specificity in this case that restricts the M8 alarm to more than a 600-km
segment of the seismic belt.

1986/10/20. On 20 October 1986, an earthquake struck the Kermadek
Islands on the background of the general activation of seismic activity in the

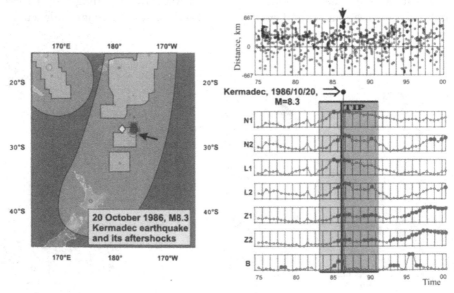

Fig. 4.7. Global testing of algorithms M8 and MSc, $M_0 = 8.0$: The 20 October 1986
Kermadek earthquake. Observe the highlighted circular areas of alarm in the first
approximation determined from algorithm M8 and the highlighted rectangular areas
of alarm in the second approximation determined from algorithm MSc. A swarm of
24 earthquakes (marked by a diamond) occurred 4 months in advance of the great
shock

region (Fig. 4.7) from the middle of 1984. This activation had an integral character related to the entire circle or even larger and could not be attributed directly to a narrow zone near the future epicenter. Seismic activity in the Tonga-Kermadek region is one of the highest on Earth. Algorithm M8 determines many TIPs that cover a large part, sometimes all, of the seismic territories considered in the test. In July 1986, MSc reduced the alarm in the second approximation to a largely extended area of intermediate and deep earthquakes in Tonga and two areas of 300–400 km in diameter in Kermadek trenches. The 1986 great earthquake and its aftershocks confirmed one of the later predictions. In the 4 months before it happened and about 150 km from its epicenter, a swarm of 24 earthquakes occurred (marked by a diamond in Fig. 4.7) of which 11 were of magnitude 5 or larger.

1989/05/23. The great 23 May 1989 Macquarie earthquake occurred in area with a very low rate of seismic activity. Enough to mention that according to [GHDB89] within the circle of radius 667 km centered on its epicenter, there were about four events of magnitude 4.5 or more per year from 1963–1988, which is far from 20 events per year required by the standard version of M8. Significant changes in activity are hard to observe in such a region;

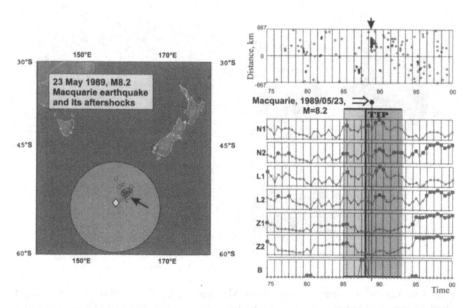

Fig. 4.8. Global testing of algorithms M8 and MSc, $M_0 = 8.0$: The 23 May 1989 Macquarie earthquake. Observe the highlighted circular areas of alarm in the first approximation determined from algorithm M8 and the highlighted rectangular areas of alarm in the second approximation found from algorithm MSc. An earthquake of magnitude 5.3 (marked by a diamond) occurred 2 weeks in advance of the great shock

nevertheless, the modified version of algorithm M8 recognized TIP in July 1988, less than a year in advance of the 1989 Macquarie earthquake. The only modification consists of using rates much smaller than $\tilde{N} = 20$ and 10 that define $\underline{M} = M_{min}(\tilde{N})$. Note that an earthquake of magnitude 5.3 did occur 2 weeks prior to and about 100 km southwest of the main shock (diamond in Fig. 4.8). Such an earthquake is very seldom there, although one could hardly draw its relation to the great Macquarie earthquake of 1989.

1993/08/08. The segment of the Bonin-Mariana deep-sea trench where the earthquake shook Guam Island on 8 August 1993 (Fig. 4.9) was already in a state of relatively high activity since 1985. In 1990, this activity got another impulse after a magnitude 7.5 event followed by a large number of aftershocks. This earthquake happened about 300 km to the northwest of the 1993 great shock epicenter (the upper diamond in Fig. 4.9). In a shorter term of about 2 months, the 1993 Guam earthquake was preceded by a magnitude 6.6 quake next to its future aftershock zone and then, 5 days prior to it, by a magnitude 5.1 (the lower diamond in Fig. 4.9) right in the epicenter area. Algorithm M8 recognized the TIP that predicted the great Guam earthquake in January

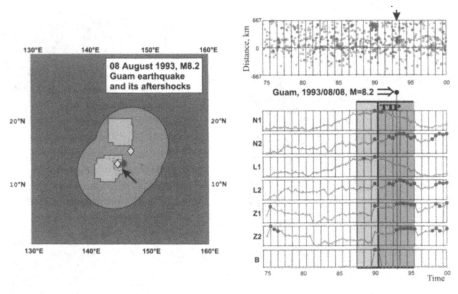

Fig. 4.9. Global testing of algorithms M8 and MSc, $M_0 = 8.0$: The 8 August 1993 Guam earthquake. Observe the highlighted circular areas of alarm in the first approximation determined by algorithm M8 and the highlighted rectangular areas of alarm in the second approximation determined by algorithm MSc. An earthquake with $M = 7.5$ (the upper diamond on the map) did occur on 5 April 1990. In a shorter term, a magnitude 6.6 event (the lower diamond) preceded the 1993 great shock within 2 months

1991. As of July 1993, the MSc reduced this alarm to two areas, each about 400 km in diameter. One of them was a correct prediction [Kos94].

1994/06/09. The great deep Bolivia earthquake [SIGB94] occurred on 6 June 1994 beneath the flat, practically aseismic territory about 600 km away from the Circumpacific coast. Its hypocenter was located on the continuation of the subducted part of the Nazca plate at the turning point of the South American coast. Formally, this earthquake occurred outside the scope of M8 testing [HKD92]; nevertheless, note that, as of January 1994, the Great Deep epicenter was less than 100 km from the area of the M8 alarm in progress (Fig. 4.10).

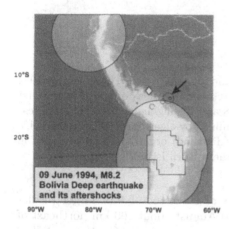

09 June 1994, M8.2
Bolivia Deep earthquake
and its aftershocks

Fig. 4.10. Global testing of algorithms M8 and MSc, $M_0 = 8.0$: The 9 June 1994 Deep Bolivia earthquake. Observe the highlighted circular areas of alarm in the first approximation determined by algorithm M8 and the highlighted rectangular areas of alarm in the second approximation determined by algorithm MSc

1994/10/04. From the end of the 1970s, the region of the southern Kuril Islands was in a steady state of activity (Fig. 4.11). The situation changed in 1990. In 1991, a swarm of earthquakes, seven of which were of magnitude 6 or more, lasted for two weeks and was followed by a magnitude 7.4 earthquake offshore Urup Island. This rise of seismic activity was first recognized in July 1992 as an M8 alarm for smaller magnitude $M_0 = 7.5$ in the region of the Eastern part of Hokkaido and the southern Kuril Islands. The first strong earthquake confirming the prediction occurred on 15 January 1993 $M = 7.6$, in Kushiro-Oki (Hokkaido) at a depth of 102 km. However, the M8 algorithm, run to predict magnitude 8.0 or larger earthquakes, determined the TIP for a larger earthquake in January 1994. The Shikotan Island earthquake that followed on 4 October 1994 fitted exactly the temporal, territorial, and magnitude ranges of the predictions (Fig. 4.11). Algorithm MSc pinpointed the location of the 4 October event by using data through the first half of 1994. The refined 210 × 160 km area of alarm coincides remarkably with the aftershock zone of the Shikotan earthquake.

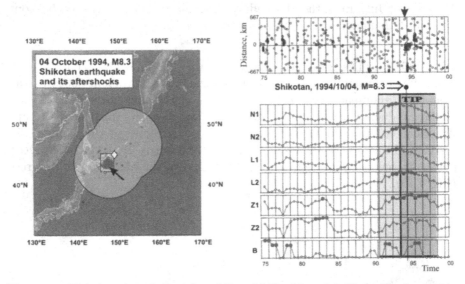

Fig. 4.11. Global testing of algorithms M8 and MSc, $M_0 = 8.0$: The 4 October 1994 Shikotan earthquake. Observe the highlighted circular areas of alarm in the first approximation determined by algorithm M8 and the highlighted rectangular areas of alarm in the second approximation determined by algorithm MSc. A remarkable swarm of seven magnitude 6 earthquakes (diamond) occurred about a month in advance of the 1994 great shock in the southern Kuril Islands

A remarkable swarm of seven quakes of magnitude 6.0 or larger (a diamond in Fig. 4.11) occurred within 14–30 August about 100 km northeast of the future aftershock zone of the 4 October main shock. On the next day, on August 31, a deeper, magnitude 6.2, earthquake outlined the opposite southwestern limit of the area, which ruptured in the 1994 great earthquake. Thus, about a month before the 8.3 main shock, two opposite sides of the M8-MSc prediction exposed rather unusual activation [KHD+96].

1995/04/07. The 7 April 1995 Samoa earthquake is somewhat different from other great earthquakes from 1985–2000. It did not produce a large number of aftershocks and can hardly be recognized on the space–time diagram of Fig. 4.12. Regardless of its large magnitude, it appears to be a common event in the seismic dynamics of the region. This characteristic, a steady high concentration of earthquakes both in space and time, might be common for Samoa and Tonga. For example, we also cannot clearly distinguish earthquakes of magnitude 7.5 or larger that occurred in 1975, 1977, 1981, and 1994 within the limits of the same circle of investigation. However, algorithm M8 managed to recognize a TIP starting from January 1994. In January 1995, the MSc algorithm outlined two areas: a larger extended area of intermediate and deep earthquakes and a 300 × 300 km area at the northern edge of the Tonga trench. The second of the refined predictions was confirmed on 7 April.

Three months in advance of the great Samoa earthquake, a remarkable swarm of 31 quakes (a diamond in Fig.4.12) was recorded on the Samoa Islands, i.e., some 250 km off its future epicenter. Although only two of them have magnitudes exceeding 5.0, it is still very unusual even in such an active region.

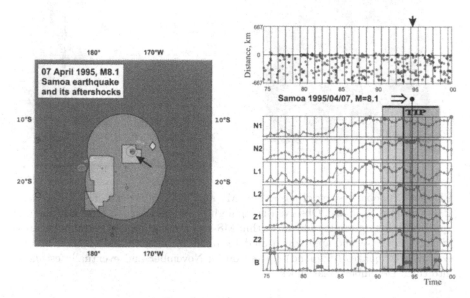

Fig. 4.12. Global testing of algorithms M8 and MSc, $M_0 = 8.0$: The 7 April 1995 Samoa earthquake. Observe the highlighted circular areas of alarm in the first approximation determined by algorithm M8 and the highlighted rectangular areas of alarm in the second approximation determined by algorithm MSc. A remarkable swarm of 31 earthquakes (diamond) occurred about 3 months in advance of the 1995 great shock

1995/12/03. The Iturup, 3 December 1995, earthquake occurred 16 months after and at about the same location as the swarm of 14–30 August described above in connection with the Shikotan earthquake. It occurred on a background of a general increase in seismic activity in the Kuril subduction zone, 200 km northeast of the 1994 Shikotan earthquake in the time period when its aftershocks continued to shake the area (Fig. 4.13). The Iturup and the preceding Shikotan earthquakes shared the same M8 alarm. However, in a year, algorithm MSc managed to change the position of the area where the great earthquake was more likely. As of July 1995, the area of the refined second approximation migrated to the northeast and expanded slightly to the size of 210×210 km [KHD+96].

The Iturup earthquake is a remarkable model example of an earthquake with a clear foreshock sequence; it started on 24 November with a magnitude

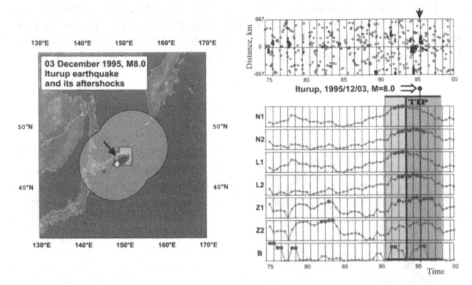

Fig. 4.13. Global testing of algorithms M8 and MSc, $M_0 = 8.0$: The 3 December 1995 Iturup earthquake. Observe the highlighted circular areas of alarm in the first approximation determined by algorithm M8 and the highlighted rectangular areas of alarm in the second approximation determined by algorithm MSc. A remarkable swarm of 64 quakes (diamond) started on 24 November and eventually escalated into the 3 December great shock

6.6 and escalated until the very main shock on 3 December (a diamond in Fig. 4.13). This swarm-like sequence consists of 64 events of magnitude 4 or above, of which 19 have $M \geq 5$ and four have $M \geq 6$.

1996/02/17. The space–time diagram of earthquake sequence dynamics in Fig. 4.14 displays certain activation through all of 1994 in the vicinity of the future source zone of the 17 February 1996 earthquake offshore of New Guinea. This activation concluded with a compact series of 24 earthquakes of magnitude 4 or more associated with a magnitude 6.1 followed by 7.1 earthquakes in March 1995. These earthquakes occurred about 400 km south of the 17 February 1996 epicenter; therefore it is hard to assert their connection with reasonable certainty.

Algorithm M8 recognized this rise of activity as precursory and declared a TIP in the region from January 1996. According to the second approximation prediction determined by algorithm MSc here, the area of a more likely occurrence of a great, magnitude 8 or more, earthquake was located in the 300×200 km area inland in New Guinea. The 17 February 1996 earthquake confirmed the M8 but missed the MSc prediction by about 200 km (Fig. 4.14).

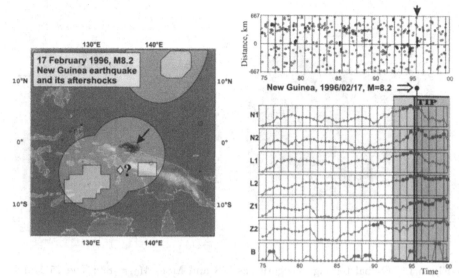

Fig. 4.14. Global testing of algorithms M8 and MSc, $M_0 = 8.0$: The 17 February 1996 New Guinea earthquake. Observe the highlighted circular areas of alarm in the first approximation determined by algorithm M8 and the highlighted rectangular areas of alarm in the second approximation determined by algorithm MSc

1998/03/25. The 25 March 1998 Balleny Sea great earthquake occurred in an exceptional location [WWL98] outside Earth's major seismic belts about 250 km away from the midoceanic ridge separating the Australian and Antarctic plates. Although such a large earthquake has never occurred in the southern high latitudes, algorithm M8, in a modification similar to that applied in the case history of the 1989 Macquarie earthquake, declared an alarm from January 1998 in a 667-km circle centered on the axis of the Indian midoceanic ridge. The seismic activity rise that resulted in a TIP is apparent on the space–time diagram in Fig. 4.15 covering the period from 1994–1997.

2000/06/04. The 4 June 2000 great earthquake offshore of southern Sumatera occurred on a high seismic activity segment of the deep oceanic trench. Nevertheless, such great seismic events in the Indian Ocean subduction zones were reported only twice in the twentieth century; they were the 1941 Andaman, $M_s = 8.1$, and the 1977 Sumbawa, $M_s = 8.0$, Islands earthquakes. The size of the seismic event in June 2000 is indirectly confirmed by the aftershock zone that extended over 200 km and by, perhaps, an associated earthquake of $M_s = 7.8$ in the Indian Ocean two weeks later.

The immediate vicinity of the 2000 Sumatera earthquake source experienced relative decay in 1996–1997 followed by a substantial rise of activity that is easily recognized on the space–time diagram in Fig. 4.16. At the same

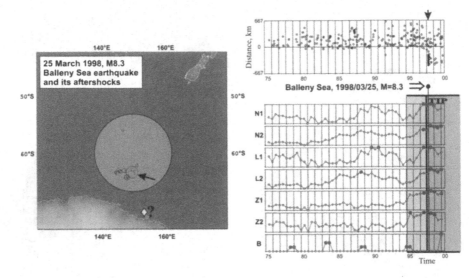

Fig. 4.15. Global testing of algorithms M8 and MSc, $M_0 = 8.0$: The 25 March 1998 Balleny Sea earthquake. Observe the highlighted circular areas of alarm in the first approximation determined by algorithm M8. The second approximation was not found due to the low seismic activity in the region

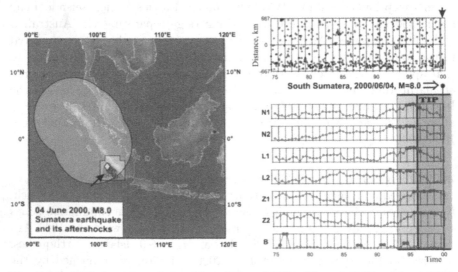

Fig. 4.16. Global testing of algorithms M8 and MSc, $M_0 = 8.0$: The 4 June 2000 Sumatera earthquake. Observe the highlighted circular areas of alarm in the first approximation determined by algorithm M8 and the highlighted rectangular areas of alarm in the second approximation determined by algorithm MSc. A foreshock of magnitude 4.7 (diamond) occurred within a day in advance of the great shock

time, in a larger territory of the M8 circle of investigation, the parameters of activity, i.e., functions N and L, reached their maximum values in the middle of 1995 and then decayed to their minimum values by the time of the great earthquake. The parameters of concentration, i.e., functions Z, did not decay from their maximum values, which means that the number of earthquakes decreased, but their energy and concentration increased. Moreover, the concentration of larger earthquakes was observed in a narrow vicinity of the coming great earthquake (see space–time diagram in Fig. 4.16). A foreshock of magnitude 4.7 (a diamond in Fig. 4.16) was recorded within a day in advance of the main shock at less than 50 km from its epicenter.

Algorithm M8 started an alarm in the middle of 1996 that was almost 4 years in advance of the 2000 Sumatera earthquake, which is different from about 1 month to 2 years in the other case histories considered. Algorithm MSc did not specify the second approximation prediction until July 1997. Since then it confirmed the 400×400-km area where the epicenter and aftershocks of the southern Sumatera earthquake occurred on 4 June 2000.

Summary for magnitude 8.0 earthquakes. Thus, our systematic analysis shows the existence of an inverse cascade in seismic activity prior to the recent great earthquakes of magnitude 8 or more. This premonitory rise of activity was first observed in intermediate-term predictions at large distances, i.e., at the scale of up to 10 sizes of the earthquake source in preparation. In many cases, detailed analysis allows distinguishing a smaller territory, down to the size of the source, where the rise of activity shows intermittent switching from a steadily high level to short periods of quiescence. On the background of such an intermediate-term rise, episodes of activity, like swarms or sequences of foreshocks, can be identified as short-term forerunners of a big earthquake. It is possible that the premonitory rise of seismic activity evolves through long, intermediate, short, and immediate, or even nucleation, phases. This is proved statistically for the long and intermediate phases and requires more data to establish this in shorter ones.

Table 4.4 summarizes the results obtained by algorithms M8 and MSc aimed at predicting great earthquakes of magnitude 8 or more. The high level of confidence is a conservative and robust estimate; it would not drop below 97% after an additional failure to predict. Being supported by case histories from regions of lower seismic activity, the results strengthen our confidence in the efficiency of the intermediate-term prediction methods, algorithms M8 and MSc.

Prediction of earthquakes with $M \geq 7.5$. A similar analysis has been performed in about 75% of the seismic regions where the catalog is sufficiently complete for an application of algorithms M8 and MSc aimed at predicting relatively smaller earthquakes of magnitude 7.5 or more. The number of circles, namely, 180, is lower than in the case of $M_0 = 8.0$, because they are smaller and some of them do not accumulate enough earthquakes reported in the catalog we use. The 147 circles of investigation represent seismic regions of the Circumpacific, and the remaining 33 circles consist of 15 from the Mediterranean, 4 from Iran, 11 from the Pamirs-Hindukush, and 3 from Burma. For the whole territory where predictions were made, M8 alarms covered, on average, 40% of the seismicity (measure μ from Sect. 4.3.4) considered, whereas MSc reduces this number to 12%. Out of 35 earthquakes of magnitude 7.5 or higher from 1985–2000, algorithm M8 predicted 24, and MSc provided a correct second approximation for 14 of them. That comes to statistical confidence of 99.9% for both approximations. However, we observe a certain decay in performance in recent years: From the total of 23 earthquakes from 1992–2000, 14 are predicted by M8 and 7 of them by MSc algorithm, which results in only 96.5 and 98.5% confidence.

How informative are the predictions by M8 and MSc? Let us compare the predictions from the previous sections with predictions based entirely on the territorial distribution of seismic activity [KRKH99]. The optimal algorithm, which expects strong earthquakes in the areas where they occur most often, is the following: The territory of alarm A_t is the same at all times and must comprise the areas of the highest seismic activity, i.e., with a high concentration of measure μ. For a selected value of $\underline{p}$, it is defined as the minimal territory in km^2 for which $p \geq \underline{p}$.

In calculations, we coarse-grain the territory under consideration into $1° \times 1°$ cells and compute μ for each of them. The performance of this algorithm from 1985–2000 is much lower than that of M8 or MSc: It predicts only two out of eight earthquakes with $M \geq 8.0$ when $\underline{p}$ is set the same as achieved by algorithm M8. Thus, with the same time–space of the alarm, algorithm M8 predicts four times more earthquakes than the optimal prediction algorithm based on the territorial distribution of earthquakes. A more detailed comparison including that for earthquakes with $M \geq 7.5$ can be found in [KRKH99].

A remark on the advance testing of M8 and MSc Since 1992. When setting up the forward test in 1992 [HKD92], the authors did not anticipate a large difference between the years of retrospection, 1985-1991, and those of real-time prediction, 1992–1997. The latter period appeared to be remarkable due to a dramatic rise in global seismic energy release. The seismic energy release rate observed during the twentieth century was changing gradually from 4.7×10^{28} erg yr^{-1} from 1907–1960 to 3.1×10^{28} erg yr^{-1} from 1961–1976 and to 2.7×10^{28} erg yr^{-1} from 1977–1993. Then, from 1994–1996 it jumped to 7.3×10^{28} erg yr^{-1}, which is slightly less than that of "the turn of the century" rate of 7.9×10^{28} erg yr^{-1} observed before 1907. Apparently, the energy release dropped to about 5×10^{28} erg yr^{-1} from 1997–2000.

One can see from Table 4.3 that the average annual number of great earthquakes with $M \geq 8.0$ increased by a factor of 2 or more during the period of real-time prediction. It was shown [KRKH99] that this rise in global seismic activity was accompanied by a specific change in the distribution of mechanisms of the largest earthquakes with $M \geq 7.5$. In the period from 1993–1997, all of them were thrusts and normal faults with a maximal plunge of principal axes above $\pi/4$, whereas prior to that the distribution of the maximal plunge was close to uniform in the range from 0 to $\pi/2$. The M8 prediction missed five out of six earthquakes with the highest possible plunge, above $7\pi/8$. The statistical significance of the prediction results for $M_0 = 7.5$ certainly improves without such earthquakes. One may even conclude that most of the earthquakes missed by M8 predictions occurred during the unusual rise of seismic energy release, have a magnitude below 7.8, and are thrust or normal fault events. Obviously, the above considerations will only help in future research and do not change the results of forward testing.

The accumulated set of successes and failures gives some indication of how the accuracy of prediction by algorithms M8 and MSc can be enhanced. At the same time, just a small part of the potentially relevant data is used in these algorithms. Sections below describe reproducible methods of extracting more information from seismic data; the current level of seismic hazard can be judged from this potentially useful information. Premonitory phenomena expressed in other data remain unexplored for the same levels of consecutive averaging, allowing for long-range correlation in active fault systems.

4.4.6   Algorithm CN

Algorithm CN [AKB+84,KKRA88,KR90] was developed in the course of retrospective analysis of seismicity patterns preceding earthquakes with $M \geq 6.5$ in California and the adjacent part of Nevada, hence its name. The essence of this algorithm can be briefly summarized as follows.

- Areas of investigation are selected in accordance with the spatial distribution of seismicity.
- Consider earthquakes with the long-term average annual number $\tilde{N} = 3$ (after eliminating aftershocks) within each area. Compared with $\tilde{N} = 20$ in algorithm M8, this implies a higher magnitude cutoff $M_{\min}$ and, therefore, relaxed requirements imposed on the completeness of catalogs. This apparent advantage comes at the cost of losing a certain degree of robustness due to lower chances of correctly identifying the current state of system dynamics.
- The sequence of earthquakes is described by nine functions (Table 4.5). Two of them, $N2$ and $N3$, are similar to N and one to B defined in Sect. 4.4.3, although with a different choice of numerical parameters. Other functions describe the following: the fraction of relatively higher magnitudes in the sequence considered G; the variations of the sequence in time,

K and Q; the value of "source energy", SIGMA; and the maximum values of "source area and diameter", S_{max} and Z_{max}.

- Following a pattern recognition routine described in [GGK+76], values of the functions are coarse-grained to distinguish "large", "medium" and "small" separated by 66 and 33 percentiles or just "large" and "small" separated by 50 percentile, i.e., the median.
- The voting of certain pairs or triplets of the discrete values of the nine functions (Table 4.6) declares (or does not declare) an alarm in the area. The combinations were found originally by applying the pattern recognition algorithm called "Subclasses" [GGK+76] to vectors (patterns) determined from the earthquake catalog of Southern California, 1938-1984, and the cutoff magnitude $M_0 = 6.4$. An alarm is declared for a certain period, $T = 1$ year, when the votes filed by traits D outscore by $\Delta = 5$ those issued by traits N.

Table 4.5. Functions describing the earthquake sequence in algorithm CN and their thresholds for discrete evaluation in Southern California

Functions	$N2(t)$	K	G	SIGMA	S_{max}	Z_{max}	$N3$	Q	B_{max}
First threshold	0	−1	0.5	36	7.9	4.1	3	0	12
Second threshold	−	1	0.67	71	14.2	4.6	5	12	24

Qualitatively, a TIP is diagnosed when earthquake clustering is high, seismic activity is also high, irregular and growing, and some quiescence preceded the increase in seismic activity. Figure 4.17 depicts examples of prediction by algorithm CN.

Algorithm CN was first tested retrospectively with prefixed parameters for the following 22 areas [KR90]: Northern and Southern California (6.4), the Gulf of California (6.6); Cocos plate margins (6.5) and adjacent to the belt of the Lesser Antillean arc (5.5); the Vrancea area of intermediate-depth earthquakes, the East Carpathians (6.4); the Pamirs (6.5); Tien Shan (6.5); Baikal Lake (6.4); Central Italy (5.6); the Caucasus (6.4); Kangra, Nepal and the Assam regions in the Himalayas (6.4); Krasnovodsk, Elbruz, and Kopet Dag regions in Turkmenia (6.4); the area of the Dead Sea Rift (5.0); the Northern and Southern Appalachians (5.0); and Brabant-Ardennes (4.5).

The earthquake catalogs available allowed to retrospectively consider time intervals from 12 to 22 years in each area, amounting to 32 years in Italy and 45 years in California. Sixty strong earthquakes occurred in all areas in the test period. Fifty (83%) of these events occurred within alarm periods, and 10 earthquakes were missed. On average, TIPs in the area considered occupied about 27% of the time, from 2 to 4 years per earthquake (except for 6 to 8 years in the southern part of the Dead Sea Rift, Kopet Dag, and Vrancea).

Table 4.6. Characteristic traits of algorithm CN

Traits D^a	N2	K		G		SIGMA		S_{max}		Z_{max}		N3		q		B_{max}	
		1	2	1	2	1	2	1	2	1	2	1	2	1	2	1	2
1	0																0
2															0		
3										0				0			0
4									0		0						
5	0										1					0	
6		1									0					0	
7	0										1					0	
8	0		0													0	
9								0		0							
10		1				0				0							
11	0	1									0						
12	0	1							0								
13		0					1										
14		0				0											
15		0		0													
16		0			1												

Traits N	N2	K		G		SIGMA		S_{max}		Z_{max}		N3		q		B_{max}	
		1	2	1	2	1	2	1	2	1	2	1	2	1	2	1	2
1								1									
2										1							1
3							1							1			1
4			1											1			1
5													0	1			1
6								1									1
7			1									1					1
8	1											1					1
9							1						0	1			
10								1						1			
11												1	0				
12								1				0					
13			1				1										
14			1	1													
15	1									1							
16		1		1				1									
17	1							1									
18	1						1										

^{a} Trait D favors declaration of an alarm, whereas trait N votes against it (obtained for Southern California). Two columns for each function stand for two thresholds, lower and upper; one threshold (and one column) is assigned to $N2$. In a given trait, 0 or 1 correspond to values above or below the threshold from Table 4.5, respectively

Southern California, M ≥ 6.4

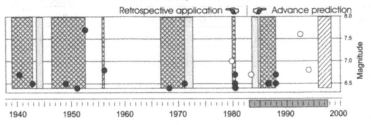

Northern California, M ≥ 6.4

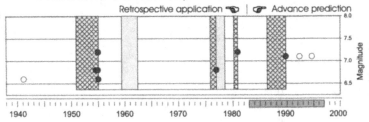

Northern Appalachians, M ≥ 5.0

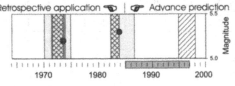

Cocos Ridge, M ≥ 6.5

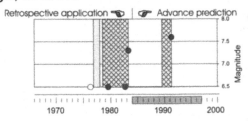

Strong earthquakes:
● predicted; ○ not predicted.

Alarms:
▨ justified; ▢ false; ▧ current.

Fig. 4.17. Some examples of algorithm CN predictions. Times of increased probability of strong earthquakes are determined for Southern and Northern California, Northern Appalachians, and Cocos Ridge (after [RN99]). Periods of advance prediction are shaded on the time axis

Advance prediction in an area was carried out for several periods, from a year in the southern Dead Sea Rift zone to about 16 years in Southern California [RN99]. Altogether, 24 strong earthquakes occurred in all regions within the test periods. Of these, 11 (46%) were predicted, and 13 were missed. Total alarm time occupied 26% of the periods considered. In 13 areas, where at least one strong earthquake occurred within the test periods, the time of alarms occupied 29.4% of the total, and it was 18.2% in the remaining nine regions where no strong earthquakes did occur.

The confidence level $\alpha = 95\%$ follows from rough statistical estimates [RN99]. However, it ignores the fact that the overall statistics are collected from different areas with different rates of seismic activity. A more rigorous and cautious estimation gives $\alpha = 91\%$, which is not very stable and may change to 96 or 81% with the next strong earthquake predicted or not [RN99]. Such a level of sensitivity is essentially due to the small sample accumulated so far and can be overcome by continuing the test. The cautious estimate presented ignores the results of monitoring in nine areas, where strong earthquakes did not occur in the test period. Therefore, the significance of predictions by algorithm CN might be higher [MDRD90].

4.4.7 Will a Subsequent Strong Earthquake Occur Soon? Algorithm SSE

Similar to CN, the algorithm SSE whose name is an abbreviation of Subsequent Strong Earthquake, is another example of applying pattern recognition methods to an earthquake prediction problem [LV92, VL94, VP93, Vor94, Vor99]. The algorithm aims at the answer to the question: Whether or not a new strong earthquake can follow one that just occurred. The answer is important for reducing the hazard caused by destabilization of buildings, lifelines, and other constructions or natural objects, like mountain slopes, glaciers, river banks, etc., after a strong earthquake has occurred. Many authors considered similar problems [Bat65, Ver69, Pro78, RJ89, Mat86, HC90].

The advance application of algorithm SSE started in 1989 demonstrates a high level of statistical significance estimated currently at more than 98%. Seventeen successful predictions have been made, including the second 1991 Rachi earthquake in Georgia (Caucasus) and three Californian earthquakes at Loma-Prieta (1989), Joshua Tree (1992), and Northridge (1994). The error score is low, consisting of two false alarms and one failure to predict.

The algorithm. Assume that a strong earthquake of magnitude M_1 has just occurred at the origin time t. The task is to predict whether or not a subsequent strong earthquake with magnitude $M \geq M_1 - a$ will occur within the time interval $(t + s, t + T)$ and the circle of radius $R(M_1)$ centered at the epicenter of the one that occurred.

Suppose that precursory symptoms similar to those revealed by algorithms M8 and CN (see above in this chapter) precede the occurrence of a subsequent strong event in the vicinity of the shock that occurred and their absence signifies that no strong earthquake will follow in a certain time interval.

Naturally, one can presume a similarity of premonitory phenomena only after rescaling which makes aftershock sequences of main shocks with different magnitudes comparable. The following scaling rules were applied in the design of algorithm SSE:

- All magnitude thresholds are derived from the magnitude of the strong earthquake, M_1, that occurred.
- The area of investigation and prediction is the circle with radius $R(M_1) = 0.03 \times 10^{0.5M_1}$ km, centered at the epicenter of the strong earthquake that occurred.
- Time constants do not scale; the period of prediction is from $t + s = t + 40$ days to $t + T = 1.5$ years.

The prediction algorithm SSE was developed in the course of retrospective analysis of 21 strong earthquakes in California with $M \geq 6.4$ [VL94]. A simple pattern recognition measure known as the Hamming distance [GZKK80], is applied to reveal a characteristic image of an earthquake followed by a subsequent one using eight functions described below. Seven of them refer to an aftershock sequence and reflect the level of aftershock activity, the expansion from the main shock, the total area of ruptures, and irregularity in the sequence. One extra function characterizes seismic activation preceding the strong earthquake that occurred.

Large values of the following five functions, accounted for in the sequence of aftershocks with magnitude equal or exceeding $M_1 - m$ during $(t + s_1, t + s_2)$, favor the occurrence of a subsequent strong earthquake.

N, the total number of aftershocks in the sequence.

S, the total area of aftershock ruptures normalized to the rupture area of the strong earthquake that occurred. Specifically, $S = \sum 10^{m_i - M_1}$, where m_i is the magnitude of the ith aftershock of the sequence.

V_{m}, the variation of magnitude value in the sequence: $V_{\mathrm{m}} = \sum |m_{i+1} - m_i|$, it adds together the absolute values of magnitude difference between subsequent aftershocks.

V_{med}, the variation of daily average magnitude in the sequence of aftershocks: $V_{\mathrm{med}} = \sum |\mu_{i+1} - \mu_i|$, where μ_i is the daily average magnitude of aftershocks during the ith day after the strong earthquake that occurred.

R_z, the deviation of the aftershock sequence from a monotonically decaying one; specifically, $R_z = (1/2) \sum (n_{i+1} - n_i + |n_{i+1} - n_i|)$, where n_i is the number of aftershocks in the interval $(t + i, t + i + s_3)$. The sum neglects negative increments of n_i.

Small values of the following three functions favor the occurrence of a subsequent strong earthquake:

V_{n}, the variation of the daily number of aftershocks in the sequence: $V_{\mathrm{n}} = \sum |n_{i+1} - n_i|$, where n_i is the number of aftershocks during the ith day.

$R_{\max}$, the largest distance between epicenters of the strong earthquake that occurred and an aftershock of the sequence divided by $R(M_1)$.

N_{for}, the number of earthquakes of magnitude $M \geq M_1 - m$ during $(t - s_1, t - s_2)$ within the distance of $1.5\,R(M_1)$.

Qualitatively, a characteristic image of an earthquake followed by a subsequent one is as follows: The activity preceding the strong earthquake that occurred is low; the number of aftershocks is large, as well as the total area of aftershock ruptures; the aftershock sequence is highly irregular in time and magnitude and decay neither monotonically nor rapidly; and the aftershocks concentrate near the main shock.

The Hamming pattern recognition procedure implies discretization of functions. Thus, for each function except V_{med}, one defines a threshold to separate its "large" and "small" values. Two thresholds separating "small", "medium", and "large" values are introduced for V_{med}. The adjustable parameters including the chosen thresholds of discretization are listed in Table 4.7.

Each function, depending on its value, "votes" for or against the *increased probability* of a subsequent strong earthquake. The rules of voting have been determined for each function by using the learning material consisting of earthquake sequences associated with 21 strong earthquakes in California and Nevada. Subsequent strong earthquakes followed six of them (type A). Seismic activity after the remaining 15 (type B) did not produce any strong event. The choice, which value of a function votes to declare an alarm, was based on comparing the occurrence of the value in sequences of different types. For example, if the "large" value of a function was observed in at least two-thirds of type A cases and at the same time in less than half of the type B cases, then the "large" value of this function votes in favor of the coming subsequent strong earthquake. Vice versa, if the "large" value was observed in at least two-thirds cases of type B and in less than half of the cases of type A, then the "large" value of this function votes against declaring an alarm. The voting of each item from the final list of eight functions was determined in a similar way.

Based on the seismicity after a strong earthquake with $M \geq M_0$ in the area considered for testing the algorithm SSE, the values of functions were calculated, and the voting was performed. If the votes *pro* outscored the votes *contra* by three (the originally adjusted threshold), then an alarm was declared. This signified that an earthquake with magnitude $M_1 - a$ or more was expected in the interval $(t + s, t + S)$ at the distance $R(M_1)$ or less from the epicenter of the strong earthquake that occurred. When the number of aftershocks was small (less than 10), the functions were not evaluated, and an alarm was not declared.

Retrospective testing. Algorithm SSE was tested retrospectively with prefixed parameters (Table 4.7) in the following eight regions (the lowest value of M_1 considered is given in parentheses): the Balkans (7.0), the Pamir and

Table 4.7. Parameters of algorithm SSE

Function	Values of parameters				Threshold values	
	m	s_1, hours	s_2, days	s_3, days		
N	3	1	10	–	24	–
S	2	1	10	–	0.1	–
V_m	3	1	40	–	0.41	–
V_{med}	3	1	40	–	0.7	2.6
R_z	3	10 days	40	10	0	–
V_n	3	1	40	–	0.98	–
R_{max}	2	–	2	–	0.23	–
N_{for}	1	5 years	90	–	2	–

Tien-Shan (6.4), the Caucasus (6.4), Iberia and Maghrib (6.0), Italy (6.0), the Baikal and Stanovoi Range (5.5), Turkmenia (5.5), and the Dead Sea Rift (5.0) [VL94,VP93]. Table 4.8 lists the results. Subsequent strong earthquakes followed 10 out of 98 strong earthquakes. Algorithm SSE missed two of the subsequent strong earthquakes, one due to a small number of aftershocks of the strong quake that occurred and one due to erroneous pattern recognition. Four alarms were false; no subsequent strong earthquake happened.

Table 4.8. Summary of retrospective testing of algorithm SSE

Region	M_0	Total $M \geq M_0$	Small number of aftershocks Single (n/e)[a]	Tested by pattern recognition		
				Total number	Single (n/e)[a]	with the next (n/e)[a]
Learning						
California	6.4	21	4/0	17	11/0	6/1
Retrospective test						
Pamir and Tien-Shan	6.4	12	4/0	8	7/1	1/0
Caucasus	6.4	5	0/0	5	5/0	0/0
Baikal and Stanovoi Range	5.5	6	4/0	2	2/1	0/0
Iberia and Maghrib	6.0	13	11/0	2	1/0	1/0
Dead Sea rift	5.0	11	10/0	1	1/0	0/0
Turkmenia	5.5	12	7/1	5	4/0	1/1
Balkans	7/0	19	7/0	12	9/1	3/0
Italy	6.0	20	9/0	11	8/1	3/0
Total retrospective test		98	52/1	46	37/4	9/1
Total		119	56/1	63	48/4	15/2

[a] n/e stands for number/error

Limitations on the algorithm. The formal definition of the algorithm permits applying it to any strong earthquake if a sufficiently complete catalog is available. However, it is necessary to consider two limitations.

First, the algorithm does not succeed in subduction zones in the Circum-pacific, where, according to [Vor99], the occurrence of a subsequent strong shock apparently does not depend on the rate of events in the aftershock sequence of an earthquake.

The completeness of the catalog at hand produces the second limitation for the magnitude of strong earthquakes M_0. Usually the threshold M_0 is chosen such that reported magnitudes $M_0 - 3$ and above are reasonably complete in the catalog. This is necessary to accurately calculate the functions used in algorithm SSE. In California, however, the value of M_0 was chosen higher than that to reduce certain errors. The tests reported in [Vor99] showed that the choice of M_0 required special investigation in each seismic region.

Advance predictions, 1989–1998. The algorithm with pre-fixed parameters (Table 4.7) has been applied to all strong earthquakes that occurred in the nine regions (Table 4.8). Table 4.9 lists the results of the advance predictions.

No earthquakes with $M \geq M_0$ occurred in the Baikal and Stanovoi Range, Turkmenia, and the Balkans. There were fifteen additional strong earthquakes that have not been tested; nine were too close in time (within 40 days) to earthquakes listed in Table 4.9; no data were available for another six earthquakes.

The advance prediction results can be summarized as follows (Table 4.10). The rate of failure to predict is low, and the rate of false alarms is considerably high compared with retrospective testing. The statistics of advance predictions by the algorithm SSE is not yet sufficient for reliable estimations; nevertheless, some preliminary calculations are quite encouraging. The statistical significance of the method, estimated by the technique proposed in [Mol97], is more than 98% [Vor99]. The relative number of failures to predict n equals 0.2; the rate of alarms τ equals 0.3 (six alarms were diagnosed after 20 strong earthquakes). Thus, the value of $n + \tau$ for advance predictions equals 0.5, low enough even for retrospective testing.

Several examples illustrate the advance predictions by algorithm SSE (Figs. 4.18–4.20).

Joshua Tree, Landers, and Northridge, Southern California. The first earthquake of the sequence occurred on 23 April 1992 near Joshua Tree and was reported as a magnitude $M = 6.3$ event. The map of its after-shocks with magnitude $M \geq 3.3$ used for prediction are shown in Fig. 4.18. The Joshua Tree earthquake produced a large number of aftershocks (54 aftershocks with $M \geq 3.3$), which caused an alarm for an earthquake with $M \geq 5.3$ in the following 1.5 years within the distance of $R(6.3) = 42$ km from

Table 4.9. Advance prediction by algorithm SSE: The results of monitoring from 1989–1998

Earthquake	M	Will a subsequent shock occur?	Outcome of prediction	Note
		California		
Loma-Prieta, 10/18/1989	7.1	NO	No shocks with $M \geq 6.1$	Success
Mendocino, 7/13/1991	6.9	NO	No shocks with $M \geq 5.9$	Success
Mendocino, 8/17/1991	7.1	NO	No shocks with $M \geq 6.1$	Success, first step
Joshua Tree, 4/23/1992	6.3	YES	Landers is predicted, $M = 7.6$	Success
Landers, 6/28/1992	7.6	YES	Northridge $M = 68$ occurred 19 days after end of alarm	False alarm
Northridge, 1/17/1994	6.8	NO	No shocks with $M \geq 5.8$	Success
Mendocino, 4/25/1992	7.1	NO	No shocks with $M \geq 6.1$	Success
Mendocino, 9/1/1994	7.1	NO	Earthquake with $M = 6.8$ occurred	Failure, first step
Mendocino, 2/19/1995	6.8	NO	No shocks with $M \geq 5.8$	Success, first step
California-Nevada border, 9/12/1994	6.3	YES	Earthquake with $M = 5.5$ occurred	Success
		Pamir and Tien Shan		
Kasakhstan, 8/19/1992	7.5	NO	No shocks with $M \geq 6.5$	Success
China, 11/19/1996	7.1	NO	No shocks with $M \geq 6.1$	Success
Iran, 5/10/1997	7.5	NO	No shocks with $M \geq 6.5$	Success
		Caucasus		
Iran, 6/20/1990	7.7	NO	No shocks with $M \geq 6.7$	Success
Rachi, 4/29/1991	7.1	YES	Earthquake with $M = 6.6$ occurred	Success
Rachi, 6/15/1991	6.6	NO	No shocks with $M \geq 5.6$	Success
Erzincan, 3/13/1992	6.8	YES	No shocks with $M \geq 5.8$	False alarm

Table 4.9. (continued)

Earthquake	M	Will a subsequent shock occur?	Outcome of prediction	Note
		Iberia and Maghrib		
Morocco, 5/26/1994	6.0	NO	No shocks with $M \geq 5.0$	Success
		Dead Sea Rift		
Gulf of Aqaba, 8/3/1993	5.8	YES	Earthquake with $M = 4.9$ occurred	Success
Gulf of Aqaba, 11/22/1995	7.3	NO	No shocks with $M \geq 6.3$	Success
		Italy		
Assisi, 9/26/1997	6.4	YES	Earthquake with $M = 5.4$ occurred	Success

Table 4.10. Summary of predictions by SSE

Prediction: Will a subsequent strong earthquake occur?	Number of predictions (total/erroneous)		
	Learning	In retrospect	In advance
NO, due to small number of aftershocks	4/0	52/1	4/1
NO, due to pattern recognition criteria	11/0	34/1	10/0
YES	6/1	12/4	6/2
Total	21/1	98/6	20/3

its epicenter. The subsequent Landers earthquake (June 28, 1992, $M = 7.6$) occurred within this distance 64 days after the Joshua Tree main shock.

The algorithm was then applied to the Landers earthquake and its aftershocks with magnitude $M \geq 4.6$ shown in Fig. 4.18. The aftershock sequence had a pretty large total area of aftershock ruptures. It was predicted [LV92] that an earthquake with $M \geq 6.6$ would occur within the distance of $R(7.6) = 199$ km and 1.5 years of the Landers main shock. This alarm expired on 28 December 1993. The Northridge earthquake (17 January 1994, $M = 6.8$) occurred within the prediction distance, but 19 days after the expiration of the alarm. This prediction counted as a false alarm in the overall statistics.

In its turn, the Northridge earthquake and its aftershocks with magnitude $M \geq 3.8$ were also tested by algorithm SSE (Fig. 4.18). Even though many aftershocks occurred (77 events with magnitude $M \geq 3.8$), the algorithm did not recognize an alarm and no earthquake with $M \geq 5.8$ occurred within the distance $R(6.8) = 75$ km and 1.5 years.

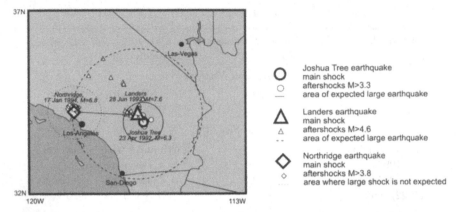

Fig. 4.18. The Joshua Tree, Landers, and Northridge earthquakes and their aftershocks

Gulf of Aqaba earthquakes from 1993–1995, Dead Sea Rift. The 3 August 1993, $M = 5.8$, Gulf of Aqaba earthquake had 171 aftershocks (Fig. 4.19). An SSE alarm was diagnosed through the analysis of this sequence. It was a successful prediction; an earthquake of magnitude 4.9 occurred 92 days after the first one.

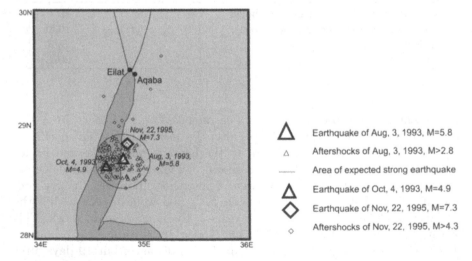

Fig. 4.19. The 1993 and 1995 Gulf of Aqaba earthquakes and their aftershocks

The largest earthquake in this region, of magnitude 7.3, occurred in the same place 2 years later (on 22 November 1995). It had only 14 aftershocks with $M \geq 4.3$ (Fig. 4.19), and no alarm was diagnosed. This prediction was also successful; no earthquake of magnitude 6.3 or more occurred within the distance $R(7.3) = 135$ km in the period of 1.5 years.

Rachi, Caucasus, Georgia earthquakes of 1991. The Rachi earthquake (29 April, 1991, $M = 7.1$) had a large aftershock sequence consisting of 77 events with magnitude $M \geq 4.1$ and a large total area of aftershock ruptures (Fig. 4.20). The diagnosed SSE alarm became another successful prediction; on 15 June 1991, an earthquake with $M = 6.6$ occurred within the distance $R(7.1) = 105$ km from the first Rachi earthquake epicenter. This second earthquake was also tested; no alarm was diagnosed for an earthquake with $M \geq 5.6$ at the distance within $R(6.6) = 59$ km in the period of 1.5 years, and no such earthquake occurred.

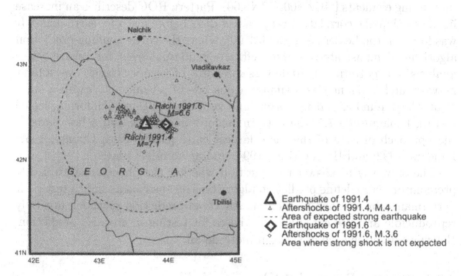

Fig. 4.20. The 1991 Rachi earthquakes and their aftershocks

The case history of the Rachi earthquake of April 1991 is significant because no subsequent strong earthquake followed any known strong earthquakes with magnitudes $M \geq 6.4$ (12 events) in the Caucasus since 1900. The Rachi earthquake produced considerably more aftershocks than a common strong earthquake in the region, and its subsequent strong earthquake produced a normal number of aftershocks.

4.5   Seismic Patterns Submitted for Testing

In the lithosphere, regarded as a system with complexity contributed by a multitude of instability mechanisms, premonitory patterns should have general features of behavior in a near-critical state [Kei90b, Tur97, NGT94, BT89, Kei92, Man83, Ben00, Ben01]. This section dwells on two such features: a specific reorganization of the space–time structure of a seismic sequence and long-range correlation between earthquakes [TNG00, STS85, SS95, BOS+98].

One premonitory phenomenon depicting a change in the spatial distribution of events prior to strong earthquakes was found in Southern California and then in the Lesser Antilles and on the Dead Sea Transform. Named "seismic reversal" (SR), it characterizes [KRS94, SGR+96, SK99] a period of opposite seismic activity in areas complementary to the bulk of long-term territorial distribution of low-magnitude earthquakes. The phenomenon happens several months prior to a strong earthquake in the area surrounding its future source. The diameter of the prediction area is of the order of 100 km.

The other two phenomena, named ROC and Accord, were designed in inspiration of evident patterns in a lattice-type "Colliding Cascades" model of interacting elements [GKZN00, GZNK00]. Pattern ROC describes an increase in the earthquake correlation range a few days before a strong earthquake. It was found in the Lesser Antilles [SZK00], where the corresponding prediction algorithm demonstrated extreme efficiency unusual even for retrospective analysis. The pattern Accord depicts spreading of seismic activity over a fault network and is defined as a simultaneous rise of seismic activity in a sufficiently large number of neighboring fault zones. In Southern California, scaled to target magnitude 7.5, the pattern Accord emerges within a few years in the approach of each of the three largest earthquakes: Kern County, 1952; Landers, 1992; and Hector Mine, 1994, and at no other time.

The only way to answer the question whether the phenomena are really premonitory is to define prediction algorithms on their basis and to test their performance in advance prediction. All three prediction methods are fully reproducible and presented as hypotheses for testing in advance prediction. Below we describe each of them in more detail.

4.5.1   Seismic Reversal (SR)

The phenomenon of Seismic Reversal (SR) can be qualitatively described as follows: Zones of relatively high seismic activity become unusually quiescent while zones of relatively low seismic activity are unusually active. This takes place a few months prior to an approaching strong earthquake within a distance of about 100 km (i.e., about ten times larger than the incipient source) from its future epicenter. Levels of low and high activity in the vicinity of each point are determined in relation to adjacent areas. As a pattern, SR is from the lower part of the intermediate-term and middle-range scale. It characterizes the combination of quiescence and the rise of seismic activity.

The phenomenon has features in common with premonitory seismic quiescence reported previously on different timescales, from tens of years ("seismic gaps" [Mog85]), to years [Wys86, WH88, WW94], and months ("doughnut" pattern [Mog85]). Rundkvist and Rotwain [RR94] reported several cases of quiescence in a structure where a large earthquake would be nucleated. Shreider [SJ90] described precursory quiescence on a timescale counted by the number of events. Various plausible physical mechanisms for premonitory quiescence have been suggested [Sch88, Sch90].

Seismic reversal is a short (order of months) and local (area diameter of the order of 100 km) transient activation of zones relatively less active on the long-term scale (order of decades) on the background of quiescence in more active zones. On a shortscale, two classes of individual events can be distinguished by analyzing long-term epicenter density in some proximity divided into "active" and "nonactive" localities. Seismic reversal is the phenomenon of an anomalously high number of epicenters in "nonactive" areas concurrent with an anomalously low number of epicenters in "active" areas.

Classification of new epicenter localities. Long-term seismicity can be represented as a map of relative levels of seismic activity. Instead of constructing high-resolution maps changing in time, we determine relative levels at the localities of recent epicenters. Each new epicenter is assigned the type of its locality, either "active" (A) or "nonactive" (N).

Consider the usual representation of an earthquake catalog: $\{t_i, \varphi_i, \lambda_i, M_i\}$, $i = 1, 2, ...; t_{i+1} \geq t_i$. Here t_i is the origin time, M_i is the magnitude, and $g_i = (\varphi_i, \lambda_i)$ is the vector of geographic coordinates of the epicenter. For each new epicenter g_j, we estimate local long-term seismicity p_j in the circle $O(g_j)$ of radius r about this epicenter. Seismicity p_j is defined as the smoothed density of epicenters in that circle, with magnitudes $\geq M_{\text{bg}}$, and in the trailing background time window, $B_j = (t_j - T_{\text{bg}} - \tau, t_j - \tau)$. The values $T_{\text{bg}}, \tau, M_{\text{bg}}$, and r are the numerical parameters of the algorithm.

Specifically, density $p_j(g)$ is defined as follows:

$$p_j(g) = \sum_k [1 - \frac{|g - g_k|}{\tau}(N_j/N_0)^{1/2}]_+ .$$

Here $[x]_+ = x$ if $x > 0$, and $[x]_+ = 0$ if $x \leq 0$; N_j is the number of epicenters with $M \geq M_{\text{bg}}$ in the circle of radius r and time window B_j; N_0 is a constant; and g_k is an epicenter in the catalog in time window B_j. We do not estimate p if the number of background events is too small, say, less than a certain threshold $N_{\min}$.

We introduce the median $p_{1/2}$ of density $p_j(g)$ within circle $O(g_j)$. When $p_j(g_j) < p_{1/2}$, the epicenter g_j is of type A; otherwise, it is of type N. Such a definition gives an approximately equal number of epicenters from the background in localities A and N, and newly occurred epicenters are expected to follow the same rule, on average.

The seismic reversal pattern is seen from the case history of the Gulf of Aqaba earthquake of 22 November 1995, $M_s = 7.3$ (NEIC). Six 3-month epicenter maps prior to the earthquake are shown in Fig. 4.21, where circles and squares stand for type A and type N epicenters, respectively. The first four 3-month maps show high activity at the southern edge of the Dead Sea Transform, which is a commonly observed premonitory intermediate-term activation. During the period from 6 to 3 months before the earthquake, this

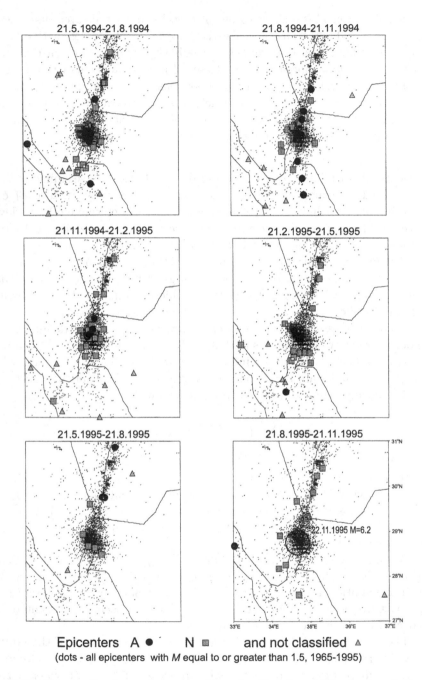

Epicenters A ●   N ▣   and not classified △
(dots - all epicenters with *M* equal to or greater than 1.5, 1965-1995)

Fig. 4.21. Consecutive 3-months maps of the classified A and N epicenters before the Gulf of Aqaba earthquake, 22.11.1995, M=7.3. The seismic reversal phenomenon is seen on the last map; no type A events occurred on the background of a large cluster of type N events during the last 3 months before the strong earthquake

activation weakened. Finally, in the last 3 months, the normally active localities became completely quiet, and an activation of normally nonactive zones occurred on this background in a wide area about the incipient epicenter. Some of those activation sites are close to, but not exactly on, the active parts of the transform.

Time−space of increased probability of strong earthquakes. Finally, we define a precursory pattern SR before a strong earthquake with $M \geq M_0$. All numerical parameters entering the description are listed in Table 4.11. Consider a function $f(t, g) = n_N/(n_A + 1)$, where n_N and n_A are the current numbers of epicenters of N and A types, respectively, with magnitude $M \geq M_c$, positioned within a circle of radius R centered at some point, which is not necessarily a reported epicenter, and having occurred in the time interval $(t - T_c\, t)$; this time window is much smaller than T_{bg}, so that f refers to the current seismicity.

We sample the territory with a rectangular grid having a constant step δ. The territory considered is filtered through a mask consisting of those grid nodes where a circle of radius r centered at the node contains at least N_{min} events with $M \geq M_{bg}$ in a time interval of length T_{bg} starting from the beginning of the catalog segment at hand, t_0.

If the values of $f(t)$ are abnormally high in a sufficiently large compact area, we regard this area and time period $(t, t + T)$ as a "Time−space of increased probability" (TSIP) of a strong earthquake with $M \geq M_0$. To find such compact areas, we consider circles of radius R centered at the nodes of the grid. Within each circle, we consider successive times, at a constant step Δt : $t_0, t_1 = t_0 + \Delta t, t_2 = t_0 + 2\Delta t, \ldots$. We look for the cases where $f(t_i, q_a) \geq F$; here q_a identifies the node, and F is a numerical threshold. In that case, an "elementary alarm" is diagnosed in the $\delta \times \delta$ square centered at node q_a. The union of such elementary alarms is an area of an alarm for the time period $(t, t + T)$.

If a strong earthquake $(M \geq M_0)$ occurs inside a TSIP, we score a *successful prediction*; otherwise, we score a *failure to predict*. The alarm is called off after the strong earthquake. If a TSIP contains no strong earthquake, we score a *false alarm*.

To eliminate false alarms caused by accidental peaks in the function f, we disregard a node when an elementary alarm is declared at less than three adjacent points, and we disregard an alarm in too small an area whose size is less than S. There is always a trade−off between false alarms and failures to predict. Retrospective results for the Antilles become better when the alarm area is extended by a narrow surrounding strip of width b; a similar condition is used in the Mendocino Scenario [KKS90].

The completeness of a catalog can be different in different areas, time periods, and magnitude ranges, owing to changes in the seismic networks, data processing routines, etc. However, algorithm SR is less affected by man−

made changes [Hab82, Hab87, Hab91, PS84], because it uses a relative number of events N and A in a rather small area.

The results obtained from algorithm SR. The algorithm was applied in three regions: in the Lesser Antilles, in Southern California, and in the Dead Sea Transform and Cyprus zone. Starting from 1 July 1995 in the Lesser Antilles region, the advance prediction was carried on. The values of the algorithm parameters for the three regions are listed in Table 4.11, and the results are summarized in Table 4.12.

Table 4.11. Values of algorithm SR parameters

Parameter		Lesser Antilles	Southern California	Dead Sea Transform
Magnitude of predicted earthquake	M_0	5	6.3	6.0
Parameters for A and N classification				
Magnitude of background earthquakes	M_{bg}	0	1.8	1.5
Duration of background time interval, years	T_{bg}	30	30	30
Radius of circles, km	r	25	20	20
Parameter for smoothing epicenter density	N_0	20	20	20
Parameter for the completeness of the catalog and for mask determination	N_{min}	10	10	10
Parameters of function f				
Magnitude of current earthquakes	M_c	0	2.3	2.0
Duration of current time interval, days	T_c	90	90	90
Radius of circles R, km	R	75	75	75
Parameters for determining TSIPs				
Step for spatial grid	δ	0.1°	0.1°	0.1°
Step for time, months	Δt	1	1	1
Threshold for function	F	5.5	5.0	5.0
Duration of alarms, months	T	12	12	12
Strip width	b	0.2°	0.2°	0
Minimum spatial size of alarm, km^2	S	5000	15000	10000

Lesser Antilles. The *retrospective* prediction in the Lesser Antilles (Fig. 4.22) covered 12.5 years from January 1984 to June 1995. The main source of data is the local catalog [IPGP01] issued by the Département des Observatoires Volcanologiques of the Istitut de Physique du Globe de Paris (IPGP). In addition, for the period prior to 1980, we also used the catalog of the Eastern Caribbean [SLT94] and NEIC/PDE Monthly Listings [PDE]. We did not introduce the limit for magnitudes: the thresholds M_{bg} and M_c were taken equal to zero. For the period starting from 1979, the catalog is almost complete for magnitudes 1.0 and higher. For many events, even

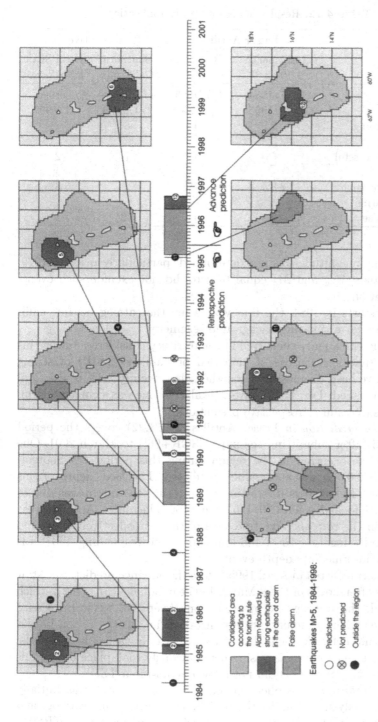

Fig. 4.22. Retrospective and real-time prediction by algorithm SR for the Lesser Antilles

Table 4.12. Results of algorithm SR application

	Lesser Antilles		Retrospective test	
	Retrospective test	Real-time test	Southern California	Dead Sea Transform
Time period	1/11/1984– 1/7/1995	1/7/1995– 1/3/2001	1/1/1968– 1/1/1999	1/1/1990– 1/1/1999
Number of strong earthquakes	7	1	8	2
Number of successful predictions	5	1	7	2
Relative space–time volume of alarms	7.3%	5.7%	9.2%	6.3%
Number of false alarms	2	1	22	1

rather large, the magnitude was not determined, particularly prior to 1979. Formal thresholds M_{bg} and M_c equal to zero did not exclude such events from consideration.

Prediction starts in 1984, the first year when the catalog became sufficiently complete for estimating background seismicity.

Seven strong earthquakes occurred in the territory considered during this period. Eight time–space areas of increased probability (TSIP) of strong earthquakes have been diagnosed. Altogether, they occupy 7.3% of the total time–space considered. Two TSIPs gave false alarms, one was a current alarm, and five were successful retrospective predictions.

The advance prediction in Lesser Antilles (Fig. 4.22) covers the period from July 1995 (after submitting the paper [SGR+96]) to March 2001. One strong earthquake occurred in the region during this period (24 September 1996, $M = 6.0$). It was successfully predicted. The focal depth of this earthquake was 138 km. It is interesting to note that the alarm was due to earthquakes with focal depths of 50–140 km in type N localities. A similar case was observed in the retrospective test. One of the retrospectively predicted earthquakes had a focal depth of 106 km, and the corresponding TSIP was also formed by intermediate depth events.

One false alarm started in April 1995 before the advance prediction test; it was due to the aftershocks of the 8 March 1995 earthquake ($M = 6.2$), which occurred outside the area covered by prediction. Its epicenter is in the sea far from the islands where seismographs are installed, i.e., at the region border where the seismic network is capable of recording only rather large events. The precursory effect we are discussing here obviously took place before this earthquake, but the thresholds did not allow diagnosing a TSIP.

Southern California. The phenomenon of seismic reversal was initially found through analysis of the Southern California earthquake catalog, and a preliminary version of algorithm SR was developed for this region [KRS94]. The final version of the algorithm was developed for the Lesser Antilles region

and yielded encouraging results; then the new algorithm was retrospectively
tested again in Southern California for the period of 31 years, 1968–1998.

The catalog of Southern California earthquakes from the NEIC Hypocenter Database CD-ROM (1932–1994) was used; for 1994 onward, it was updated by the Southern California Earthquake Center (SCEC) catalog, which
is available on the web site http://scec.gps.caltech.edu. Since seismicity in
California is higher than that in the Antilles, all magnitude thresholds of
algorithm SR were changed (see Table 4.11); the parameters r, S, and F
were also adapted for the region.

Seven out of eight strong earthquakes were retrospectively predicted and
TSIPs occupied altogether 9.2% of the total time–space considered. However,
there are 22 false alarms. The period of advance prediction (2 years) is yet
too short for conclusions, but its results are less encouraging than in the
Lesser Antilles: there were two alarms, both of them false, and one strong
earthquake was missed (1999 Hector Mine, $M = 7.3$).

Dead Sea Transform and Cyprus. The retrospective test of algorithm SR
was performed in this region for the period from 1 January 1990 to 1 January
1999. The local catalog of the Geophysical Institute of Israel (GII), Holon,
Israel, was used. The geological structure and the seismicity of the region has
some traits similar to those of California [ARS+92]. Accordingly, the same

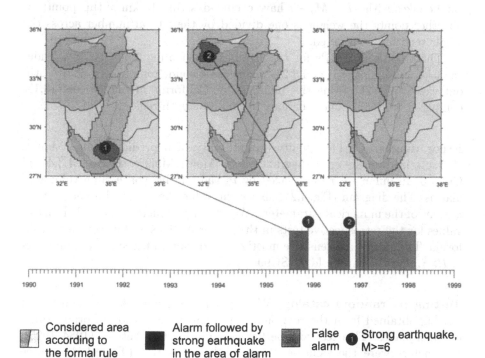

Fig. 4.23. Retrospective and real-time prediction by algorithm SR for the Dead
Sea Transform and Cyprus region

values of parameters were taken, except for the magnitude thresholds. They
were reduced by 0.3 (see Table 4.11). This change makes the corresponding
numbers of earthquakes per unit time and unit area approximately the same
as those in California. This transfer of parameters gives a reasonable retrospective result (Fig. 4.23); both strong earthquakes in the region in the
considered time interval were successfully predicted. The total time–space
volume of alarms is only 6.3%, and only one false alarm was diagnosed.

Measure of space. As pointed out in Sect. 4.3.4, the relative time–space
volume of TSIPs, ν, can be underestimated. In standard tests, a mask is
introduced to limit the area considered. In addition, we performed two tests
where a certain weight was assigned to each point on the map. This weight
depended on the level of background seismicity about the point (measure μ
from Sect. 4.3.4).

We consider two null hypotheses: (1) strong earthquakes ($M \geq M_0$) have
the same spatial distribution as the background seismicity ($M \geq M_{rmbg}$),
and (2) strong earthquakes that can occur only in places where earthquakes
with $M \geq M_0 - 1$ have already occurred. In the first case the weight of a point
is proportional to the number of events in the background seismicity within
25 km of the point. In the second case, the weight of a point is zero if no
earthquakes with $M \geq M_0 - 1$ have occurred within 10 km of this point; for
all other points the weight is one divided by their total number across the
whole territory considered.

The volume of TSIPs for the Antilles (1984–June 1995) is 7.6% under
the first null hypothesis and 8.2% under the second, compared to 7.3% without weighting. The respective numbers for California are 11.9% and 13.1%
compared to 12.1%. The difference is small for both regions.

Error diagrams. The quality and stability of a prediction algorithm is
characterized by the error diagram (n, τ) [Mol91, Mol97]; see Sect. 4.3.4 and
Chap. 5. In our case, τ is replaced by ν, the relative space–time volume of
alarms. The diagram (Fig. 4.24) shows the values (n, ν) for different combinations of the numerical parameters. We varied parameters in a wide range of
values for the retrospective tests in the Lesser Antilles and in Southern California. The results are sensitive mostly to variation in the spatial parameters
r, R, S, and the threshold F [SK99].

Testing by random catalog. We applied algorithm SR to a randomized
catalog obtained from the real one by random permutation of origin times
of earthquakes. This test was performed only for the Antilles (1984–1995).
Parameters were the same as those in the main test (Table 4.11). Three
versions of the randomized catalog were tested, yielding three, three, and
five TSIPs with the respective relative volumes 3.7%, 4.2%, and 5.8%, and

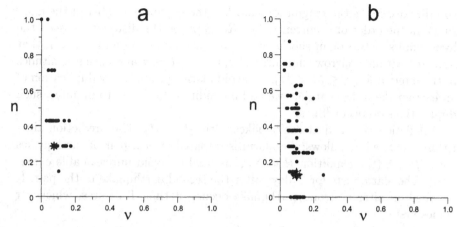

Fig. 4.24. Error diagram for the stability test of algorithm SR in the Lesser Antilles (**a**) and Southern California (**b**). n is the number of failures to predict divided by the total number of strong earthquakes; ν is the total time–space volume of alarms divided by the total volume considered. *Stars* stand for the main results (the values of the adjustable parameters are in Table 4.11); *dots* mark different values of the parameters (only one parameter is varied each time)

one, zero, and two "successful predictions" of a total seven. The number of TSIPs is much smaller than that in the main result. The number of successes drops dramatically. The testing on such a shuffled catalog indicates that the main result is by no means random.

4.5.2   Premonitory Increase of the Correlation Range (Pattern ROC)

There is growing evidence that the correlation length increases at the approach of an extreme event in a dynamic system [TNG00, STS85, SS95, BOS+98]. The analysis of simulated seismicity obtained from the colliding cascade model suggested the way to estimate the correlation range. The first attempt to use of this estimation was performed on seismic data in the Lesser Antilles [SZK00]; it is described below.

Algorithm ROC. Precursor ROC, which stands for "range of correlation," represents a nearly simultaneous occurrence of two earthquakes at large distances from each other. A. Prozorov [Pro75] introduced a termless precursor, which immediately follows a major earthquake, but well apart from it. He concluded that "long-range aftershocks" mark the location of a future major earthquake. The ROC is a short-term seismic pattern of the middle-range spatial uncertainty.

We consider a sequence of main shocks $\{t_k, \boldsymbol{g}_k, M_k\}$, $k = 1, 2, ...,$ where t_k is the earthquake occurrence time, $\boldsymbol{g}_k$ is the vector of the hypocenter coordinates, M_k is the magnitude, and k is the sequence number of the main shock in the order of occurrence. Let $R(i, j | \tau)$ be the distance between the hypocenters of two main shocks with sequence numbers i and j, $i < j$, that occurred within a narrow time interval τ, i.e., $t_j - t_i \leq \tau$ and have magnitudes in the interval $M_{min} \leq M < M_0$. To avoid alarms produced by duplication of an earthquake in the catalog, we add a condition $t_j - t_i \geq 10$ minutes; such duplications do occur [She92].

Prediction is aimed at earthquakes with $M \geq M_0$. The prediction algorithm is defined as follows: An alarm is declared after a pair of earthquakes with $R \geq \Delta$ (the condition $R < \Delta_{max}$ is used to avoid unreasonable extension). The alarm lasts for T days after the second earthquake in the pair. It is called off after a strong earthquake occurs or time T expires, whichever comes first.

Performance of algorithm ROC. The occurrence of a pattern prior to an event that one aims to predict is not sufficient to ensure satisfactory prediction based on this pattern alone. For example, most precursors can concentrate before one of strong events, leaving others unpredicted. To evaluate the performance of precursor ROC, we apply the algorithm defined above with the following numerical parameters: targeted earthquakes are those with magnitude $M_0 = 5.5$ or greater, duration of alarm $T = 40$ days, time window $\tau = 3$ days, threshold for declaration of alarm $\Delta = 150$ km, $\Delta_{max} = 300$ km, and $M_{min} = 3.8$.

A total of 13 short-term alarms is determined; three strong earthquakes occurred within these alarms, one strong earthquake is missed, and 10 alarms are false. The total duration of alarms is 10% of the interval considered. Although the total duration of alarms is low, the number of false alarms is rather large. The number of false alarms can be drastically reduced if ROC delivers the second approximation of a longer term prediction algorithm in the time intervals already determined as dangerous.

Consecutive application of algorithms SR and ROC. Earthquake prediction research knows just a few but very successful examples of reproducible consecutive predictions with increasing accuracy. Among them is the prediction of the Haicheng earthquake in China, 1975, which went through four stages from long-term to immediate [MFZ+90]; the same holds for the increase of the territorial accuracy of intermediate-term prediction by the algorithm "Mendocino Scenario" [KKS90].

The comparison of alarms determined by SR and ROC (lower and middle rows in Fig. 4.25) suggests regarding ROC as the second approximation of SR and declaring an alarm in two steps: first, define intermediate-term alarms by algorithm SR; second, during the alarms achieved, determine a short-term alarm by algorithm ROC.

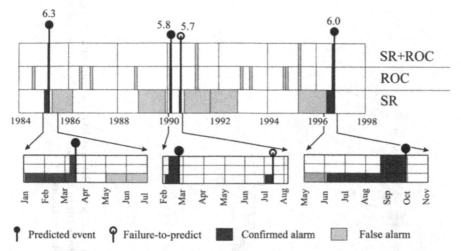

Predicted event   Failure-to-predict   ■ Confirmed alarm   □ False alarm

Fig. 4.25. Retrospective prediction combining both algorithm SR and precursor ROC. Lower, middle, and upper rows show alarms issued by SR, ROC, and SR and ROC combined, respectively

Alarms determined in this way are shown in Fig. 4.25. Figure 4.26 shows how short-term prediction by two algorithms was unraveled in space and time. Arrows in Fig. 4.26 connect the first and second earthquakes in the pairs that generated alarms; both lie close to the area covered by the SR alarm. We see that three out of four strong earthquakes occurred during a short-term alarm. Two alarms are false. The total duration of alarms is 3% of the time considered. An area of alarm occupies 15–25% of the territory, and all alarms together occupy 0.5% of the time–space considered. Any such scores would be a great success in advance prediction. One should remember, however, that adjustable parameters of the precursor were not determined *a priori* on a model and were data-fitted on observations considered. Accordingly, this study merely formulates a hypothesis to be tested in advance prediction.

Error diagrams. We tested the stability of the precursor ROC considered independently (Fig. 4.27) and on top the SR alarms (Fig. 4.28). In the second case, only the parameters of the ROC were varied because we used alarms already issued by SR [SK99]. Figure 4.28 also shows the results of a randomized prediction. A combination of adjustable parameters listed at the bottom of Fig. 4.28 is randomly selected. For each combination, we know the number N (4 to 40) of alarms and their duration T. We distribute, randomly, the same number of alarms of the same duration. The scores of (n, τ) and (n, f), where f is the rate of false alarms, are shown by gray dots in Fig. 4.28. Naturally, random predictions give a much inferior score; just a few of them overlap with predictions from the combined algorithm.

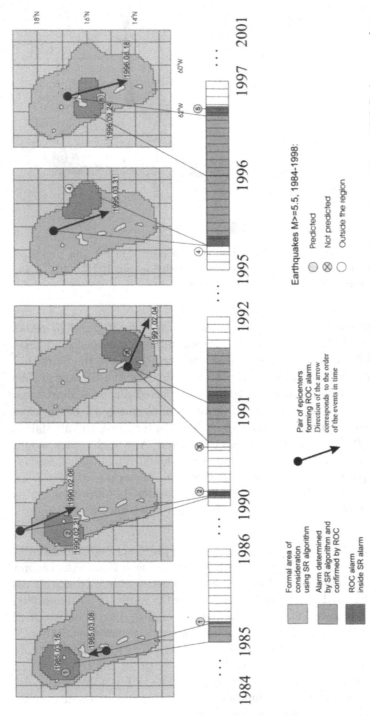

Fig. 4.26. Transition to short term prediction in time and space. Year intervals without joint SR and ROC alarms and strong earthquakes are omitted on the figure. Pairs of events forming ROC precursor seem to occur in the same part of the region. False alarms determined by both SR and ROC were initiated, probably, by preceding strong earthquakes

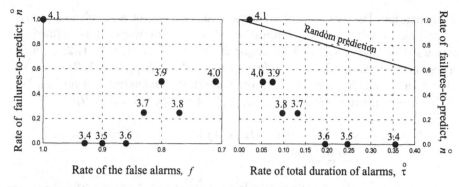

Fig. 4.27. Error diagrams for precursor ROC. Points on the diagrams are labeled by threshold values M_{min}. Other parameters are fixed as follows: $\tau = 3$ days, $\Delta = 150$ km, $\Delta_{max} = 1000$ km, and $T = 40$ days

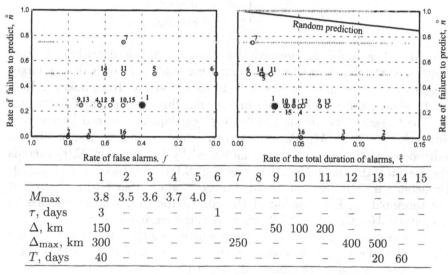

	1	2	3	4	5	6	7	8	9	10	11	12	13	14	15
M_{max}	3.8	3.5	3.6	3.7	4.0	–	–	–	–	–	–	–	–	–	–
τ, days	3	–	–	–	–	1	–	–	–	–	–	–	–	–	–
Δ, km	150	–	–	–	–	–	–	50	100	200	–	–	–	–	
Δ_{max}, km	300	–	–	–	–	–	250	–	–	–	–	400	500	–	–
T, days	40	–	–	–	–	–	–	–	–	–	–	–	20	60	

Fig. 4.28. Error diagrams for joint prediction by SR and ROC

4.5.3   Premonitory Spreading of Seismicity Across the Network of Faults: Pattern Accord

Another manifestation of long-range correlation in seismicity is the premonitory spreading of activity. The seismicity pattern Accord represents this phenomenon. It is defined as a simultaneous rise of seismic activity in a sufficiently large number of fault zones. Similar to pattern ROC, pattern Accord was found first in a synthetic catalog generated by the colliding cascade model [GKZN00, GZNK00], and then observed in the seismicity of Southern California [ZKA02].

Definitions. Pattern Accord consists, qualitatively, of a simultaneous rise in seismic activity in several branches of a fault network. It is defined as follows. Consider a network of major faults $F = \{Fi\}$, $i = 1, 2, ..., n$ in a region R. It is divided into *subregions* R_i, $R = \{R_i, \}$, $i = 1, 2, \cdots, n$ so that a subregion R_i contains F_i (more generally, a *subnetwork* of faults). We define the measure of seismic activity $\Sigma_i(t)$ for each subregion R_i:

$$\Sigma_i(t, s, \lambda) = \sum_{\substack{k:\, M_1 \leq M_2;\\ E_k \in R_i}} H(t - t_k, s, \lambda)\, 10^{Bm_k}, \qquad i = 1, \cdots, n,$$

where the summation is taken over all main shocks E_k from the ith subregion with magnitudes m_k, $M_1 \leq m_k \leq M_2$; t_k is the origin time of the kth earthquake; the function H is defined as follows:

$$H(t, s, \lambda) = \begin{cases} 10^{t/\lambda}, & 0 \leq t \leq s, \\ 0 & \text{otherwise} \end{cases}$$

M_1, M_2, B, s, and λ are numerical parameters of the algorithm. A finite positive λ introduces an effective attenuating "memory" by assigning larger weight to more recent earthquakes. A. Khokhlov and V. Kossobokov [KK94] used a similar concept of memory in a seismic flux description of seismicity.

The targets of prediction are "strong" earthquakes defined by the condition $M^0 > M \geq M_0$. Pattern Accord is determined by the number of subregions where seismic activity measured by the function $\Sigma_i(t)$ exceeds the common threshold Σ_0: $A(t) = \#\{i : \Sigma_i(t) \geq \Sigma_i(t_0), \quad i = 1, ..., n\}$. By definition, $A(t)$ takes on positive integer values from 1 to n. Note that the threshold Σ_0 is a fraction of the contribution from an earthquake of magnitude M_0: $\Sigma_0 = 10^{BM_0}/R$, where R is the parameter of the algorithm.

An alarm is declared from t to $t + \Delta$ whenever $A(t) \geq C_A$. The threshold C_A is determined from the condition that $A(t)$ is lower that C_A during $q\%$ of the time. Δ and q are numerical parameters of the algorithm. A strong earthquake terminates an alarm before its expiration.

Application in Southern California. The prediction algorithm was applied retrospectively to simulate actual advance prediction; at time t, the algorithm uses information on seismicity only prior to t. First, the three largest Southern California earthquakes were targets of prediction and then the earthquakes with magnitudes from 6.5 to 7.3 were considered.

The division of Southern California into seven subregions (Fig. 4.29) is based on fault orientation, subjective grouping, and the direction of slip, along with tectonic settings. The fault data were taken from [Jen77, Jen94]. In general, the division of Southern California into subregions is natural. Northern regions are separated from southern ones by the Transverse Ranges, where east–west striking left-lateral faults dominate. The relatively stable

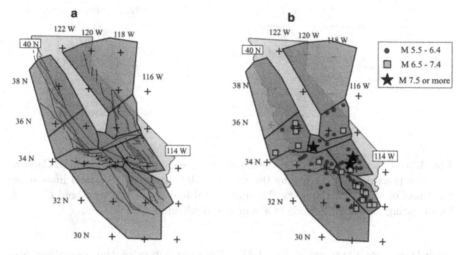

Fig. 4.29. The fault network of Southern California. (a) Faults and subregions (fault zones). (b) Subregions and seismicity, 1928–2000

Sierra Nevada-Great Valley block and the Peninsular Ranges separate the eastern areas.

The authors analyzed the data from the USGS-NEIC earthquake catalog [GHDB94], from 1 January 1928 to 30 January 2000. The catalog was preprocessed by the same rules as in other prediction algorithms of that kind (see Sect. 4.3.1). Figure 4.29b displays the distribution of main shocks.

First, pattern Accord and the largest earthquakes in Southern California were juxtaposed. During the time considered, three such earthquakes occurred: Kern County, $M = 7.7$, 21 July 1952; Landers, $M = 7.6$, 28 June 1992; and Hector Mine, $M = 7.4$, 16 October 1999. The values of the parameters of the algorithm are listed in the first row of Table 4.13. Figure 4.30 shows that $A(t)$ exceeds the value 5 in the approach and a few years after the targeted earthquakes. This suggests the choice $C_A = 5$ and the corresponding value of $q = 80\%$. With this choice of parameters, each strong earthquake is preceded by an alarm, and there are no false alarms. The total duration of alarms is 29 years, i.e., 41% of the time considered.

To study the stability of this result, numerous experiments were performed with adjustable parameters varied independently. The results are summed

Table 4.13. Numerical parameters of the algorithm Accord

M_0	M^0	M_1	M_2	s, years	λ, years	Δ, years	B	R	q, %
7.5	∞	5.0	7.4	12	∞	3	0.9	21	80
6.5	7.5	5.0	6.4	2	∞	2	0.9	25	90

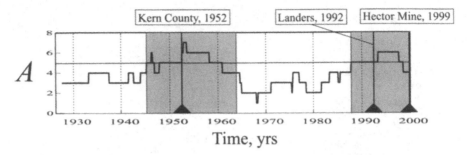

Fig. 4.30. Precursor Accord and the three largest earthquakes in Southern California. The precursor is depicted by the function $A(t)$ (see text). Vertical lines show the times of strong earthquakes. The horizontal line shows the threshold $C_A = 5$ for declaring an alarm. Periods of alarm are highlighted

up in the error diagram in Fig. 4.31. The parameters of the algorithm are interdependent. For example, reducing the "memory" λ, one has to reduce the threshold Σ_0 to produce the same prediction score. The stars in Fig. 4.31 correspond to predictions with R calculated from the empirical relation $R = \lambda^{0.55} 10^{3.5}$; other parameters are varied independently.

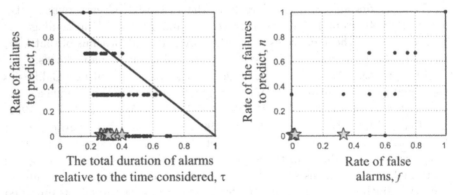

Fig. 4.31. Error diagram for retrospective prediction of three largest earthquakes in Southern California, 1928–2000. Heavy dots and stars correspond to independent variation of parameters in the empirical relation between R and λ, respectively (see text)

Pattern Accord was tested in a similar way before 11 earthquakes of magnitudes from 6.5 to 7.3 occurred during the time period considered. Two of them, the Colorado River Delta of 31 December 1934, $M = 7.0$, and the Superstition Hills of 24 November 1987, $M = 6.7$, are not considered in the statistics, because they follow strong earthquakes within the too short time of 29 and 12 hours, respectively. Figure 4.32 shows the function $A(t)$ and alarms determined with values of parameters given in the second row of Table 4.13.

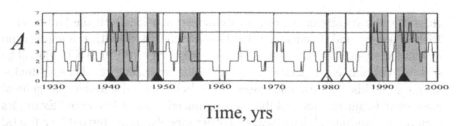

Time, yrs

Fig. 4.32. Precursor Accord and earthquakes with magnitudes from 6.5 to 7.3 in Southern California. "Predicted" and "unpredicted" earthquakes are marked by filled and empty triangles, respectively. Other notations are the same as in Fig. 4.30

Six out of nine strong earthquakes are "predicted" with no false alarms; three earthquakes are missed. The total duration of alarms is 20 years, i.e., 28% of the time considered.

The stability test (Fig. 4.33) shows again that the quality of prediction remains acceptable for some special combinations of parameters and also for a wide domain in the space of parameters. It is encouraging that combinations leading to high scores are not scattered randomly across the parameter space but form a cluster.

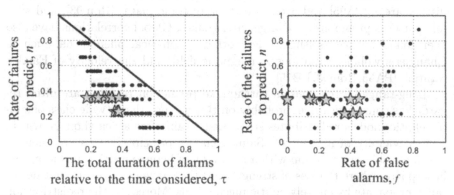

Fig. 4.33. Error diagram for the retrospective prediction of strong ($6.5 \leq M < 7.5$) earthquakes in Southern California, 1928–2000. The parameters of the algorithm are varied independently. Heavy *dots* and *stars* correspond to broad and narrowed ranges of the variation, respectively

4.6 Discussion

It is widely accepted that the number of intermediate size earthquakes in a region increases prior to the characteristic earthquake for the region [KM64, Var89, Kei90b, BV93, BOS+98]. This precursory activation, it has been shown, exhibits power-law scaling and occurs over a region about ten times larger than the rupture size of the characteristic earthquake. Both the Gutenberg–Richter scaling and precursory activation are consistent with the application of condensed-matter theory to regional seismicity [BT89, KKTM00]. The scaling of small earthquakes in a region is equivalent to thermal fluctuations in solids, liquids, and gases. This behavior falls under the general class of phenomena that exhibit self-organized critical behavior. Examples include the sandpile, slider-block, and other models, characterized by a fractal frequency–area distribution of avalanches [ALM82, Ben00, Ben01, BT89, NS90, NGT94, SCL97, STS85, TNG00, Yam98]. These models can be explained in terms of an inverse cascade involving the coalescence of metastable clusters [GZNK00, GKZN00].

The characteristic earthquake appears to behave as a second-order phase transition [KKTM00]. As the phase change is approached, the scale of the region over which the correlated activity occurs (the correlation length) increases. In terms of earthquakes, this correlated region is the region of precursory activation.

One prominent, yet controversial, feature of seismic activity is the correlation of earthquake occurrences at long distances far exceeding source dimensions. This correlation is expressed in many ways: simultaneous change of seismic activity within large regions [Mog68]; migration of earthquakes along fault zones [Mog68, VS83, MFZ+90]; alternate rise of seismicity in distant areas [PA95] and even in distant tectonic plates [Rom93]; and seismic patterns premonitory of large earthquakes. Global correlations have also been found between seismicity and other geophysical phenomena, such as Chandler's wobble, variations in magnetic field, and the velocity of Earth's rotation [PA95, Rom93, PB75].

Long-range correlations in earthquake prediction research are often regarded as counterintuitive, probably on the grounds that a wide class of simple elastic models redistributes stress and strain after an earthquake within a small vicinity of its source, "Saint-Venant's principle". This argument is not applicable to a medium with a hierarchy of heterogeneities, including the lithosphere, where the loss of strength and the change of stress do not necessarily propagate by entirely elastic mechanisms. Moreover, the redistribution of stress is not necessarily relevant to this argument because earthquakes involved in long-range correlations may not trigger each other but reflect an underlying large-scale process such as microfluctuations in the movement of tectonic plates [PA95] and perturbation of the ductile layer beneath the seismically active zone [Aki96]. Accordingly, there is no reason to look for premonitory phenomena only near an incipient fault break.

Earthquake prediction studies report many evidences that long-range correlation between earthquakes is reflected in some phenomena precursory to large earthquakes. One of the examples is the remarkably successful prediction of the Haicheng earthquake in China, 1975 [MFZ+90]. In its long-term stage, the prediction was made by extrapolating migration of seismicity across distances of about 10^3 km. In the timescale of years, many premonitory seismic patterns are formed within areas of a linear size 10 times larger than the dimensions of the source of an incipient large earthquake [Kei90b]; this estimation is validated by advance earthquake prediction [KRKH99, Vor99]. F. Press and C. Allen [PA95] found that on the timescale of tens of years, this size may even reach about five times larger: earthquakes of magnitude 6 in Parkfield, California, are preceded by an increase of seismic activity in the Grand Basins and/or the Gulf of California. Such large distances over which seismicity is correlated are explained well by microfluctuations in the movement of tectonic plates [PA95] or by the interaction of crustal blocks [SVP99, GKJ96]. Different models explaining long-range interaction between earthquakes are naturally divided into two classes: models rooted in statistical physics, such as renormalization models originated by Allegre et al. [ALM82] and [NS90], and models based on specific local mechanisms [GBB97, HRM+93, RG83, SKL92]. We emphasize that these models are not contradictory but complementary.

The algorithms presented in this chapter use seismic activation and the growing correlation of earthquakes at the approach of the Big one. On average, the algorithms predict about 80% of large characteristic earthquakes in a given region with alarms occupying from 20 to 30% of space–time. This could be done on the basis of earthquake catalogs routinely available in the majority of regions. With more complete catalogs, the areas of alarm may be substantially reduced in the second approximation at the cost of additional failures to predict. There are serious limitations in this performance. The areas covered by reliable alarms are large (especially in the first approximation), and many of them will inevitably expire without a strong earthquake. Nevertheless, considerable damage may be prevented by knowledgeable use of such predictions when their formulation is timely and specific.

In conclusion, we comment on the recently revived discussions: To what extent are earthquakes predictable? No current theory of dynamics of seismic process can answer this question. Inevitably, a negative statement that asserts a nontrivial limitation on predictability is merely a conjecture. On the other hand, by forward testing a reproducible prediction method we can unequivocally establish a certain degree of predictability of earthquakes. The results described here did confirm a positive statement on the predictability of earthquakes on the intermediate-term scale. Furthermore, it appears that premonitory activation evolves through long-, intermediate-, short-, and immediate-term phases.

The accumulated set of successes and failures gives some indications how the accuracy of prediction algorithms could be enhanced. At the same time, just a small part of potentially relevant data is used in these algorithms. Premonitory phenomena expressed in other data remain yet unexplored. The results described here suggest that the approach deserves further testing and development.

5 Earthquake Prediction Strategies: A Theoretical Analysis

G.M. Molchan

The earthquake "prediction problem" is examined from the "standpoint" of decision theory. The "problem" stated in these terms has an exact formulation as optimization of the goal function γ. The two types of γ discussed are at the research phase and application phase of prediction incorporating economics. In both cases, we find the structure of prediction strategies that optimize γ. The error diagram and specific techniques used in prediction practice are discussed.

5.1 Introduction

Earthquake prediction is a very difficult scientific and socioeconomic problem. It is no wonder therefore that its history is rife with fits of optimism and disappointment [Gel97]. The 131:3 (1997) issue of the Geophysical Journal International contains a collection of papers by opponents of earthquake prediction; they assert that "earthquakes cannot be predicted" [GJK97]. The assertion is practically unobjectionable, when prediction is understood to mean *dynamic prediction*, i.e., a deterministic (100%) localization of a future large event in a sufficiently narrow space–time window. Dynamic prediction is usually related to the study of dynamic systems; its limitations are due to dynamic chaos or strong instability of paths in the system's phase space. Dynamic chaos is typical of dissipative dynamic systems of large dimensions; these concepts are transferred wholesale to lithosphere dynamics. No exact equations are available for the seismic process, so such a transfer does not provide evidence from which to judge the horizons (space–time scale) of prediction; what it does is to warn of possible difficulties in the way of dynamic prediction.

The restriction of prediction to a deterministic one has unfortunately cast doubt on the existence of precursors [Gel97]. This is quite logical, when deterministic precursors are involved. The denial of stochastic precursors is based on the fact that most of them have low statistical significance [Kag97]. In this connection, we wish to make the following remark. The most complete and long (50–70 years) observations used for prediction are earthquake catalogs.

A large event in regions like California is considered one that occurs once every 7 years, on average. For this reason, possible amounts of observation available for analysis of regional precursors can be 7 to 10 large events at most, which is certainly not sufficient for statistical inference about significance. The situation is changed, when a precursor can be extended to a set of regions without substantial addition to the parameters involved. It is in this manner that statistical significance was proved for precursors such as a "burst of aftershocks" [MDRD90], a prediction algorithm, or the collective precursor M8 [KRKH99], as well as the Vorobieva algorithm for predicting a second large event [Vor99]. These facts in the statistical analysis of seismicity should not be undervalued, or overvalued, considering that a precursor, even though significant, may be of little use in prediction. To take an instance, a large sample of newborn infants will reveal that the frequencies with which boys and girls are born are unequal, but the difference is so small as to be of little practical value in predicting the sex of an infant about to be born.

Going beyond the limits of deterministic prediction, the question posed by Kagan [Kag97]: "Are earthquakes predictable?" can be restated more constructively. Forecasts are always feasible; the essence of the matter is in the relative efficiency of forecasts. The assertion that earthquakes are unpredictable is in itself a form of prediction; it is worded so as to influence public opinion and implies that earthquake prediction is entirely unreasonable economically. The problem therefore needs comprehensive analysis, including its economic aspect. Little has unfortunately been done toward that goal. Below we are going to discuss a possible formulation of the prediction problem as a whole and to provide a theoretical analysis. This will enable us to grasp the meaning of empirical activities in earthquake prediction and to understand those tasks that face prediction specialists, on the one hand, and economists and decision makers, on the other.

It is high time to address the problem of interaction between economists and geophysicists. This is demonstrated by contemporary prediction practice. Today, there are a number of prediction techniques (algorithms) that are not of very high quality. The number of such algorithms grows with time, thus complicating the situation. As a matter of fact, the algorithms are decision functions that call (or do not call) an alert for each time unit at a given point or in a region. In this situation, two nonequivalent techniques can (and do) lead to contradictory alert decisions. It is impossible to avoid this difficulty by choosing the "best" method at the research phase without interaction with the consumer of the forecasts. The best method does not exist because the quality of stochastic prediction is characterized by a vector quantity (see below) which obviously cannot be ordered in a linear manner. On the other hand, decision makers have to understand the objective principles on which prediction algorithms are based. These principles are vague because they are not fully realized by the authors themselves or else are based on artificial efficiency criteria that have little relevance to reality. As a result, decision makers deal with a set of ready-made (and a priori contradictory) decisions, and the authors of the algorithms do not know whether their results are geared to applications.

A goal function is defined here for theoretical analysis of the prediction problem as a whole. This may be a forecast-related loss function or the reciprocal utility function, in particular, expected prevented loss. As a result, the multiplicity of forecasts, instead of being an obstacle, becomes a promising basis from which to select the best decision. Below we investigate two models of the loss function. One model is important for most practical prediction algorithms; it is useful during the research phase of prediction (just where we are at present), and the other is a crude simulation of prediction economics. In both of these cases, we will find the forecast structure that optimizes the loss function under very general conditions on the information $J(t)$ at time t. We will show that the optimal forecast in simple cases is based on a hazard function $r(t)$, i.e., on the conditional (given $J(t)$) occurrence rate of large earthquakes. That fact supports and refines the standpoint of many investigators (see [Ver78], [Aki81]) that estimation of the hazard function $r(t)$ is the principal problem facing an earthquake prediction expert. When more complicated cases are dealt with, one also needs, besides $r(t)$, the transitional probabilities for states of $J(t)$ in adjacent time intervals. The study of the second statistic (transitional probabilities for J) is still in its infancy. The information necessary for this accumulates in the form of detailed descriptions of earthquake preparation processes and theories that summarize these observations.

The plan of this chapter is as follows. Section 5.2 considers the simplest type of forecasts with two alert states, yes/no. In the special case in which $J(t)$ is the time elapsed from the last large event, we give a detailed study of forecast stability.

Section 5.3 presents an analysis involving an arbitrary number of alert states. Optimization of mean forecast-related losses yields a Bellman-type equation. This part of the chapter helps clarify which statistics are useful in the problem of prediction as a whole. For convenience of reference, all technical proofs have been relegated to the Appendix.

5.2   Prediction Involving Two Types of Alert

5.2.1   The Error Diagram

Recent discussions of earthquake prediction make increasing appeal to the language and general properties of dissipative dynamic systems. When dealing with a dynamic system, one is usually interested in the amount of deviation of the system's position from the true one at a given time. The value of a deviation is measured in a suitable metric on the phase space and is a measure of prediction performance. For example, the classical Kolmogoroff–Wiener problem is concerned with the prediction of a random time series $x(t)$

based on observations available with some time delay τ by the time $t - \tau$. Prediction performance is given by a single value, namely, the relative rms error $E|x(t) - \hat{x}_\tau(t)|^2/E|x(t)|^2$, where $\hat{x}_\tau$ is the forecast of x and E denotes mathematical expectation or averaging over the paths of x.

Speaking in the language of dynamic systems, one can treat large seismic events as anomalous states (disasters) in lithosphere dynamics; we are interested first of all in the positions and times τ_n of the disasters. To take an instance, when a time series x is discussed, the disasters may be at random times when x goes beyond a critical level (a limiting acceptable load in a physical system). Dealing with a general dynamic system, the τ_n are the times at which the path of the system occurs in a selected region of phase space. In application to a seismic process, this may be a spatial zone G and a magnitude range $M > M_0$. It goes without saying that the Lyapunoff exponents of a dynamic system, which can be used to judge the horizons of dynamic prediction, tell us practically nothing as to whether disasters can be predicted. In addition, a forecast of disaster time that is very accurate, but delayed $\hat{\tau}_n > \tau_n$, can be unacceptable in practical terms, because a forecast of τ_n has to be made in advance. For this reason, the metrical proximity of $\hat{\tau}_n$ and τ_n is useless in disaster prediction.

The performance of an earthquake prediction technique actually requires at least two quantities rather than a single one to characterize it: the rate of failures to predict and the relative alert time τ. These can be given precise meaning, when more definitions have been formulated. Here and below, we will discuss only the temporal behavior of large earthquakes; in other words, forecasts will concern events of magnitude $M > M_0$ in a given area G. The sequence of large events is considered as a random point process $dN(t)$ [$N(t)$ is the number of events in the interval $(0, t)$] of finite rate $\lambda > 0$, i.e., $EdN(t) = \lambda dt$.

Let $J(t)$ be the information available at time t for predicting events in the point process $dN(t)$. In practice, $J(t)$ may include an earthquake catalog for the region containing G, data on physical fields, and observations of precursors. Any type of information is relevant to a constant moving interval of the form $(t - t_i, t - \tau_i)$, where τ_i is the delay of the ith data type. The simplest case is where the observer uses information $J(t)$ and makes the decision $\pi(t)$: calling ($\pi = 1$) or not calling ($\pi = 0$) an alert in the time interval $(t, t + \delta)$, where δ is some time unit. This may be equal to the time increment at which the information is updated. An event is considered to have been predicted, when it occurred during an alert period, and is a failure to predict otherwise. The set of decisions $\{\pi(t)\} = \pi$ is called the prediction *strategy*. In practice, the strategy is defined by the method or by the prediction algorithm. It is useful to consider a class of strategies π where decisions can be made with some probabilities, i.e., after an additional test of the coin-tossing type with outcome probabilities $(p, 1 - p)$ depending on $J(t)$. In practice, deterministic solutions are usually preferred, where $p = 0$ or $p = 1$. Discrete-time strategies were defined above just to simplify the discussion. For this reason, the point process $dN(t)$ will be considered on the lattice $Z_\delta = \{\delta k,\ k = 0, \pm 1, \pm 2, ...\}$ as well, assuming that no more than one event can occur in the interval $(t, t + \delta)$:

$$\mathrm{Prob}\{\delta N(t) = N(t + \delta) - N(t) > 1\} = 0\ , \quad t \in Z_\delta\ .$$

We now define quantities to characterize the predictive properties of a strategy π in the interval $(0, T)$. These are the relative number of failures to predict

$$\hat{n}_\pi = \sum_{0 < t < T} (1 - \pi(t)) \cdot \delta N(t)/N(T)\ , \tag{5.1}$$

and the relative alert time

$$\hat{\tau}_\pi = \sum_{0 < t < T} \pi(t)\delta/T\ . \tag{5.2}$$

Without loss of generality, we can assume that the information flow $J(t)$ on which the prediction strategy is based is a multivariate random process. Suppose that the process $[\delta N(t), J(t)]$, $t \in Z_\delta$, is stationary and ergodic. The application of the individual ergodic theorem [Bil65] to (5.1), (5.2) will give the result that $\hat{n}_\pi$ and $\hat{\tau}_\pi$ converge to the respective constants n_π and τ_π with a probability of one. These limits define two long-term prediction errors for the strategy π, *the rate of failures to predict* n_π and *the relative alert time* τ_π.

Representation of a strategy π by a pair of numbers n and τ gives a subset $\mathcal{E}(J)$ of the square $[0, 1] \times [0, 1]$ which depends on information flow J (Fig. 5.1). It turns out that this set admits an effective description. The key observation is that any two strategies π_1 and π_2 of the type considered can be combined into a new strategy that independently uses π_1 or π_2 with probabilities q and $1 - q$ in each time interval δ. This leads to a mixture of parameters $(n, \tau)_i$ of the original strategies with the same weights q and $1 - q$. Hence the error set $\mathcal{E} = \{(n, \tau)_\pi\}$ corresponding to various strategies is convex, if these strategies are based on the same information $J(t)$. Now note that the error set $\mathcal{E}$ contains points $(1, 0)$ and $(0, 1)$ and, by convexity of $\mathcal{E}$, the diagonal $n + \tau = 1$. The first point stands for the widespread *optimistic strategy* in which an alert is never declared. The second point corresponds to the total *pessimistic strategy* in which a continuous alert is kept. Points on the diagonal $n + \tau = 1$ correspond to the strategy of a *random guess* in which an alert is declared with probability p independent of $J(t)$.

Trivial strategies on the diagonal $n + \tau = 1$ are not infrequent in applications. The prediction experiments in Parkfield, U.S. and Tokay, Japan, focused on recording a large event in small spatial regions during long time intervals [Gel97]. The implementation of these experiments proceeded in accordance with the pessimistic strategy. Seismic hazard assessment is based

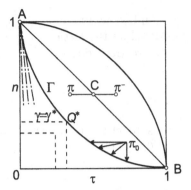

Fig. 5.1. Error set $\mathcal{E}(J)$ for prediction strategies based on a fixed type of information J. Point A corresponds to an optimistic strategy, point B to a pessimistic strategy, and the interval AB corresponds to strategies of random guess. C is the center of symmetry of $\mathcal{E}(J)$. π and π^- are a strategy and its antipodal strategy. Γ is the error diagram of optimal strategies. *Arrows* indicate a better forecast relative to the strategy π_0. *Dashed lines* are contours of the loss function $\gamma = \max(n, \tau)$. Q^* are errors of the minimax strategy, $n = \tau$. *Dash-dotted* lines are contours of the loss function $\gamma = \tau/(1 - n)$

on the Poisson seismicity model and a long-term rate $\lambda(g, M)$ of events that occur at point g with magnitude M. This is equivalent to the random guess strategy with probability $p = \lambda(g, M)\delta g\, \delta M\, \delta t$. The same strategy with the probability of correct guess $p = \int\limits_G \int\limits_M^{\infty} \lambda(g, m)\, dg\, dm$ simulates the attitude of an opponent of prediction who rejects the very possibility of predicting large events occurring in G.

The set $\mathcal{E}$ has the center of symmetry $(1/2, 1/2)$, because every prediction corresponds to the antipodal prediction π^- where an alert and a nonalert swap places and errors (n, τ) are replaced by $(1 - n, 1 - \tau)$. Therefore, all points of $\mathcal{E}$ above the diagonal $n + \tau = 1$ correspond to strategies constructed by rejecting nontrivial strategies with $n + \tau < 1$.

Let us show that only strategies at the lower boundary Γ of the set $\mathcal{E}$ are important. The boundary Γ connects the points $(1, 0)$ and $(0, 1)$. It is monotonic and concave due to the properties of $\mathcal{E}$. So far, there is no strategy with errors $(0, 0)$, i.e., an *ideal strategy* that guarantees a 100% prediction of large events in G with no alerts at all. Consequently, Γ does not contain $(0, 0)$. The points of Γ are incomparable, i.e., if $\tau_1 < \tau_2$, then $n_1 \geq n_2$. For any point $(n, \tau) \in \mathcal{E}$, there exists another point $(n_1, \tau_1) \in \Gamma$ where $n_1 < n, \tau_1 < \tau$, which corresponds to a better prediction. Therefore, the total error set $\mathcal{E}$ contains a minimum set of best and incomparable strategies. The number of these strategies is infinite, and they are described by an *error diagram* Γ (Fig. 5.1).

To be able to compare strategies, we choose some one-dimensional characteristic $\gamma = \gamma(n, \tau)$ that is a function of (n, τ). We will call it a *loss function* if

γ increases in each argument. Typical examples of γ that have been employed at the research phase of prediction are functions of the form $\gamma_1 = n + \tau$, $\gamma_2 = \max(n, \tau)$, $\gamma_3 = \tau/(1 - n)$, $\gamma_4 = n/\theta(\tau_0 - \tau)$, or $\gamma_5 = \tau/\theta(n_0 - n)$, where $\theta(x) = 1$ for $x > 0$ and $\theta(x) = 0$ otherwise. The strategy π^* will be called γ-optimal, when it minimizes $\gamma(n_\pi, \tau_\pi)$. For example, the optimal strategy minimizes the mean prediction error $(n+\tau)/2$ in the γ_1 case and optimizes the number of successes at a given level of alert time in the γ_4 case. Knowing the Γ diagram, one can easily find the errors of the γ-optimal strategy graphically. We assume that the sets of levels of γ, $A_u = \{(n, \tau) : \gamma(n, \tau) < u\}$, are convex for any level u. The sets A_u increase with increasing u. Obviously, there is a critical level u_* where A_u and $\mathcal{E}(J)$ touch each other. Since A_u and $\mathcal{E}(J)$ are convex, this will be a single point (the regular case) or a line segment (not a typical case). The point of contact $Q^* = (n^*, \tau^*)$ will determine the γ-optimal errors. By construction, it belongs to the Γ error diagram.

It is easily seen that any point of the Γ curve can be made γ-optimal by a suitable choice of the loss function. This can be demonstrated as follows. The Γ curve is always on one side of its tangent. Let $a(n-n^*)+b(\tau-\tau^*) = 0$ be the equation of the tangent to Γ at the point (n^*, τ^*). Then a and b have the same sign, since Γ is decreasing along the n axis, and $f = |a|n + |b|\tau$ is the desired loss function for the point (n^*, τ^*).

To sum up, if there is no ideal forecast with zero errors, the Γ curve consists of a continuum of points corresponding to γ-optimal strategies. The absence of a universal prediction strategy for a given information flow was far from being quickly grasped in prediction practice.

When the information flow is updated, $J \subset J'$, the set $\mathcal{E}(J)$ expands; for this reason, the diagram $\Gamma(J')$ will be below (to be more accurate, not higher than) $\Gamma(J)$. The $\Gamma(J)$ curve can be regarded as characterizing the limiting capability of information J in the prediction of large events in region G. The paradox here consists of the fact that the Γ curve always includes the end points $(0, 1)$ and $(1, 0)$ corresponding to the trivial strategies of an optimist and a pessimist. These ignore all information and become optimal with a special choice of the loss function. For example, looking from the economic point of view, there is no sense in predicting seismic events where there is no threat to the economy and population.

Below we consider some examples of the loss function.

Example 1. Publications dealing with prediction frequently employ the quantity $(1 - n)/\tau$ both (a) as a measure of prediction efficiency [Gus76] and (b) as the "probability gain" for a large event, given observations of a precursor with predictive parameters (n, τ) [Aki81]. The inverse of this, $\gamma = \tau/(1 - n)$, is the loss function. Its isolines form a beam of the straight lines $1 - n = c\tau$ centered at $P = (1, 0)$ (Fig. 5.1). Since $P \in \Gamma(J)$, the γ-optimal strategy is the trivial strategy of the optimist who never calls an alert. This strategy has the best efficiency for any information J, hence the quantity $(1-n)/\tau$ cannot be used to compare different prediction techniques.

It can, however, be meaningful because the quantity $\log[(1-n)/\tau]$ is identical to the Shannon information on a large event acquired by the observer from observing a single precursor (see below).

Example 2. Let $\gamma = \max(\alpha n, \beta \tau)$; its isolines form rectangles with vertices on the straight line $\alpha n = \beta \tau$ (Fig. 5.1). Consequently, the γ-optimal error point Q^* lies at the intersection of the line $\alpha n = \beta \tau$ and Γ. The γ-optimal strategy will be defined as the *minimax strategy with weights* (α, β) (or simply *the minimax strategy* in the case of equal weights), because $\gamma(n^*, \tau^*) = \min_{\pi} \max(\alpha n_\pi, \beta \tau_\pi)$. Parameter fitting for prediction algorithms frequently gives the final results of retrospective prediction that have approximately equal errors: $n \simeq \tau$. This relation is automatically obtained, when the function $\gamma = \max(n, \tau)$ is optimized, because $n^* = \tau^*$ for the minimax strategy. Even though the authors of prediction algorithms are not always capable of clearly stating the goal principles underlying their respective algorithms, the retrospective results show that they try to optimize losses of the form $\max(n, \tau)$.

5.2.2 The Optimal Prediction Strategy

We will try to determine the structure of the optimal prediction strategies based on the information flow $J(t)$. To do this, we define the *hazard function* $r(t)$, which is the conditional (with respect to the information $J(t)$) rate of predicted events:

$$r(t) = \text{Prob}\{\text{an event occurs in } (t, t+\delta)|J(t)\}/\delta .$$

The symbol λ above stands for the unconditional rate of large events, i.e., $E\delta N(t)/\delta = \lambda$.

The statement that follows provides a description of the optimal strategy with errors $(n^*, \tau^*) \in \Gamma$. The description is not unique.

Statement 1. *If the flow $\{n(t), J(t)\}$ is stationary and ergodic, then there exists a threshold r^* depending on the loss function γ such that the optimal prediction strategy declares an alert every time when $r(t) > r^*$. In rare cases in which the relation $r(t) = r^*$ has a nonzero probability, an alert is selected with some probability p^*. If Q^* is the point where the isoline $\gamma = \gamma^*$ touches the error curve Γ, then the threshold r^* is expressed in terms of a derivative common to Γ and lines $\gamma = \gamma^*$ at Q^*:*

$$r^* = -\lambda \frac{dn}{d\tau}(Q^*) . \tag{5.3}$$

If one of the curves is not differentiable at Q^*, then the derivative is the slope of any straight line tangent to Γ at Q^*.

The proof of the above statement and the description of the parameter p^* are given in the Appendix. The statement is remarkable in that it holds for a very broad class of the processes $\{n(t), J(t)\}$. Nevertheless, the proof is quite elementary and reduces to the classical Neyman–Pearson lemma in the statistical theory of hypothesis testing. The generality results from the choice of the class of loss functions that involve only two prediction characteristics, n and τ. It is just statistics such as these that are considered at the research phase of prediction. This is not at all sufficient from the practical point of view. For example, the goal function of the form $\gamma(n, \tau)$ ignores the rate of points where the strategy changes state. Frequent changes from alert to non-alert make for lower trust in the prediction involved. This circumstance is well known from practical forecasts of aftershocks, where the population begins to ignore seismologists' warnings when they frequently call short-lived alerts.

We continue with our discussion of examples.

Example 3. Consider $\gamma = \alpha \lambda n + \beta \tau$. This loss function can be given an economic meaning, even though a naive one. Let α be the mean loss prevented by successful prediction. The use of a strategy with errors (n, τ) will fail to predict λn events per unit time. Consequently, $\alpha \lambda n$ will give the loss per unit time resulting from failures to predict. The quantity $\beta \tau$ gives the loss due to alerts per unit time, when β denotes the cost of maintaining the state of alert per unit time. It follows that γ gives the loss per unit time. The example of a loss that is linear in n and τ is important in that one can find the optimal strategy for it without knowing the relevant error diagram Γ. This can be demonstrated as follows. The isolines $\gamma = c$ form a set of parallel straight lines with the slope $dn/d\tau = -\beta(\alpha\lambda)^{-1}$. The latter determines the slope of the tangent to Γ at the point Q^*. Therefore, the use of (5.3) gives the optimal γ-strategy as

$$\pi(t) = \begin{cases} 1 & r(t) > \beta/\alpha \\ 0 & r(t) < \beta/\alpha \,. \end{cases} \tag{5.4}$$

We have assumed that $P\{r(t) = \beta/\alpha\} = 0$, which generally holds.

Relation (5.4) is highly important for the analysis of the earthquake prediction problem as a whole. In the present case, it separates into two independent problems. One (seismological) reduces to estimation of the hazard function $r(t)$, and the other (economic) problem is to estimate the economic parameter β/α. Nevertheless, it is important during the research phase of prediction to know at least the order of β/α, since one can hardly hope to get stable estimates of r in the entire range of values.

It is difficult to estimate β/α because the above approach to the economic part is oversimplified. For this reason, we will modify our approach considering the normalized linear losses

$$D = \frac{\alpha \lambda n_\pi + \beta \tau_\pi}{0.5(\alpha\lambda + \beta)} \,.$$

Here, the denominator gives the loss guaranteed by the choice of the strategy of random guess with probability $1/2$. Since β/α is an unknown quantity, let us find the optimal strategy for which D is equal to $\max_{\alpha,\beta} \min_{\pi} D$.

To do this, we note that the tangent to Γ at a point (n_π, τ_π) meets the diagonal $n = \tau$ at the point $(D/2, D/2)$. Since Γ is concave, the intersection lies on the diagonal between $(0, 0)$ and $(a, a) \in \Gamma$. The latter point corresponds to the errors of the minimax strategy. Consequently, $\max_{\alpha,\beta} \min_\pi D$ is reached at the minimax strategy. In other words, the minimax strategy is D-optimal for the least favorable value of β/α. This circumstance may argue for the choice of $\max(n, \tau)$ as the goal function during the research phase. If the denominator in D is based on a random guess with probability p, i.e., is equal to $\alpha\lambda q + \beta p$, $q + p = 1$, then the D-optimal strategy for the least favorable value of β/α is the minimax strategy with weights (p, q). The proof remains the same, if we use the line $np = \tau q$ instead of the diagonal $n = \tau$.

5.2.3   Prediction of the Characteristic Earthquake

Characteristic earthquakes, it is thought, are the largest earthquakes occurring on fixed individual segments of tectonic faults. Their occurrence, it is thought, is quasi-periodic. For this reason, the characteristic events are predicted using a single statistic, namely, the time u elapsed since the last event [AE91]. Since the information flow $J(t)$ is so simple, the hazard function $r(t)$ can be found explicitly. In that case, it is convenient to consider $r(t)$, not as a function of time, but as a function, r_u, of state or elapsed time, $J(t) = u$. Let τ be the interevent time with distribution F. Then

$$r_u = \text{Prob}\{\tau \in (u, u + \delta)|\tau \geq u\}/\delta \simeq F'(u)[1 - F(u)]^{-1}, \ u > 0 \ , \quad (5.5)$$

when $\delta \ll 1$. That equality can be inverted:

$$1 - F(x) = \exp(-\int_0^x r_u \, du) \ . \quad (5.6)$$

The regularity requirement $\text{Prob}\{r(t) = \text{constant}\} = 0$ in Statement 1 means that F should not be locally exponential, i.e., there should be no intervals Δ where $1 - F(x) = \exp(-\lambda x)$.

If the regularity requirement holds, the γ-optimal strategy is defined as a function of state $J(t) = u \geq 0$, as follows:

$$\pi(u) = 1 \quad \text{if} \quad F'(u)/[1 - F(u)] > r^* \ . \quad (5.7)$$

The time of a large subsequent event is the stopping time when the strategy is terminated. A new prediction cycle starts at τ. The prediction errors are given in the regular case by

$$n_\pi = \int_0^\infty [r_u < r^*] \, dF(u) \ ,$$

$$\tau_\pi = \lambda \int_0^\infty dF(s)(\int_0^s [r_u > r^*] \, du) \ ,$$

with the notation $[C] = 1$ if the statement C is true and 0 otherwise; $\lambda^{-1} = E\tau = \int_0^\infty s\, dF(s)$, and the threshold r^* is found by minimizing the loss function $\gamma = \gamma(n, \tau)$.

The exponential distribution $F(x) = 1 - \exp(-\lambda x)$ is relevant to an irregular situation. In that case, $r_u \equiv \lambda$ and $\text{Prob}(r_u \equiv \lambda) = 1$, so that the optimal strategy and its errors are computed differently when $r^* = \lambda$ (see the Appendix). Then, $n + \tau = 1$, as was to be expected in that case, whereas the optimal strategy $\pi(u) = 1$, $u \in A$ with $n_\pi = n_0$ proceeds by selecting any set A on the semiaxis $u \geq 0$; the only requirement is that $\int_A dF(u) = 1 - n_0$.

In terms of $r(t)$, the requirement of quasi-periodicity for large events can be treated as unbounded growth of r_u for large u, $r_u \uparrow \infty$, as $u \uparrow \infty$. Quantitative verification of the characteristic earthquake concept showed [KJ91] that large events can occur in pairs at certain locations (clustering). The effect can be incorporated by assuming that $r_u \uparrow \infty$, as $u \to 0$. Thus, it appears that one possible model of F for interevent time distribution could be (5.6) with a U-type r_u. For example, $r_u = au^{\alpha-1} + bu^\beta$ with the exponents $0 < \alpha < 1$ and $\beta > 0$ and the constants $a > 0$, $b > 0$.

Distributions with power-law tails have become very popular in physical applications, i.e., $1 - F(x) \simeq cx^{-a}$, $x \gg 1$. In that case, r_u decreases like a/u, $u \gg 1$ for any $a > 0$; this acquires a paradoxical meaning in the context of characteristic earthquakes, namely, the larger the elapsed time, the safer the relevant fault segment. Unfortunately, we do not possess a large enough database to come to a definite conclusion about the behavior of the tail of F. For this reason, a physical interpretation of r_u may happen to be meaningful for inferences about the tail of $F(x)$.

The optimal alert times for a U-shaped hazard function r_u make up two intervals: $(0, u_1)$ and (u_2, ∞). The first alert interval is a response to the clustering of large events, and the second is in agreement with the quasi-periodicity of large events. The second interval is called the period of positive aging in engineering practice to describe endurance failures of materials.

The appearance of an infinite alert interval in the optimal strategy is not as obvious a fact. Consider an example, a distribution involving a well-expressed peak in the density F' (examples are presented in the section that follows). The neighborhood of the peak defines the most probable or typical values of τ. One feels, therefore, that an alert should be called in this neighborhood. The rule (5.7) shows that such a choice is not invariably the optimal one. Empirical forecasting also shows a tendency to reduce the duration of alerts. The first examples of infinite alerts were in Fedotov's prediction for Kamchatka [FSB+77] and in Prozorov's method of long-range aftershocks [PR72].

220   G.M. Molchan

5.2.4 Stability of the Minimax Strategy

Consider three types of interevent time distributions traditionally popular in applications in general and in the prediction of characteristic earthquakes in particular:

The lognormal distribution for which τ has the following representation: $\tau = m \cdot \exp(a\xi - a^2/2)$ where ξ is the standard Gaussian variable and $m = E\tau$. In this model, $r_u \simeq a^{-2} \ln u/u$, $u \gg 1$.

The gamma distribution with density $F'(x) = cx^{\alpha-1} \exp(-x\alpha/m)$, $\alpha > 0$; here, r_u is asymptotically constant, $r_u \simeq \alpha/m$, $u \gg 1$; $m = E\tau$.

The Weibull distribution for which $r_u = cu^{\alpha-1}$, $\alpha > 0$.

The behavior of the hazard function r_u, $u > 0$, for these models is presented schematically in Fig. 5.2. It depends on the parameter $I = \sigma/m$ well known in statistics as the coefficient of variation. Here, $m = E\tau$ is the mean and σ^2 the variance of τ. $I = 0$ for a periodic sequence of large events, and $I = 1$ for a purely random or Poisson sequence for which $F(x) = 1 - \exp(\lambda x)$. Hence, the intermediate values $I \in (0,1)$ can be treated as a measure of quasi-periodicity for large earthquakes. Data for segments of the San Andreas fault [Nis89] show that the coefficient of variation I is in the range from 0.25 to 0.6. The hazard function r_u for Gamma and Weibull distributions varies monotonically as u is increasing. Consequently, the optimal alert set $A = \{u : r_u > r^*\}$ on the semiaxis $u > 0$ consists of a single interval. When $I < 1$, r_u is increasing; $A = (k_\gamma m, \infty)$ where k_γ is a dimensionless threshold depending on the loss function γ, and r_u is decreasing when $I > 1$, so $A = (0, \bar{k}_\gamma m)$ is bounded. When $I < 1$, alerts are called in accordance with the quasi-periodicity of large events and are not bounded from above, whereas alerts for $I > 1$ incorporate possible clustering of events, hence are called immediately after large events.

It is easy to see that if w_1 and w_2 are independent and obey the Weibull distribution with parameters $0 < \alpha_1 < 1$ and $\alpha_2 > 1$, respectively, then $\min(w_1, w_2)$ corresponds to the U-shaped hazard function $r_u = c_1 u^{\alpha_1-1} + c_2 u^{\alpha_2-1}$ discussed above.

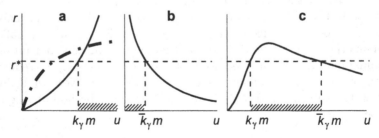

Fig. 5.2. A sketch of hazard function r_u for three types of interevent time distribution F: (a) Weibull and Gamma, with the coefficient of variation $I < 1$; (b) the same for $I > 1$; (c) lognormal distribution. *Dashed intervals* show the alert set A where $r_u > r^*$

The alert set $A = \{u : r_u > r^*\}$ for the lognormal distribution is merely formally different from the preceding models. Here, the alert set $A = (k_\gamma m, \bar{k}_\gamma m)$ is bounded away from zero and ∞. However, the threshold $\bar{k}$ is large enough for $I < 0.6$ (see below), and this allows replacing A a semi-infinite interval with very slightly different errors (n, τ).

In the above three examples of F, the hazard function has entirely different asymptotics for large u and $I < 1$: $O((\ln u)/u)$, $O(1)$, and $O(u^\beta)$, where $\beta = \alpha - 1 > 0$ for the lognormal, Gamma, and Weibull distributions, respectively. In other words, r_u is very sensitive to the choice of the F model, and statistical data do not tell us which of the tails of F is the right one. One asks how stable the γ-optimal strategy is under these conditions. We will examine the issue for the minimax strategy with $\gamma = \max(n, \tau)$.

The thresholds k_γ of the mimimax strategy for the three models of F considered here are shown in Fig. 5.3 a and b. They practically coincide for Gamma and Weibull when $I < 1$, differing by 0.01 or 1.5% at most. Practically coincident with these is the lower threshold k_γ for a lognormal distribution for $I < 0.6$; the deviation is $|\Delta k_\gamma| < 0.03$. The upper threshold $\bar{k}_\gamma$ in the same range $I < 0.6$ is large, $\bar{k}_\gamma > 3.3$, and makes little contribution to prediction errors. Therefore, it is natural in the lognormal model to consider a slightly simpler γ-optimal strategy involving only a lower alert threshold $\tilde{k}_\gamma$. (The value of $\tilde{k}_\gamma$ is again found from the condition $n = \tau$.) The threshold

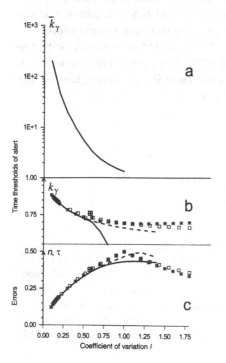

Fig. 5.3. Parameters of the minimax strategy for four models of interevent time distribution F: Weibull (*squares*), Gamma (*asterisks*), lognormal (*solid line* for the case of two alert thresholds and (*dashed line* for the case of a single one), and uniform (crosses inside squares, two points with $I = 0.58$). The following variables are plotted: (**a**) threshold $\bar{k}_\gamma$ for lognormal distribution versus the coefficient of variation I; (**b**) dimensionless alert threshold k_γ versus I; (**c**) prediction errors, $n = \tau$, versus I

is consistent with k_γ for Gamma and Weibull distributions in the interval $J < 1$: $|\Delta k_\gamma| < 0.03$.

For the practically interesting range $I = 0.25$–0.6, the thresholds k_γ and $\tilde{k}_\gamma$ are generally very stable: 0.75 ± 0.05. Note that the alert set $A = ((3 - \sqrt{5})m, \infty) \simeq (0.76m, \infty)$ is optimal for the rectangular distribution, $F(u) = u/(2m)$, $0 < u < 2m$, with $I = 3^{-1/2} \simeq 0.58$, whereas $(3 - \sqrt{5})m$ is the golden section(!) of the interval $(0, 2m)$. These results point to a stable structure of the minimax prediction strategy.

The second type of stability is relevant for prediction results, i.e., for the values of the loss $\gamma = \max(n, \tau)$; see Fig. 5.3 c. The values of γ for optimal strategies involving a single alert threshold differ within 0.02 or 3%. As a result, we can set up a *very simple prediction strategy for the case $I \in (0.25, 0.6)$: given a large event, wait three-quarters of the recurrence time, $m = E\tau$, and then call an alert to be canceled only by the next large event. This guarantees that the errors $n \simeq \tau \leq 0.35$*, provided of course that the characteristic earthquake concept is valid. We note for comparison that the well-known CN and M8 prediction algorithms provide the mean error $(n + \tau)/2 \simeq 0.4$ in a more complex situation. This space–time prediction involves events with cutoff magnitudes $M \geq 6.4$ [RN99] and $M \geq 7.5$ [KRKH99]. The situation with the $M \geq 8$ prediction is much better, $(n + \tau)/2 \simeq 0.2$ [KRKH99]. The fact that the estimates of $n + \tau$ are comparable for simple and complex prediction algorithms leaves room for the hope that a space–time prediction with the error level $(n + \tau)/2$ about 0.3–0.4, where τ is the space–time alert rate can be attained using comparatively simple tools.

Figure 5.4 presents the error diagram Γ for the three F models with the coefficient of variation $I = 0.25$, 0.5, and 0.75. The curves are consistent in the area $n > \tau$. This indicates that the prediction, its structure, and results are similar for a broad class of loss functions.

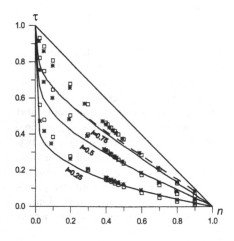

Fig. 5.4. Error diagrams Γ for three distributions $F(x)$: Weibull (*squares*), Gamma (*asterisks*), lognormal (*solid line* for the case of two alert thresholds and *dashed line* for the case of a single one), and for three values of the coefficient of variation I : $0.25, 0.5$, and 0.75

5.2.5   Prediction on the San Andreas Fault

Nishenko and Buland [NB87] analyzed 53 intervals τ_i in 14 regions with high seismicity. Out of three types of distributions $F(x)$, normal, Weibull, and log-normal, preference was given to the third type. The Working Group on California Earthquake Probabilities [WGCE88] applied the lognormal model to predict the largest earthquakes on the San Andreas fault. Table 5.1 contains data on eight of the fault segments. The WGCEP forecast was constructed in a more prosaic way than described above: the conditional probability of a large event during the future $\Delta = 30$ years was computed for each segment irrespective of the recurrence time:

$$R_\Delta = \mathrm{Prob}\left\{n(t_0, t_0 + \Delta) > 0 \mid \tau \geq u\right\} = \frac{F(u + \Delta) - F(u)}{1 - F(u)},$$

where t_0 is the year 1988 and u is the time since the last event on the segment. It follows from Table 5.1 that $R_\Delta > 0.9$ for the Parkfield (P) region and $R_\Delta < 0.4$ for the other segments. The circumstance was one of the arguments in favor of the Parkfield experiment. Concentrated observation in the region was to record the precursory process of a large earthquake.

Table 5.1. Conditional probability R_Δ of major earthquakes on segments of the San Andreas fault, 1988–2018[a,b]

Fault segment	L	M	t_0	m	R_Δ	I	t_1	k	$\bar{k}$	$n = \tau$
San Francisco Peninsula	90	7	1906	196	0.38	0.42	2032	0.75	8.8	0.30
Santa Cruz Mountains	35	6.5	1906	136	0.44	0.43	2007	0.74	8.4	0.30
Parkfield	30	6	1966	21	0.9	0.24	1983	0.81	37	0.19
Cholame	55	7	1857	159	0.3	0.53	1972	0.72	4.8	0.34
Carrizo	145	8	1857	296	0.1	0.31	2082	0.76	13	0.28
Mojave	100	7.5	1857	162	0.3	0.41	1979	0.75	9	0.29
San Bernardino Mountains	100	7.5	1812	198	0.2	0.60	1951	0.7	3.4	0.36
Coachella Valley	100	7.5	1680	256	0.4	0.30	1880	0.78	19	0.25

[a] L, length (km); M, expected magnitude; t_0, date of most recent event; m, expected recurrence time (yr); R_Δ, test; I, coefficient of variation; t_1, start of minimax alert; k and $\bar{k}$ are dimensionless time alert thresholds; $n = \tau$, minimax errors

[b] The data (L, M, t_0, m, R_Δ) are from [WGCEP]

The use of R_Δ as a test for comparison among the segments will pronounce the P region hazardous practically at any time. $R_\Delta < 0.45$ for all segments except Parkfield for any elapsed time u. At the same time, $R_\Delta > 0.9$ for segment P with any $0 < u < 800$ years. Moreover, the R_Δ value in Parkfield can reach the level of 0.5 only in the impossible case of quiescence for 4000 yr. Otherwise, the R_Δ-test causes a steady alert in P, but no alerts in other segments.

To resolve this paradox, let us measure the time on each segment in its own units of recurrence time m. Then we shall have identical recurrence times $E\tau = 1$ but different forecast intervals Δ/m. We have a large value $\Delta/m = 1.5$ for segment P and small ones for the other segments: $\Delta/m = 0.1 \div 0.2$. Calculation of R_Δ involves increments of F at intervals of Δ/m. Since the means have been made equal to one and $I < 1$, the increment will be small for $\Delta/m = o(1)$ and large enough for $\Delta/m > 1$. The conclusion is only slightly affected by the times elapsed.

Let us apply the minimax strategy to the segments. The alert set on the u axis is determined by the set $A = (km, \bar{k}m)$. From Table 5.1, it follows that $\bar{k}m$ varies within the range 700–5000 yr. When $\bar{k}$ has been replaced with ∞, that will affect the errors of the minimax strategy by 0.001 at most. The main conclusion for the minimax strategy is as follows: If the parameters of the F distribution are accurate, then five of eight fault segments should be in a state of alert by 1988, in particular, the region P since 1983 and the Coacella Valley (CV) since 1880(!), although according to R_Δ, an alert is impossible in CV. However, the higher of the mean number of predicted events $\Delta m^{-1}(1 - n)$ for the time period considered, 1988–2018, is for region P.

As a matter of fact, the conclusion is a neater form of argumentation in favor of the Parkfield experiment. The situation becomes less obvious, when it is assumed that the loss function has the form $\gamma = \alpha m^{-1} n + \beta L\tau$, where L is the length of the segment concerned, $m = E\tau$, and α and β are independent of the fault. Thresholds l_1 and l_2 then exist such that the optimal loss γ on segment P is the least when $\beta/\alpha < l_1$ and is the greatest among all the segments considered when $\beta/\alpha > l_2$ [Mol90].

5.3   Prediction with Multiphase Alerts

The prediction model considered above is sufficiently general and yields a simple optimal strategy. It clearly divides the domain of activity into two parts: one is the province of geophysics (estimation of the hazard function); the other is related to economics (for example, estimation of the loss ratio β/α.) In the case of the linear loss function, the process $(n(t), J(t))$ can even be nonstationary. The rejection of stationarity leads to a dependence of the threshold r^* on time. We will see below that the resulting prediction is optimal under linear losses per unit time both on the interval δ (local optimality) and on the entire time axis (global optimality).

The simplest model considered is suitable for many types of practical forecasts that involve only two alert states, i.e., where an alert is declared or called off. However, a real alert must be multiphase as a rule, because different degrees of hazard require different systems of protective measures [Sad86]. Hence we modify the prediction model by introducing multiphase alerts and generalized linear losses.

Let us assume that an observer can select any alert from a given set of alerts $(A_0, A_1, ..., A_m)$ using the information $J(t)$. The cancellation of an alert is included in the set; it is A_0. We also assume that every alert A_i requires the cost β_i per unit time and that α_i is the loss prevented per successful prediction. In particular, $\alpha_0 = \beta_0 = 0$ for A_0.

We assume that any change in alerts leads to loss c_{ij}, $0 \leq c_{ij} < \infty$. The case $c_{ij} = \infty$ means that the change $A_i \to A_j$ is forbidden. For example, the population can be evacuated only after overcoming transportation problems. Whereas some of the protective measures require an ordering of corresponding alert types, other protective measures can be carried out in parallel. A set of such parallel measures is considered a single measure in our model.

Nonzero c_{ij} values result in stability of alert sequences because they prevent fast alternation of alerts. However, the introduction of c_{ij} complicates the problem; locally optimal decisions are not globally optimal in this case.

Denote by z_t the losses associated with the decision $\pi(t)$. Let us consider the total losses associated with the prediction strategy on the semiaxis $t > 0$ relative to the initial moment $t = 0$ with time factor ρ:

$$Z_\pi = \sum_{k \geq 0} z_{k\delta} \exp(-\rho \cdot k\delta) = \sum_{k \geq 0} z_{k\delta} \theta^k , \tag{5.8}$$

where $\theta = \exp(-\rho\delta)$. Z_π is called discounted loss in the theory of optimal control (see [How71], [Ros70]). In practical problems, the factor ρ can stand for the efficiency of capital investments. Mathematically, the introduction of ρ allows us to consider the problem on a finite interval of order $1/\rho$, escaping difficulties due to boundary effects when stationary prediction methods are studied.

The loss function now is the mean total discounted loss, i.e., the prediction goal is the minimization of

$$S = EZ_\pi . \tag{5.9}$$

Consider the case in which the change of alert types does not lead to additional losses, i.e., $c_{ij} = 0$ over all i and j.

Statement 2. Let $c_{ij} = 0$, $i, j = 1, 2, ..., m$, and let $\pi(t)$ depend only on the stationary information sequence $J(t)$. Then the optimal strategy is such that

$$\pi(t) = A_{j^*} , \tag{5.10}$$

where the subscript j^* realizes the minimum

$$\min_j [\beta_j - \alpha_j r(t)] = S[r(t)] \tag{5.11}$$

for the current value of the hazard function $r(t)$.

Remark 1. The function $S[r]$ in equation (5.11) is the convex polygonal envelope of the system of straight lines $y = \beta_j - \alpha_j r$ (Fig. 5.5a). Let $P_0(0, 0)$,

$P_1(r_1, y_1), ..., P_k(r_k, y_k)$ be the vertices of the polygon $S(r)$ that are ordered in r, $0 < r_1 < ... < r_k < r_{k+1} = \infty$, and let $j(n)$ be the number of the straight lines $y = \beta_j - \alpha_j r$ with the pair of vertices P_n and P_{n+1}. The prediction strategy (5.10), (5.11) means that there exist $k \leq m$ hazard levels $r(t) : \{r_i\}$ such that the alert with $j^* = j(n)$ is always declared in the interval $r(t) \in (r_n, r_{n+1})$ (see Fig. 5.5b). A number of alerts $\{A_i\}$ can be cost-ineffective; such is the alert A_3 in Fig. 5.5a.

Remark 2. The quantity (5.11) defines the minimum conditional mean loss per unit time in the interval $(t, t+\delta)$ under the given information $J(t)$. Hence the strategy (5.10), (5.11) is optimal at once globally and locally. It does not depend on the time factor ρ and is the generalization of the prediction strategy for two-phase alerts with the linear loss function discussed above.

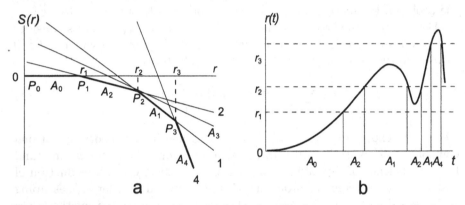

Fig. 5.5. Optimal strategy with multiphase alert for the case $c_{ij} = 0$. (a) Optimal mean loss per unit time $s(r)$ as a function of hazard level r; $s(r)$ is the envelope of straight lines $\beta_i - \alpha_i r$ indexed by the type of alert A_i; $\{P_\alpha\}$ are vertices of the envelope; $\{r_\alpha\}$ are hazard levels for the change of alerts $\{A_i\}$; here, the alert A_3 is not cost-effective. (b) Hazard function $r(t)$ and optimal alert $A_{j(t)}$ as functions of time under the conditions of (a)

To study the general case we introduce the following notion. We say that the process $\{\delta N(t), J(t)\}$ has an M property if the information sequence is a Markov process, i.e.,

$$\text{Prob}\{J(t + \delta) = v | J(t) = u; J(s), \forall s < t\} =$$
$$\text{Prob}\{J(t + \delta) = v | J(t) = u\} = P_{uv} , \tag{5.12}$$

and if

$$E\{\delta N(t) | J(s), s \leq t\} = E\{\delta N(t) | J(t)\} = r(t)\delta . \tag{5.13}$$

Conditions (5.12) and (5.13) hold when information $J(t)$ contains all past data on the predictors up to the moment t and all prehistory of the process

$dN(\cdot)$. In other words, the past $\{J(s), \delta N(s), s < t\}$ is measurable relative to $J(t)$. If the sequence $\delta N(\cdot)$ and a physical process $x(t)$ used to predict $\delta N(t)$ have a finite memory, i.e., a finite correlation interval Δ, then (5.12) and (5.13) are true for the information sequence $J(t) = \{\delta N(s), x(s), t - \Delta < s < t\}$; its dimensionality is less compared with the case of infinite memory.

Adding new requirements to the description of the process $\{n(t), J(t)\}$, we may extend the class of decisions; $\pi(t)$ can depend on $J(t)$ and past decisions $\pi(s), s < t$. This dependence may be stochastic.

Statement 3. Assume that $\{\delta N(t), J(t)\}$ is stationary, has the M property, and that decisions $\pi(t)$ depend on $J(t)$ and $\{\pi(s), s < t\}$. Then

(a) The optimal strategy minimizing (5.9) exists and can be chosen to be stationary, i.e., decisions $\pi(t)$ depend only on t in terms of the current state of information $J(t) = u$ and the current state of alert $\pi(t - \delta) = A_i$.

(b) Minimal mean loss (5.9), $S^*(u, i)$, under initial conditions $J(0) = u$ and $\pi(-\delta) = i$, is the solution of

$$S(u, i) = \min_j [c_{i,j} + \beta_j \delta - r_u \delta \alpha_j + \theta \sum_v P_{uv} S(v, j)], \qquad (5.14)$$

where $r_u = r(t)$ under the condition $J(t) = u$. We assume for simplicity that the set of states $J(t)$ is countable.

(c) If $\theta \in (0, 1)$ or $\rho > 0$, then equation (5.14) has a unique solution. This solution can be found by an iterative procedure

$$S^{(0)}(u, i) = 0, \qquad S^{(n+1)} = T_\theta S^{(n)}, \qquad (5.15)$$

where T_θ is the operator defined by the right-hand side of (5.14) with the domain of functions $f(u, i)$. The error of the nth iteration is

$$|S^{(n)} - S^*| < \theta^n (1 - \theta)^{-1} L,$$

where

$$L = \max_{ij : c_{ij} < \infty} |c_{ij} + b_j \delta + \alpha_j|.$$

d) Under conditions $J(t) = u$ and $\pi(t - \delta) = A_j$, the optimal decision is $\pi(t) = A_{j^*}$, where the subscript $j^* = j(u, i)$ minimizes the right-hand side of (5.14).

Remark 1. Equation (5.14) is of the Bellman type in the theory of optimal control ([How71], [Ros70]). The specific feature of our case is that the control parameter j enters the loss function rather than the transition matrix $[P_{uv}]$.

Remark 2. The recurrence (5.15) leads to the set of functions $S_k = S^{(N-k)}$, $k = 0, ..., n$, which are optimal mean discounted losses in the intervals $(k\delta, N\delta)$, $k = 0, 1, ..., N - 1$. The sequence of subscripts $j_k^*(u, i)$ minimizing (5.14) with

$S = S_k$ defines the sequence of optimal decisions in intervals $(k\delta, (k+1)\delta)$ under information states $\{J(t), \pi(t - \delta)\}$, $t = k\delta$.

The algorithm described is also suitable for optimizing of total losses in the interval $(0, N\delta)$ when $\theta = 1$, i.e., without the time factor ρ. Unfortunately, the optimal prediction strategy for a finite time interval is nonstationary when $[c_{ij}] \neq 0$.

Let us consider two examples.

Renewal process. Consider the prediction of characteristic earthquakes again. Suppose that interevent intervals are independent and have the distribution $F(x)$; the information is the time $J(t) = u$ that elapsed from the last event. This model satisfies conditions (5.12) and (5.13). The hazard function is defined by (5.5), and the transition matrix P_{uv} is such that only two transitions from state u are possible, one to $u + \delta$ with the probability $1 - r_u \delta$ (no events) and the other to zero with the probability $r_u \delta$ (an earthquake has occurred). Therefore, equation (5.14) takes the form

$$S(u, i) = \min_j \left\{ c_{ij} + \beta_j \delta - \alpha_j r_u \delta + \theta \cdot \left[S(u + \delta, j)(1 - r_u \delta) + S(0, j) r_u \delta \right] \right\} .$$

Cyclic Poisson process. To describe a sequence of catastrophic events, Vere-Jones [Ver78] used the model of a Poisson process with periodic rate, $\lambda(t) = \lambda(t + T)$. Clearly, the information takes the form $J(t) = t(\text{mod} T)$. Therefore, conditions (5.12) and (5.13) are true. Though this model is nonstationary, Statement 3 still holds. Equation (5.14) takes the form

$$S(u, i) = \min_j [c_{ij} + \beta_j \delta - \alpha_j \lambda(u)\delta + \theta S(u + \delta, j)] .$$

We also add the obvious condition of periodicity $S(u, i) = S(u + T, i)$ and $' = N\delta$.

Despite the simplicity of these examples, the optimal prediction cannot be obtained in an explicit form if $[c_{ij}] \neq 0$, even in the case of two-phase alerts. The case $[c_{ij}] \neq 0$ involves the hazard function r_u, as well as the matrix of transitional probabilities for information states $J(t)$ in successive time intervals. The practical estimation of this matrix P_{uv} is complicated, and the problem has not yet been formulated. Difficulties in estimating $[P_{uv}]$ depend on the type of information sequence $J(t)$ and on detailing the phase of its states. The results of the previous section show that it is sufficient to use an information phase space of low dimension.

Optimization of mean loss rate. The limiting case of the minimization problem (5.8), (5.9), as $\rho \to 0$ $(\theta \to 1)$, stands for the situation in which the loss function takes the form of total expected losses per unit time, i.e.,

$$\gamma_\pi = \lim_{n \to \infty} \inf E \frac{z_1 + \ldots + z_n}{n\delta} .$$

In the case of a two-phase alert,

$$\gamma = \alpha \lambda n + \beta \tau + c\nu , \qquad (5.16)$$

where ν is the number of transitions from the state $\pi = 0$ (nonalert) to $\pi = 1$ (alert) per unit time. The cost of such a transition is determined by the parameter c. An interpretation of γ has been given above for $c = 0$. An analysis of (5.16) can be found in [MK92] and the general case in [Mol92].

5.4 Statistical Problems

5.4.1 The Performance of Prediction Algorithms

Intermediate-term prediction techniques recently developed actually address the theoretical problem of whether earthquakes are predictable. Therefore, the techniques mostly reduce to the simplest two-phase alert, as characterized by the errors (n, τ). The positive answer to that question will be found in the proof that the error diagram Γ is significantly different from the straight line $n + \tau = 1$. The diagram can be estimated by the lower bound of the convex hull of points $(n, \tau)_A$ relevant to different prediction algorithms $\{A\}$ based on the same data set, the same prediction domain, and the magnitude range of large events.

The above idea can be used for comparisons among algorithms. Most algorithms involve internal parameters that are subsequently held fixed in an arbitrary manner. By varying the essential parameters θ of an algorithm A, one gets an error set $(n, \tau)_\theta$. Considering again the lower bound of its convex hull, one arrives at the error curve Γ_A representing the predictive power of the algorithm based on the data set chosen. Suppose that the curves Γ_A for two algorithms (Fig. 5.6) intersect at an intermediate point (the end points are always the same). Let Γ_{A_1} and Γ_{A_2} have a common tangent of slope $-p$ (Fig. 5.6). When the goal is a linear loss of the $\gamma = an + b\tau$ type, then it follows from Fig. 5.6 that A_1 is to be preferred when $a/b > p$ and A_2 otherwise $(a/b < p)$.

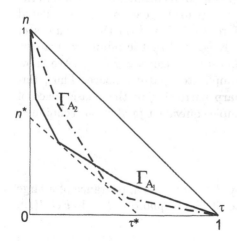

Fig. 5.6. Comparison of algorithms by error diagrams Γ_A. The *solid line* and *dash-dotted* line are error diagrams for two algorithms. The line (n^*, τ^*) is the common tangent for these diagrams

5.4.2   Estimation of (n, τ)

Statistical estimators of (n, τ) are generally unstable owing to the short history of a forward forecast. When errors (n, τ) refer to a time–space forecast, the value of τ measures the relative space–time occupied by alerts. Unjustified extension of the space by adding aseismic areas can make τ as small as one likes. One way out of this difficulty is to collect estimates of (n, τ) for prediction algorithms having a common prediction space and a common magnitude range of large events. Anomalies in the (n, τ) estimates become evident in the (n, τ) diagram.

Empirical estimates of the prediction errors, $(\hat{n}, \hat{\tau})$, admit theoretical analysis of the simplest case of predicting a renewal process, to be discussed in Sects. 5.2.4 and 5.3. Molchan [Mol90] proved asymptotic normality for the estimates $(\hat{n}, \hat{\tau})$ on large time intervals and found asymptotic expressions for the mean and covariance matrix of the $(\hat{n}, \hat{\tau})$ vector. These results were used to forecast large events in a synthetic earthquake catalog as follows. A catalog was generated from a block model of lithosphere dynamics [GLR90]; it contained 92 large events with $M \geq 6.4$ for a period of 720 years. The frequency of large events was made consistent with the Southern California seismisity rate. A minimax strategy was used for prediction, as described in 5.2.4 for the characteristic earthquake. The forecast was based on the empirical interevent time distribution. The normalized estimates $\hat{n}, \hat{\tau}$ based on three nonintersecting time intervals of 240 years each are

$$(\hat{n} - E\hat{n})/\sigma(\hat{n}) = -0.94, \ 0.55, \ 0.92 \ ,$$

$$(\hat{\tau} - E\hat{\tau})/\sigma(\hat{\tau}) = -0.51, \ 1.34, \ 0.16 \ ,$$

where $\sigma^2(\xi)$ is the variance of ξ. These deviations of $\hat{n}, \hat{\tau}$ from the theory for the renewal process model are well within Gaussian error limits. This is an argument in favor of the weak interdependence of recurrence intervals τ_i in the block model. [GLR90] provide forecast results for large events in a simulated model using the CN algorithm. The forecast is based on the dynamics of low magnitude seismicity which is totally ignored by the minimax strategy. Nevertheless, the forecast results are almost identical: $n \simeq \tau = 0.4$. We know that the CN algorithm uses a very complicated pattern space. Thus, this example indicates possibilities in a sharp narrowing of the phase space of a dynamical system for constructing simple equivalent prediction techniques.

5.5   Estimation of $r(t)$

The amount of information provided by $J(t)$ as to the appearance of a large event in the interval δt is given by the Shannon quantity $I = \ln PG[J(t)]$, where

$$PG[J(t)] = \text{Prob} \left\{ \delta N(t) = 1 \mid J(t) \right\} / \text{Prob} \left\{ \delta N(t) = 1 \right\} = r(t)/\lambda \ .$$

Aki [Aki81] has called $PG(J)$ *the probability gain* and proposed the following prediction program based on a combination of precursors:
- choose simple, weakly correlated, and sufficiently informative precursors $A = \{A_1, ..., A_k, ...\}$,
- estimate the quantity $PG(A_k)$ for each,
- find $PG[A(t)]$ for the whole set of precursors $A(t) = (A_{j_1}, ...A_{j_k})$ observed up to the time t from the relation

$$PG[A(t)] = q_t PG(A_{j_1}) \cdots PG(A_{j_k}) . \tag{5.17}$$

The factor q_t is related to precursors that have not been observed by time t. The above program was implemented for the Caucasus region [SCZ$^+$91]. Unfortunately, this method of estimating $r(t)$ seems oversimplified. The equality (5.17) means that the precursors $A_1, ..., A_n$ are conditionally independent with respect to the event $\{\delta N(t) = 1\}$. That is impossible physically, when the $\{A_k\}$ are really precursors, even though independent. Consider a formal example: suppose that ξ_1 and ξ_2 are independent random variables and a large event occurs, when $|\xi_1 + \xi_2 - 1| < \varepsilon$. Hence $\xi_1 + \xi_2 \simeq 1$ in the conditional situation.

Vere–Jones [Ver78], Utsu [Uts77], and Aki [Aki81] considered estimating of $r(t)$ the foremost task of a prediction specialist. This is true only in part (see Statement 3). The estimate of $r(t)$ in the entire range of values is unstable. This has been demonstrated by using the simple problem of a characteristic earthquake prediction. Stability is possible in an "academic" forecast involving two alert states (Sect. 5.2.4) because exact knowledge of $r(t)$ is needed in the neighborhood of a fixed level. The number of such levels in a realistic situation has to be greater than two or even infinitely many (Statements 2 and 3). For this reason, special importance will be attached to the maximum possible reduction of dimensionality for the prediction functionals. The best solution to that problem today is provided by the $M8$ algorithm.

An important example of estimating $r(\cdot)$ is statistical modeling of earthquake catalogs dating back to Hawkes [Haw71] and Kagan [Kag73]. We are speaking of the so-called self-exciting model in which events $x =$(time t, magnitude M, location g) divide into "main" and "offspring." The main events make a Poisson process with the rate $\lambda_0(x)$; once occurring, any event x_i will generate offspring events that form a Poisson process with the rate $\lambda(x|x_i)$ (note that $\int \lambda(x|x_i)\, dx \leq \rho < 1$). Reproduction of any event (a main or offspring) occurs once and independently of one another. Consequently, the probability of finding an event on the interval dx is $r(x)\, dx$ with

$$r(x) = \lambda_0(x) + \sum_{t(x_i) < t(x)} \lambda(x|x_i) , \tag{5.18}$$

where $t(x)$ is the time coordinate of the point x.

The offspring events with a common main event as progenitor x_0 play the part of aftershocks of x_0, and offspring events for x can be considered as direct aftershocks of x. The properties of aftershocks are well known: the Gutenberg–Richter law for the magnitudes and the Omori law for the time decay. The properties of direct aftershocks are not known, and this causes difficulties and ambiguities in the parameterization of $\lambda(x|x_i)$ (compare the solutions of this problem in [KK92] and [KJ91]).

Model (5.18) is essentially designed for short-term prediction of aftershocks [JK99]. However, Ogata [Oga88] and Khokhlov and Kossobokov [KK92] found that the hazard function $r(x)$ can be useful for predicting of large events. With this end in view, one should, in contradiction of Statement 1, call alerts using two thresholds: $r(x) < c$ and $r(x) > r^*$. The paradox lies in the fact that (5.18) responds to a seismicity increase by higher values of $r(x)$ and to quiescence by lower values. The successes achieved in the analysis of seismicity during the last decade have confirmed the predictive value of patterns like quiescence and seismicity increases. Thus, it appears that the successes in the prediction of large earthquakes using (5.18) with two thresholds will, paradoxically enough, provide a statistical argument against using the model for describing earthquake catalogs.

5.5.1 Comments

The above discussion is based on this author's work ([Mol90, Mol91, Mol92]) which arose from an attempt to understand empirical earthquake forecasting practice. Discussions with V.I. Keilis-Borok. A. Prozorov, and Y. Kagan during this research have been very fruitful.

The papers by De Maré [DeM80], Lindgren [Lin85], and Ellis [Ell85], which unfortunately became known too late to me, are directly related to Statement 1. G. Lindgren seems to have been the first to grasp the difference between the conventional forecast of a time series $x(t)$ and the forecast of level crossing time for $x(t)$ (forecast of a catastrophe). De Maré and Lindgren studied the problem of optimal prediction of "catastrophes" for the Gaussian stochastic process where alerts are instantaneous. Our error diagram shows such situations as points $(n, \tau = 0)$, which is typical of short-term prediction. This means that the Γ diagram can be supplemented with a similar curve in an infinitely small neighborhood of $\tau = 0$ where a different τ scale is used.

The study by Ellis is directly related to optimization of earthquake prediction; he found the structure of the optimal prediction for linear loss $\gamma = \alpha n + \beta \tau$. We investigated the general case of loss functions $\gamma(n, \tau)$ using the error diagram, which is very helpful practically. Consideration of the forecast problem involving any number of alert types is a new element for earthquake prediction theory as well.

5.6   Appendix

Proof of Statement 1. The prediction errors (n, τ) have the form

$$n_\pi = E[1 - \pi(t)]\delta N(t)/E\delta N(t) = E_1[1 - \pi(t)] , \qquad (A1)$$

$$\tau_\pi = E\pi(t) = E_0\pi(t) . \qquad (A2)$$

Here, E_i denotes averaging over the measures $\mathcal{P}_i$. If $P(\cdot)$ is the original measure, then $\mathcal{P}_0$ is a narrowing of P on the events of $J(t)$, and $\mathcal{P}_1$ is a conditional measure on $J(t)$ given a large event in the interval $\delta t = (t, t + \delta)$. Formally,

$$\mathcal{P}_1(\omega) = \text{Prob} \{\omega, \delta N(t) = 1\} / \text{Prob} \{\delta N(t) = 1\} ,$$

where ω is an event related to $J(t)$.

Relations (A1, A2) have the following interpretation. There is an observation $J(t)$ and two hypotheses of its distribution: $\mathcal{P}_0$ (hypothesis H_0) and $\mathcal{P}_1$ (hypothesis H_1). The decision rule $\pi = 1$ ($\pi = 0$) classifies $J(t)$ as belonging to $H_1(H_0)$ and gives rise to errors of two kinds: n_π and τ_π. If one of these errors is fixed, say $\tau < \tau_0$, and the other is to be minimized, then the optimal decision must have the Neyman-Pearson structure [Leh60], i.e., the form

$$\text{Prob}\{\pi_t(J(t)) = 1\} = \begin{cases} 1 & L_t > c \\ p & L_t = c \\ 0 & L_t < c, \end{cases} \qquad (A3)$$

where $c > 0$ and $p \in [0, 1]$ are deterministic constants and

$$L_t = \mathcal{P}_1(dJ)/\mathcal{P}_0(dJ) = \text{Prob} \{dN(t) = 1, dJ\} / [\text{Prob} \{\delta N(t) = 1\} \cdot \text{Prob}(dJ)]$$

is the likelihood ratio. By definition $\text{Prob} \{\delta N(t) = 1, dJ\} / \text{Prob}(dJ) = r(t)\delta$ where $r(t)$ is the hazard function and $\text{Prob} \{\delta N(t) = 1\} = \lambda\delta$. Therefore,

$$L_t = r(t)/\lambda . \qquad (A4)$$

The parameters (c, p) in (A3) are found from the requirements $\tau \leq \tau_0$ and $n = \min$. Remember that $P(r_t = \lambda c) = 0$ if $p = 0$.

Summing up, (A3) imposes the structure of the γ-optimal strategy for $\gamma = n/\theta(\tau_0 - \tau)$. Here, $\theta(x) = 1$ when $x > 0$ and zero when $x < 0$. The optimal errors (n^*, τ^*) are given by the intersection of Γ and the line $\tau = \tau_0$. Varying τ_0 in the range $0 < \tau_0 < 1$ will yield all optimal strategies described by the curve Γ. It follows that substitution of (A3) in (A1, A2) gives a parameterized representation of Γ. The use of (A4) then yields

$$n = \lambda^{-1}E\, r(t)[r(t) < \lambda c] + (1 - p)c \cdot \text{Prob}\{r(t) = \lambda c\} , \qquad (A5)$$

$$\tau = E\,[r(t) > \lambda c] + p \cdot \text{Prob}\{r(t) = \lambda c\} , \qquad (A6)$$

where $[C]$ is the logical function, $[C] = 1$ or 0. Since Γ is convex, it is differentiable everywhere, except at a countable set of points at most; one-sided derivatives defined everywhere. In virtue of (A5), (A6), are at points where Γ is smooth,

$$dn/d\tau = -c , \tag{A7}$$

provided $\text{Prob}\{r(t) = \lambda c\} = 0$. At critical points where $\text{Prob}\{r(t) = \lambda c\} \neq 0$, n and τ are linear functions of p; therefore, they correspond to linear segments of Γ with a slope of $-c$. The curve Γ is thus parameterized by its derivative (A7) and by one more parameter $p \in (0,1)$ on its linear segments. Hence, one gets a complete description of the γ-optimal strategy in terms of Γ. Let $Q^* = (n^*, \tau^*)$ be the γ-optimal errors. Then, $c = -dn/d\tau(Q^*)$ in (A3) where the right-side derivative has been taken for the sake of definiteness. If τ^* belongs to the linear segment $[\tau_0, \tau_1]$ of Γ, then $p = (\tau^* - \tau_0)/(\tau_1 - \tau_0)$.

Proof of Statement 2. Let $\pi(t) = A_j$; then the loss due to forecast δN in the interval $(t, t + \delta)$ is

$$Z_t = \beta_j \delta - \alpha_j \delta N(t) .$$

Here, α_j has the minus sign, because α_j denotes prevented loss. The decision $\pi(t)$ depends on $J(t)$, so the conditional mean is

$$E[Z_t | J(t)] = \sum_j (\beta_j \delta - \alpha_j r(t)\delta)[\pi(t) = A_j] \geq s(r(t))\delta , \tag{A8}$$

where $s(r) = \min_j (\beta_j - \alpha_j r)$. The equality in (A8) is attained at the decision $\pi^*(t)$ described in Statement 2. Averaging (A8) yields

$$E\{E[Z_t | J(t)]\} = EZ_t \geq Es[r(t)]\delta .$$

The inequality becomes an equality at the strategy $\pi^* = \{\pi^*(t)\}$ by virtue of the property of $\pi^*(t)$ mentioned above. The optimal decision $\pi^*(t)$ in the interval $(t, t + \delta)$ does not affect the loss in other intervals because it has been assumed that $\pi(t)$ depends only on $J(t)$. For this reason, the total mean weighted loss $\sum z_{t_k} \theta^k$, $t_k = k \cdot \delta$ on the semiaxis $t \geq 0$ or the same loss with equal weights, $\theta = 1$, on $0 < t < T_0$ is minimal, if each decision $\pi(t)$ is locally optimal.

Proof of Statement 3. Let $Z(t)$ be the discounted earthquake losses for the interval (t, ∞):

$$Z(t) = \sum_{k \geq 0} z_{t+k\delta} \theta^k = z_t + \theta Z(t + \delta) .$$

The loss z_t in the interval δt is the sum of the loss c_{ij} due to an alert transition $A_i \rightarrow A_j$ and the loss $\beta_j \delta$ for maintenance of the alert, minus prevented loss

α_j, if a large event has occurred in the interval δt. Consequently,

$$|z_t| < \max_{i,j:c_{ij}<\infty} |c_{ij} + \beta_j\delta + \alpha_j| = L \, , \qquad \text{(A9)}$$

so that

$$|Z(t)| < L(1-\theta)^{-1} \, . \qquad \text{(A10)}$$

The conditional mean loss $z(t)$ under the condition $J(t) = u$, $\pi(t-\delta) = i$ at decision $\pi(t) = j$ is

$$c_{ij} + \beta_j\delta - \alpha_j E\{\delta N(t)|J(t) = u, \, \pi(t) = j, \, \pi(t-\delta) = i\} \, .$$

The M property (5.12, 5.13) yields the following relation for conditional means:

$$E\{\delta N(t)|J(t)\} = E\{\delta N(t)|\mathcal{F}_t\} = E\{\delta N(t)|\mathcal{F}_t, \mathcal{W}_t\} \, , \qquad \text{(A11)}$$

where $\mathcal{F}_t = \{J(s), s \leq t\}$ is the past of the process $J(t)$, and $\mathcal{W}_t$ is that of the additional experiments $\{\mathcal{W}(s), s \leq t\}$ which have been involved in making the decisions $\{\pi(s), s \leq t\}$. From the inclusions

$$\{J(t)\} \subset \{J(t), \pi(t), \pi(t-\delta)\} \subset \{J(s), \mathcal{W}(s) \, s \leq t\}$$

combined with (A11), it follows that

$$E\{\delta N(t)|J(t)\} = E\{\delta N(t)|J(t), \pi(t), \pi(t-\delta)\} \, .$$

For this reason,

$$E\{z_t|J(t) = u, \, \pi(t-\delta) = i, \, \pi(t) = j\} = c_{ij} + \beta_j\delta - r_u \cdot \delta \cdot \alpha_j \, , \qquad \text{(A12)}$$

where r_u is the hazard function $r(t)$ in terms of the state of information $J(t) = u$ at time t. Similarly,

$$E\{Z(t+\delta)|J(t), \, \pi(t-\delta), \, \pi(t)\} =$$
$$E[E\{Z(t+\delta)|J(t+\delta), \, \pi(t), \, J(t), \, \pi(t-\delta)\}|J(t), \, \pi(t-\delta), \, \pi(t)] \, . \text{(A13)}$$

By virtue of the M property,

$$E\{Z(t+\delta)|\mathcal{F}_{t+\delta}, \mathcal{W}_{t+\delta}\} = E\{Z(t+\delta)|J(t+\delta), \, \pi(t)\} \, ,$$

so that (A13) can be continued:

$$\text{(A13)} = E[E\{Z(t+\delta)|J(t+\delta), \, \pi(t)\}|J(t), \, \pi(t-\delta), \, \pi(t)] \, . \qquad \text{(A14)}$$

Let

$$S_k(u, i) = E\{Z(k\delta)|\pi((k-1)\delta) = i, \, J(k\delta) = u\}$$

be the optimal mean discounted loss for the interval $(k\delta, \infty)$ with the initial conditions $J = u$, $\pi = i$. Taking into account (A9, A12–A14), one finds that S_k and S_{k+1} must be related through

$$S_k(u, i) = \min_j [c_{ij} + \beta_j \delta - r_u \cdot \delta \cdot \alpha_j + \theta \sum_v P_{uv} S_{k+1}(v, j)] . \quad \text{(A15)}$$

This can be seen as follows. The bracketed expression gives the mean discounted loss for the interval $(k\delta, \infty)$ under the initial conditions $J = u$, $\pi = i$ at decision $\pi_{k\delta} = j$. The loss on $(k\delta, \infty)$ will be minimal, if it is minimal on $((k + 1)\delta, \infty)$ under any initial conditions $J((k + 1)\delta) = v$, $\pi(k\delta) = j$, and if the decision $\pi k\delta = j^*$ makes (A15) the minimum at time $k\delta$. The above statement is the Bellman principle expressed by (A15).

Represent (A15) in the operator form: $S_k = T_\theta S_{k+1}$, and consider the space H of functions $f(u, i)$ on the set of states $J(t), \pi(t)$ with the norm $\|f\| = \sup |f| \le L(1 - \theta)^{-1}$. The constant L is given by (A9). The choice of the threshold $L(1 - \theta)^{-1}$ follows from (A10). In that case,

  (a) T_θ maps H onto itself;
  (b) T_θ is a contraction operator:

$$\|T_\theta f_1 - T_\theta f_2\| \le \theta \|f_1 - f_2\|, \qquad f_1 \in H .$$

The property (a) follows from (A15):

$$|T_\theta f| < L + \theta \|f\| \le L + \theta(1 - \theta)^{-1} = L(1 - \theta)^{-1} .$$

We take into account here that $r_n \delta = E(\delta N | J) < 1$. The property (b) is a corollary of the inequality

$$|\min_{1 \le i \le n} x_i - \min_{1 \le i \le n} y_i| \le \max_{1 \le i \le n} |x_i - y_i| .$$

The $n = 2$ case of this inequality is simple and yields the general case

$$|\min_i x_i - \min_i y_i| = |x_\alpha - y_\beta| =$$
$$|\min_{i=\alpha,\beta} x_i - \min_{i=\alpha,\beta} y_i| \le \max_{i=\alpha,\beta} |x_i - y_i| \le \max_i |x_i - y_i| .$$

The space H is closed if the set of states $J(t)$ is discrete. It follows from the general theory of contraction operators [Yos65] that T_θ has a single fixed point in H: $T_\theta S^* = S^*$, which can be found by iteration, $T_\theta^n S \to S^*$. Moreover,

$$|T_\theta^n S - S^*| < \theta^N (1 - \theta)^{-1} \|T_\theta S - S\| , \quad \text{(A16)}$$

where S is any initial approximation. Setting $S = 0$, one derives item (c) in Statement 3.

Let $S_\pi(u, i)$ be the mean discounted loss in $(0, \infty)$ resulting from the strategy π. We are going to show that $S_\pi(u, i) \ge S^*(u, i)$.

The demonstration is as follows. Suppose that π_n is part of the strategy π on $(n\delta, \infty)$. By virtue of (A16), one can choose $n = n(\varepsilon)$ to make

$$|T_\theta^n S - S^*| < \varepsilon .$$

From the definition of T_θ, it follows that $T_\theta^n S_{\pi_n} \leq S_\pi$; hence,

$$S^* - \varepsilon \leq T_\theta^n S_{\pi_n} \leq S_\pi .$$

Since i is arbitrary, $S^* \leq S_\pi$.

Let us consider the strategy π^* for which $\pi^*(t) = A_{j*}$, given $J(t) = u$, $\pi(t - \delta) = i$, where $j^* = j(u, i)$ is the index that makes

$$\psi(j) = c_{ij} + \beta_i \delta - r_u \cdot \delta \cdot \alpha_j + \theta \sum_v P_{uv} S^*(v, j)$$

the minimum. If the argument j is not defined uniquely, then it can be sampled from a probability distribution $p_{ui}(\cdot)$ defined on the set of these j^*. Suppose that the strategy π^* causes a mean loss $S(u, i)$ on $(0, \infty)$. Since the strategy is stationary,

$$S(u, i) \leq z(j^*|u, i) + \theta \sum_v P_{uv} S(v, j^*) ,$$

where $z(j^*|u, i)$ is the mean loss on δt (see A12 with $j = j^*$). This inequality becomes an equality for $S^*(u, i)$. Subtraction of the two yields

$$0 < S(u, i) - S^*(u, i) \leq \theta \sum_v P_{uv} \{S(v, j^*) - S^*(v, j^*)\} .$$

Hence, $|S - S^*| \leq \theta \|S - S^*\|$. This gives $S = S^*$. The items (a, b, d) in Statement 3 are proved.

6 Recognition of Earthquake-Prone Areas

A. Gorshkov, V. Kossobokov and A. Soloviev

6.1 Introduction

This chapter describes a systematic approach to resolving the problem where may large earthquakes occur at present. The approach is based on the pattern recognition technique developed for studies involving small samples and initially bypasses a number of low level approximations, like "large quakes occur in seismic regions, where tectonic movement is evident at recent ages."

As shown in Chap. 1 and numerous publications, the supporting medium of earthquakes is a hierarchical dynamic system of lithospheric blocks and their boundaries, faults. The driving forces arising from thermal and gravitational convection in the mantle of Earth govern movements in this medium. Some parts of the system are apparently active, and some are not. Each earthquake manifests a catastrophe, i.e., an abrupt change of system characteristics that results in sharp displacements accompanied by radiation of seismic waves. Some of the system catastrophes occur without earthquakes. A large earthquake involves and affects a large domain of the system and therefore cannot happen anytime and anywhere due to natural on-path obstacles emerging constantly in a process of movement. In the long run, the system self-adjusts and organizes its traffic at all levels of hierarchy from major tectonic plates and seismic belts to grains of rock and microcracks; it also develops specific patterns that accompany extremes in both space and time. A snapshot of this self-organization, which delivers a global map of earthquake epicenters, is very indicative and thought provoking. Enough to say that patterns of worldwide epicenter distribution allowed A. Wegener [Weg15] to propose the hypothesis of drifting continents long before possible mechanisms and driving forces were discovered. As already mentioned in the introduction to Chap. 4, spatial patterns of seismic distribution may help to detect temporal ones, for example, precursors of large catastrophic earthquakes. Naturally, tectonic movements leave traces, which accumulate on a timescale of tens of thousand years or longer and provide geographic, geologic, gravitational, and magnetic evidence of the intensity of driving forces, their directivity, and dating. This evidence, termless in a common sense, both apparent and masked, requires analysis and interpretation before it is used in favor of a conclusion about present day, specifically, seismic activity.

Can we do better than predicting large earthquakes in places, earthquake-prone areas, where they are already known? Can we recognize an earthquake-prone area in advance of a large quake striking it? The positive answer to these questions is supported by vast investigations initiated in the 1970s by I.M. Gelfand and V.I. Keilis-Borok [GGI+72b, GGI+72c, GGK+76] whose results were confirmed by large earthquakes that occurred after the publications.

In general, the recipe for resolving these questions is as follows: Taking a closer look at a region with multitudinous manifestations of seismic activity accumulated in specific structures and physical parameters, (i) make a judgment and choose potential earthquake-prone areas, (ii) check their association with large earthquakes, (iii) apply pattern recognition algorithms to determine an unambiguous classification of potential earthquake-prone areas into "dangerous" and "non dangerous", (iv) perform an exhaustive test of the classification obtained in a number of control tests on data available, and (v) wait until new data confirm or disprove the results.

Following the recipe, we first describe a general target set in terms of pattern recognition (Sect. 6.2) and the basics of formalized morphostructural zoning (Sect. 6.3) essential for an adequate choice of recognition objects. Then, we present a sample of specifics and some accumulated experience in Sect. 6.4 and finally conclude with a summary of seismic reality after publication confirming the methodology presented here.

6.2   Unraveling Earthquake-Prone Areas as a Pattern Recognition Problem

A zero approximation in predicting large catastrophic earthquakes is resolving a question where such events can happen. The question is highly important for knowledgeable seismic hazard and risk assessment. A simple answer that large earthquakes can happen only in places of a seismic region where smaller magnitude quakes were registered is, on one hand, supported by the Gutenberg–Richter law, on the other hand, contradicts seismological practice, which has many case histories of "surprises" in seismic zoning based exclusively on the seismic record (for an extreme example, see Sect. 4.4.5 describing the 1998 Balleny Sea earthquake [WWL98]).

Pattern recognition suggests another approach to resolving the question. Places already marked by large seismic events might have a similar portrayal that can be used to identify sites, which did not yet explicitly show up as earthquake-prone. Such a portrayal requires several assumptions. First, a potential earthquake-prone area must be associated with a natural object of recognition represented as a vector in a parameter space. Second, most of the reported earthquakes must associate with potential earthquake-prone areas, so that the hypothesis of association is acceptable with a high level of confidence. In this case one can try applying a pattern recognition algorithm to unravel a generalized portrayal of earthquake-prone areas by comparing objects associated and not associated with known large earthquakes.

Usually, strong earthquakes with magnitude $M \geq M_0$ associate with morphostructural nodes, specific structures that are formed about intersections of fault zones (see the next section of this chapter). This is true for the majority of strong earthquakes. The nodes are described as objects of recognition by the characteristics of topographical, geologic, geomorphological, and other geophysical data; such descriptions are vectors whose components are the values of these characteristics.

Thus, the pattern recognition problem consists of assigning the vectors to two classes: vectors **D** (historically stands for "Dangerous") representing nodes where earthquakes with $M \geq M_0$ can occur and vectors **N** (stands for "Not dangerous") describing nodes where only earthquakes with $M < M_0$ can happen. The classification is known for some objects of recognition on the basis of the seismic activity observed in the region. These objects form a training set of vectors that belong to known classes. The training set consists of vectors $\mathbf{D}_0$ and $\mathbf{N}_0$ that represent, respectively, the nodes where strong earthquakes occurred and the nodes that are far from the known epicenters of such earthquakes. A pattern recognition algorithm provides a classification of the vector space into **D** and **N**. Earthquake-prone areas were determined in this way in several seismic regions worldwide.

The important stage of a pattern recognition study consists of control tests necessary to ensure the reliability of the classification obtained, especially for small samples $\mathbf{D}_0$ and $\mathbf{N}_0$. The results of such tests usually illustrate, although do not prove, either reliability or non-randomness. The proof in a strict statistical sense can be achieved when a number of strong earthquakes confirms the results of pattern recognition classification. We compare and present the pattern recognition classification versus the strong earthquakes that occurred in the corresponding regions after publications.

Formulation of the problem and main stages of its investigation. Consider a selected magnitude cutoff M_0 that defines large earthquakes in the region under study. Roughly speaking, the problem of determining earthquake-prone areas aims at separating places of potential earthquakes into two parts, **D** where earthquakes with magnitude $M \geq M_0$ can happen and **N** where earthquakes with magnitude $M \geq M_0$ are impossible.

The first question arising in a strict formulation of the pattern recognition problem is how to select the region and magnitude cutoff M_0. The experience accumulated in [GGI+72a, GGZ+73, GGZ+74a, GGZ+74b, ZRS75, GGK+76, GZKK78, ZK80, CKO+80, GK81, Kos83, GS84, GZS84, CGG+85, WGG+86, GNRS87, GGK+87, GZRT91, BCF+92, BRC+94, GKPS00] suggests the following heuristic criteria.

- The number of large earthquakes with $M \geq M_0$ in the region should be at least 10–20.

- The circles centered at epicenters of reported earthquakes with $M \geq M_0$ that have radii about the size of their source should not cover all of the region (otherwise, the problem has a trivial solution where the whole region is **D**).
- The region has to be tectonically uniform in sense of the similarity of possible causes of earthquakes with $M \geq M_0$.

These criteria establish certain limitations on the size of the region and the threshold M_0. For instance, $M_0 = 5.0$–6.0 implies the linear size of a region of the order of hundreds kilometers, whereas for $M_0 = 7.0$–7.5 this size should be larger than a thousand kilometers. $M_0 = 8.0$ requires a region tens of thousands kilometers long. These limitations were met in practice, for example, in Italy, $M_0 = 6.0$ [CKO+80]; California, $M_0 = 6.5$ [GGK+76]; South America and Kamchatka, $M_0 = 7.75$ [GZS84], and the whole Circumpacific, $M_0 = 8.0$ [GZKK78]. The experience accumulated in a decade confirmed that pattern recognition methods might reliably distinguish earthquake-prone areas on different scales of lithospheric block hierarchy and in different seismic and tectonic environments [GGI+72a, GGZ+73, GGZ+74a, GGZ+74b, ZRS75, GGK+76, GZKK78, ZK80, CKO+80, GK81, Kos83, GS84, GZS84, CGG+85, WGG+86, GNRS87, GGK+87].

Following is the description of methods applicable to partitioning a finite set of objects **W** into two classes. When selecting the region and threshold magnitude M_0, it is necessary to define the objects of recognition $\mathbf{w} \in \mathbf{W}$.

Gelfand et al. [GGI+72b, GGI+72c] were the first who applied pattern recognition methods to determine earthquake-prone areas in the Pamirs and Tien Shan. Since then, several important improvements of such a determination have been developed, including a broader choice of natural objects for recognition. In general, one may consider three types of objects in a study of earthquake-prone areas: planar areas, segments of linear structures, and points.

In 1972, Gelfand et al. [GGI+72b, GGI+72c] used planar morphostructural nodes of the Pamirs and Tien Shan as candidates for earthquake-prone places. At that time, even a formal definition of this structure that permits reproducible identification did not exist and was the subject of further analysis by geomorphologists and mathematicians [AGG+77] (a brief definition of a morphostructural node is given in Sect. 6.2 below). However, because most fractured areas are characterized by multidirectional intensive tectonic movements, nodes essentially attract epicenters of large earthquakes. The fact that most earthquakes with $M \geq M_0$ in a region originate within nodes is a necessary precondition for using them as objects of recognition. Ranzman [Ran79] formulated the geomorphological basis that favors this precondition. Gvishiani and Soloviev [GS81] suggested a statistical method for testing it in practice, even when the boundaries of nodes are not defined precisely.

In planar nodes, pattern recognition algorithms classify each morphostructural node in the region either as a **D** node, which is prone to earthquakes with $M \geq M_0$, or as a **N** node, where strong earthquakes are not possible. Such a classification determines the area **D** as the union of all **D** nodes in the region and the area **N** as the union of all **N** nodes. The remaining territories of the region complementary to the nodes are not assumed to be dangerous (they are rejected with a certain level of confidence by preconditioning strong earthquake - node association).

This natural choice of objects entails a difficult problem of outlining the boundaries of morphostructural nodes. When the difficulty is overwhelming, one may try substituting the nodes with intersections of morphostructural lineaments as done in [GGZ+74b]. Tracing lineaments and their intersections is a much easier task for a geomorphologist (see Sect. 6.3) that essentially delivers similar (though less complete) information on the most fractured places of multidirectional intensive tectonic movements. That is why intersections of morphostructural lineaments were commonly used for determining of earthquake-prone areas [GGZ+74b, ZRS75, GGK+76, ZK80, CKO+80, GS84, GZS84, CGG+85, WGG+86, GNRS87, GGK+87]. The necessary precondition of using nodes as recognition objects is transferred to a hypothesis that epicenters of strong earthquakes originate near intersections of morphostructural lineaments [GGZ+74b]. This hypothesis is likely to be confirmed in a region if the following two conditions are valid: (1) the distance from all accurately determined epicenters of earthquakes with $M \geq M_0$ to the nearest intersection does not exceed a predefined distance ρ; (2) the area covered by circles of radius ρ centered at all intersections is a small part of the total area of the region. A statistical justification of the hypothesis can be obtained by using the following algorithm [GS81].

Let N be the number of earthquakes with $M \geq M_0$ in the region considered; $N(\rho)$ denotes the number of epicenters whose distance to the nearest intersection of lineaments is ρ or less, and the function $p(\rho) = N(\rho)/N$ is the empirical distribution of the distance to the nearest intersection. By randomizing a position of intersections, we may judge the validity of their association with strong earthquake epicenters. Each intersection is put at random in one of the uniformly distributed locations within the circle of radius R centered at the intersection. The radius R must be not too large, so that the area covered by the circles does not considerably exceed the area of the region. The empirical distribution of the distance from epicenters of strong earthquakes to the nearest intersection averaged over a large enough sample of random positions of intersections, $\overline{p(\rho)}$, characterizes the association at random. If $p(\rho)$ is much larger than $\overline{p(\rho)}$ for a sufficiently large range of ρ (the difference between $p(\rho)$ and $\overline{p(\rho)}$ is divided by the standard deviation of the random empirical distribution function [GS81]), then the validity of the intersection - epicenter association is confirmed, and the original intersections of morphostructural lineaments can be used as point objects in a pattern recognition study. Usually, point objects are considered with their circular vicinities of the same certain radius, which makes them analogous to the nodes. The circular vicinities may overlap, thus creating another complication for a classification that was overcome by designing a special algorithm [GGZ+74b].

Pattern recognition algorithms assign the vectors that describe intersections of lineaments to two classes: class $\mathbf{D}$ of intersections having vicinities prone to earthquakes with $M \geq M_0$ ($\mathbf{D}$ intersections) and class $\mathbf{N}$. The classification of vectors determines the preimage of area $\mathbf{D}$ as the union of all vicinities of $\mathbf{D}$ intersections. The area $\mathbf{N}$ is the complement of area $\mathbf{D}$ in the union of all vicinities of intersections. It is assumed that the remaining territories of the region complementary to all vicinities of intersections are not dangerous.

Usually, earthquakes are associated with segments of faults that they rupture. Therefore linear objects of recognition, like segments of active faults or fault zones, may seem most natural to many seismologists ([GGK+76] remains an excellent demonstration of how the problem is viewed differently). Pattern recognition algorithms divide linked linear objects into two classes that have about the same length (the length of a class is the total length of its linear objects): $\mathbf{D}$ segments capable of originating an earthquakes with $M \geq M_0$ and $\mathbf{N}$ segments that are not.

Segments of linear structures were used as objects for recognition of earthquake-prone areas in California [GGK+76], where the basic linear structure was the San-Andreas fault, in the whole linear structure of the Circumpacific seismic belt [GZKK78], and in the Western Alps [WGG+86].

The usage of pattern recognition algorithms with learning necessitates an a priori selection of the training set $\mathbf{W}_0$, which is the union of two subsets that do not overlap: the training set $\mathbf{D}_0$ from class $\mathbf{D}$ and the training set $\mathbf{N}_0$ from class $\mathbf{N}$. Such a selection of $\mathbf{W}_0 = (\mathbf{D}_0, \mathbf{N}_0)$ depends on the types of objects for recognition. In the case of planar objects, all of those, including known epicenters of earthquakes with $M \geq M_0$, form $\mathbf{D}_0$, whereas the subset $\mathbf{N}_0$ consists of all remaining objects from $\mathbf{W}$, $\mathbf{N}_0 = \mathbf{W} \setminus \mathbf{D}_0$, or those of such objects that do not contain known epicenters of earthquakes with $M \geq M_0 - \delta$ (where $\delta > 0$ is usually 0.5 or about this value). It is necessary to emphasize that $\mathbf{N}_0$ is not a "pure" training set in the sense that some of its members belong to class $\mathbf{D}$. In the first case, where $\mathbf{N}_0 = \mathbf{W} \setminus \mathbf{D}_0$, the problem consists of distinguishing samples that spoil the purity of $\mathbf{N}_0$. Such a fuzzy type of learning highlights a specific difficulty in locating possible earthquake-prone areas by pattern recognition techniques.

It is natural to require the condition $\mathbf{D}_0 \subseteq \mathbf{D}$, where $\mathbf{D}$ denotes the vectors classified as belonging to class $\mathbf{D}$. In other words, all places of strong earthquakes that are known should be recognized. When $\mathbf{D}_0$ contains many vectors a part of it can be excluded from the training set and reserved to verify the reliability of the decision rule obtained.

When recognition objects are points, the training set $\mathbf{D}_0$ is assembled from those that are situated at a distance not exceeding a certain fixed value r from the reported epicenters of earthquakes with $M \geq M_0$. The choice of r must satisfy the condition that the distance from most (practically all) of the well located epicenters of strong earthquakes in the region to the nearest recognition point is less than r. Naturally r scales with M_0. For instance, Zhidkov and Kosobokov [ZK80] used r = 40 km for $M_0 = 6.5$ in the eastern part of Central Asia; Gvishiani and Soloviev [GS84] chose $r = 100$ km for $M_0 = 7.75$ on the Pacific coast of South America . The training set $\mathbf{N}_0$ consists of either all remaining points or those of them that are at a distance r_1 $(r_1 \geq r)$ or longer from the epicenters of earthquakes with $M \geq M_0 - \delta$ $(\delta > 0)$. In this case, the training set $\mathbf{N}_0$ can also contain points that are potentially from class $\mathbf{D}$.

There is a certain difficulty when recognition objects are points; one epicenter can be attributed to several objects if its distance to each of them is r or less. In such a case, the training set $\mathbf{D}_0$ may have some objects from class $\mathbf{N}$. A special algorithm called CLUSTERS has been designed to resolve the problem [GGK$^+$76]. In case of ambiguity, the condition that $\mathbf{D}_0 \subseteq \mathbf{D}$ is changed for another natural one: each epicenter of an earthquake with $M \geq M_0$ has a point $\mathbf{D}$ object at a distance r or less.

When recognition objects are linear segments, the training set $\mathbf{D}_0$ assembles those containing a projection of an epicenter of a strong earthquake. The training set $\mathbf{N}_0$ is either $\mathbf{N}_0 = \mathbf{W} \setminus \mathbf{D}_0$ or contains segments from $\mathbf{W}$ that are not neighbors of $\mathbf{D}_0$. Another way to form $\mathbf{N}_0$ is to exclude those segments from $\mathbf{W} \setminus \mathbf{D}_0$ that contain a projection of an epicenter of an earthquake with $M \geq M_0 - \delta$ (where $\delta > 0$ is a parameter). As a rule, there is a unique projection of an epicenter that does not create ambiguity in selecting $\mathbf{D}_0$; therefore, it is natural to require that $\mathbf{D}_0 \subseteq \mathbf{D}$.

Pattern recognition algorithms operate with vectors of characteristics representing natural recognition objects. As far as earthquake-prone areas are considered, it appears natural to use the characteristics describing, either directly or indirectly, the intensity of recent tectonic activity at the locality of each object. The accumulated experience in recognizing earthquake-prone areas has established the following characteristics as typical:

- a multitude of characteristics describing topography;
- characteristics describing the complexity of geomorphological and neotectonic network of structures;
- characteristics describing gravitational field anomalies.

In principle, all available information related directly or indirectly to the level of seismic activity can be used to characterize objects. The only necessary precondition for a characteristic is the availability of uniform measurements across the entire region under consideration. After measuring selected characteristics for all objects $\mathbf{w} \in \mathbf{W}$, they are converted to vectors

$\mathbf{w}^i = \{w_1^i, w_2^i, ..., w_m^i, \}$, $i = 1, 2, ..., n$, where m is the total number of characteristics, n is the total number of objects in $\mathbf{W}$, and w_k^i is the value of the kth characteristic measured for the ith object. Some pattern recognition algorithms (specifically, Cora-3 and CLUSTERS) work in a binary vector space. Their application requires a transformation of vectors that describe natural recognition objects into binary ones. A specific transformation, so-called coding of characteristics is described in Sect. 6.2.1.

Given the training sets of vectors $\mathbf{W}_0 = (\mathbf{D}_0, \mathbf{N}_0)$, a pattern recognition algorithm determines a classification

$$\mathbf{W} = \mathbf{D} \cup \mathbf{N}, \tag{6.1}$$

where $\mathbf{D}$ and $\mathbf{N}$ are vectors of classes $\mathbf{D}$ and $\mathbf{N}$, respectively. Sect. 6.2.1 describes pattern recognition algorithms used in practice to determine earthquake-prone areas.

As pointed above, the resulting classification should satisfy certain conditions, like $\mathbf{D}_0 \subseteq \mathbf{D}$ for area objects. To avoid a trivial solution when all places considered belong to $\mathbf{D}$, the following condition is usually introduced:

$$|\mathbf{D}| \leq \beta |\mathbf{W}|, \tag{6.2}$$

where $|\mathbf{D}|$ and $|\mathbf{W}|$ stand for the numbers of objects in sets $\mathbf{D}$ and $\mathbf{W}$, respectively, and β, $0 < \beta < 1$, is a real constant, which sets an a priori upper bound for the fraction of $\mathbf{D}$ vectors in $\mathbf{W}$. The value and justification of β must result from an expert evaluation of geological, seismological, and other available information on the region.

The quality and reliability of a classification can be verified in control tests. If successful, such test favors the classification that actually divides the region into earthquake-prone areas and areas where earthquakes with $M \geq M_0$ are not likely. Usually, pattern recognition of earthquake-prone areas involves a small sample of natural objects whose size does not allow reserving a control set for verification. Nevertheless, certain verification of the classification can be achieved by a comprehensive analysis of the result and additional information that was not used initially, of which the most important are data on epicenters of large earthquakes, e.g., noninstrumental, either historical or paleoseismological. Section 6.2.2 describes some the typical control tests and other methods for evaluating the reliability of small sample statistics.

Classifications that are not satisfactory and have no meaningful interpretation are usually not reported. To get a satisfactory classification, a researcher can perform several cycles of trial and error through the following stages of recognition:

- definition of the region under study and the magnitude cutoff attributed to earthquake-prone areas;
- choice of the natural recognition objects;

- selection of the training set $\mathbf{W}_0 = \mathbf{D}_0 \cup \mathbf{N}_0$;
- description of objects as vectors;
- classification of vector space $\mathbf{W} = \mathbf{D} \cup \mathbf{N}$ by a pattern recognition algorithm;
- evaluation of the reliability of classification from control tests;
- interpretation of the classification $\mathbf{W} = \mathbf{D} \cup \mathbf{N}$ as a division of the region into earthquake-prone and other areas;
- generalization of geological and geomorphological interpretation of classification and the rules used to obtain it.

6.2.1 Parameterization of Recognition Patterns

After measuring the values of the characteristics, the natural objects are represented as vectors, $\mathbf{W} = \{\boldsymbol{w}^i\}$, $i = 1, 2, ..., n$, where n is the total number of vectors. In a vector $\mathbf{w}^i = (w_1^i, w_2^i, ..., w_m^i)$, m is the number of characteristics, and w_k^i is the value of the kth characteristic.

The pattern recognition algorithms that were commonly used in [GGI+72a, GGZ+73, GGZ+74a, GGZ+74b, ZRS75, GGK+76, GZKK78, ZK80, CKO+80, GK81, Kos83, GS84, GZS84, CGG+85, WGG+86, GNRS87, GGK+87, GZRT91, BCF+92, BRC+94, GKPS00] operate in a binary vector space. Some characteristics initially have only two values and need no additional discretization. In general, the value of such a characteristic answers the question whether the object has a certain property. The affirmative answer sets the value to one. If the answer is no, the value equals zero.

Some characteristics are integers, for example, the number of lineaments in the node. Such characteristics are initially discrete and need no additional discretization but a binary coding that unites integers into intervals. Similar integration of values into certain intervals is necessary to represent other numerical (real-valued) characteristics, like the distance to the nearest morphostructural intersection, in a binary form.

Let us consider a numerical characteristic w_k and assume its values $\{w_k^i\}$ sorted in ascending order from the minimum $w_k^1 = x_k^0$ to the maximum $w_k^n = x_k^f$ ($x_k^f > x_k^0$). A discretization of w_k that inherits the natural ordering of a numerical characteristic requires establishing the thresholds $x_k^0 < x_k^1 < x_k^2 < \cdots < x_k^{r_k} < x_k^f$ (where $r_k + 1$ is the number of discretization intervals); thus, the discrete image of w_k^i equals j if $x_k^{j-1} < w_k^i \leq x_k^i$ (where $x_k^{r_k+1} = x_k^f$). The thresholds can be introduced either arbitrarily on the basis of various natural considerations or objectively using the empirical distribution of w_k. In the latter case, the thresholds are determined to equalize the numbers of vectors of the same discrete value j, $j = 1, 2, ..., r_k + 1$. This type of discretization is rather objective and automatic and requires from an expert just the number of discrete values $r_k + 1$. For example, two such values imply distinguishing "small" and "large" through the median of the characteristic.

A discretization can be used in the form of a nonlinear normalization factor when applied independently to the vectors representing locations in different seismic and/or tectonic environments. For instance, Caputo et al. [CKO$^+$80] used objective discretization to normalize evidently different altitudes in Italy and Sicily.

The classification potency of w_k can be estimated in the following way. Let us compute the fractions of w_k^i that take on the same discrete value j in $\mathbf{D}_0$ and $\mathbf{N}_0$, p_k^j and q_k^j, respectively, and set

$$R_k = \max_{1 \leq j \leq r_k+1} |p_k^j - q_k^j| \, .$$

This value characterizes deviation of the empirical distribution functions of the two classes from each other. The larger R_k, the better w_k. In the practice of small sample analysis, the characteristic w_k is usually assumed satisfactory if $R_k > 15\%$. If it is not so, the discretization of w_k is not promising and can be excluded from further analysis (unless its combination with other characteristics is satisfactory). Note that in many cases, objective discretization with $r_k > 2$ can obscure the analysis due to a small number of samples per value. For the same reason, it is advisable to group some values of naturally discrete characteristics when $r_k > 2$.

A binary vector space delivers a natural presentation of a questionnaire, one for a positive and zero for a negative answer to a certain question. The coding transforms discrete vectors $\mathbf{w}$ into their binary representations $\mathbf{s}$. The number of components s_k describing a characteristic w_k depends on the number of its discrete values w_k, as well as on the type of coding. Let us consider two types of coding for discrete components w_k.

The *I type* coding (I stands for "impulse") converts $r_k + 1$ discrete values of a component w_k into $r_k + 1$ components of a binary vector. Let us denote these components by δ_1, δ_2, ..., δ_{r_k+1}. The value δ_j signifies the answer to the question whether $w_k = j$. When the answer is positive, $\delta_1 = 0, \dots, \delta_{j-1} - 0$, $\delta_j = 1$, $\delta_{j+1} - 0$, ..., $\delta_{r_k+1} = 0$.

The *S type* coding (S stands for "stair") presumes natural ordering of the characteristic and converts $r_k + 1$ discrete values of w_k into r_k components of a binary vector. Let us denote these components by δ_1, δ_2, ..., δ_{r_k}. The value of δ_j signifies the answer to the question whether $wk \leq j$. When the answer is positive, $\delta_1 = 0, \dots, \delta_{j-1} = 0$, $\delta_j = 1$, $\delta_{j+1} = 1, \dots, \delta_{r_k} = 1$.

Let us illustrate the difference between I and S type coding. If $r_k = 2$ and $x_k^0 < w_k^i \leq x_k^1$, I coding results in 100, whereas S coding yields 11. When $x_k^1 < w_k^i \leq x_k^2$, the coding results in 010 (I type) and 01 (S type). Finally, when $x_k^2 < w_k^i \leq x_k^3$, the coding produces 001 (I type) and 00 (S type).

The discretization and coding transform recognition objects $\mathbf{W} = \{\mathbf{w}_i\}$, $i = 1, 2, \dots, n$, into binary vectors in a binary vector space of a certain dimension l. Thus, the initial problem is converted to classification of a finite set of l-dimensional binary vectors. It allows applying general pattern recognition methods and validation techniques.

To simplify formulas, we shall use the same notation as before discretization and coding. Note that the l-dimensional binary vectors representing different natural objects can coincide. Pattern recognition algorithms are applied to classify binary vectors of the set $\mathbf{W} = \{\mathbf{w}_i\}$ using training sets $\mathbf{W}_0 = \mathbf{D}_0 \cup \mathbf{N}_0$, where $i = 1, 2, ..., n$; $\mathbf{w}^i = (w^i_1, w^i_2, ..., w^i_j, ..., w^i_l)$, and each w^i_j takes on values zero or one.

A basic set of classification algorithms with learning This section describes some basic pattern recognition algorithms widely used for locating earthquake-prone areas: algorithm CORA-3 [Bon67, GGK$^+$76], its modification CLUSTERS [GGZ$^+$74b, GGK$^+$76], and algorithms HAMMING and HAMMING-1 [GK81, Kos83].

The simplest of them is HAMMING. It is applied in two stages. In the first stage, the algorithm computes the sums of w^i_j over $\mathbf{D}_0$ and $\mathbf{N}_0$, $q_D(j)$ and $q_N(j)$, respectively, for each component $\mathbf{w}_j$, $j = 1, 2, ..., l$. Then, it computes the values

$$\alpha_D(j) = \frac{q_D(j) + 1}{n_1 + 2}, \qquad \alpha_N(j) = \frac{q_N(j) + 1}{n_2 + 2},$$

where $n_1 = |\mathbf{D}_0|$ and $n_2 = |\mathbf{N}_0|$.

The binary vector $\mathcal{K} = (\kappa_1, \kappa_2, ..., \kappa_l)$ called the kernel of class $\mathbf{D}$ is determined as follows:

$$\kappa_j = \begin{cases} 1 & \text{if} \alpha_D(j) \geq \alpha_N(j), \\ 0 & \text{if} \alpha_D(j) < \alpha_N(j). \end{cases}$$

The calculation of kernel $\mathbf{K}$, whose components are more typical of $\mathbf{D}_0$ than of $\mathbf{N}_0$, completes the first stage.

In the second stage, the algorithm computes Hamming's distance

$$\rho_i = \sum_{j=1}^{l} |w^i_j - \kappa_j| \tag{6.3}$$

for each vector $\mathbf{w}_i \in \mathbf{W}$. The classification is defined as follows:

$$\mathbf{D} = \{\mathbf{w}_i | \, \rho_i \leq \mathcal{R}\}, \qquad \mathbf{N} = \{\mathbf{w}_i | \, \rho > \mathcal{R}\},$$

where R is a parameter of the algorithm.

The condition $\mathbf{D}_0 \subseteq \mathbf{D}$ is satisfied if R equals or exceeds the maximum value of ρ_i over $\mathbf{D}_0$,

$$\mathcal{R} \geq \max_{\mathbf{D}_0} \rho_i. \tag{6.4}$$

If the number $|\mathbf{D}|$ satisfies (6.2), the classification $\mathbf{W} = \mathbf{D} \cup \mathbf{N}$ obtained is a satisfactory solution of the pattern recognition problem.

HAMMING-1 provides more perfection compared to HAMMING. It operates with the generalized Hamming's distance

$$\rho_i = \sum_{j=1}^{l} \xi_j |w_j^i - \kappa_j| \,, \tag{6.5}$$

instead of (6.3). Weights $\xi_j > 0$ are parameters of the algorithm. They can be assigned arbitrarily or computed from objective considerations that reduce the danger of self-deception; for example,

$$\xi_j = \frac{|\alpha_D(j) - \alpha_N(j)|}{\max_j |\alpha_D(j) - \alpha_N(j)|} \,, \tag{6.6}$$

where the maximum is taken over all components.

Since 1972 [GGI+72a], the algorithm CORA-3 has been widely used for solving geophysical problems including determination of earthquake-prone areas. This powerful pattern recognition algorithm also operates in two stages. In the learning stage, the algorithm determines characteristic traits for classes $\mathbf{D}$ and $\mathbf{N}$ using vectors from $\mathbf{D}_0$ and $\mathbf{N}_0$.

A matrix $\mathbf{A}$,

$$\mathbf{A} = \left\| \begin{matrix} j_1 & j_2 & j_3 \\ \delta_1 & \delta_2 & \delta_3 \end{matrix} \right\| \,,$$

denotes a trait, where j_1, j_2, and j_3, $1 \leq j_1 \leq j_2 \leq j_3 \leq l$, are the numbers of binary vector components and δ_1, δ_2, and δ_3 are their binary values. We say that a binary vector $\mathbf{w}_i = (w_1^i, w_2^i, ..., w_l^i)$ has the trait $\mathbf{A}$ if $w_{j_1}^i = \delta_1$, $w_{j_2}^i = \delta_2$, $w_{j_3}^i = \delta_3$.

Let $\mathbf{W}' \subseteq \mathbf{W}$. Denote the number of vectors $\mathbf{w}_i \in \mathbf{W}'$ that have trait $\mathbf{A}$ by $K(\mathbf{W}', \mathbf{A})$. The algorithm has four free parameters k_1, $\overline{k}_1$, k_2, and $\overline{k}_2$, which are nonnegative integers used to define characteristic traits of the two classes. Trait $\mathbf{A}$ is a characteristic trait of class $\mathbf{D}$ if $K(\mathbf{D}_0, \mathbf{A}) \geq k_1$ and $K(\mathbf{N}_0, \mathbf{A}) \leq \overline{k}_1$. A trait $\mathbf{A}$ is a characteristic trait of class $\mathbf{N}$ if $K(\mathbf{N}_0, \mathbf{A}) \geq k_2$ and $K(\mathbf{D}_0, \mathbf{A}) \leq \overline{k}_2$.

The number of characteristic traits can be rather large. Some of them occur on the same vectors from training sets. The algorithm distinguishes such cases and does not include all characteristic traits in the final list. Specifically, denote by $\Omega(\mathbf{A})$ a subset of $\mathbf{W}$ such that $\mathbf{w}_i \in \Omega(\mathbf{A})$ has trait $\mathbf{A}$. Let $\mathbf{A}_1$ and $\mathbf{A}_2$ be two characteristic traits of class $\mathbf{D}$. Trait $\mathbf{A}_1$ is *weaker* than trait $\mathbf{A}_2$ (or $\mathbf{A}_2$ is *stronger* than $\mathbf{A}_1$) if

$$\Omega(\mathbf{A}_1) \cap \mathbf{D}_0 \subset \Omega(\mathbf{A}_2) \cap \mathbf{D}_0 \quad \text{and} [\Omega(\mathbf{A}_2) \cap \mathbf{D}_0] \setminus [\Omega(\mathbf{A}_1) \cap \mathbf{D}_0] \neq \oslash \,.$$

This condition means that all vectors from $\mathbf{D}_0$ that have $\mathbf{A}_1$ also possess $\mathbf{A}_2$; at the same time there is at least one vector from $\mathbf{D}_0$, which has trait $\mathbf{A}_2$, and does not have $\mathbf{A}_1$.

A similar inclusion is valid for characteristic traits of class $\mathbf{N}$: Let $\mathbf{A}_1$ and $\mathbf{A}_2$ be two characteristic traits of class $\mathbf{N}$. Then trait $\mathbf{A}_1$ is weaker than trait $\mathbf{A}_2$ if

$$\Omega(\mathbf{A}_1) \cap \mathbf{N}_0 \subset \Omega(\mathbf{A}_2) \cap \mathbf{N}_0 \quad \text{and} (\Omega[\mathbf{A}_2] \cap \mathbf{N}_0] \setminus [\Omega(\mathbf{A}_1) \cap \mathbf{N}_0] \neq \emptyset .$$

Two characteristic traits $\mathbf{A}_1$ and $\mathbf{A}_2$ of class $\mathbf{D}$ are called *equivalent* if they are found on the same vectors of $\mathbf{D}_0$, i.e., $\Omega(\mathbf{A}_1 \cap \mathbf{D}_0) = \Omega(\mathbf{A}_2 \cap \mathbf{D}_0)$. Similarly, characteristic traits $\mathbf{A}_1$ and $\mathbf{A}_2$ of class $\mathbf{N}$ are called equivalent if $\Omega(\mathbf{A}_1 \cap \mathbf{N}_0) = \Omega(\mathbf{A}_2 \cap \mathbf{N}_0)$.

The algorithm excludes from the list of characteristic traits those that are weaker or equivalent to a selected trait.

Thus, the learning stage results in the final list of p_D and p_N characteristic traits of classes $\mathbf{D}$ and $\mathbf{N}$, respectively. Any member of this list does not have weaker or equivalent members.

In the second stage, the algorithm performs voting and classification using the final list of characteristic traits. For each vector $\mathbf{w}_i \in \mathbf{W}$, it calculates the number n_D^i of its characteristic traits from class $\mathbf{D}$, the number n_N^i of those from class N, and the difference $\Delta_i = n_D^i - n_N^i$ called voting.

The classification is defined as follows:

$$\mathbf{D} = \{\mathbf{w}^i \,|\, \Delta_i \geq \Delta\} \quad \text{and} \quad \mathbf{N} = \{\mathbf{w}^i \,|\, \Delta_i < \Delta\} ,$$

where Δ is a parameter of the algorithm. The condition $\mathbf{D}_0 \subseteq \mathbf{D}$ is satisfied if Δ does not exceed the minimum value of Δ_1 on $\mathbf{D}_0$,

$$\Delta \leq \min_{\mathbf{D}_0} \Delta_i . \tag{6.7}$$

If the number $|\mathbf{D}|$ satisfies (6.2), the classification $\mathbf{W} = \mathbf{D} \cup \mathbf{N}$ obtained is a satisfactory solution of the pattern recognition problem.

Algorithm CLUSTERS is a modification of CORA-3 [GGZ+74a,GGZ+74b, GGK+76]. It is useful when there is ambiguity in attributing a phenomenon to a recognition object, e.g., in the classification of earthquake-prone areas when natural recognition objects are points and $\mathbf{D}_0$ consists of those located at a distance r or less from epicenters of reported earthquakes with $M \geq M_0$. Let K be the number of the epicenters. Each epicenter defines a subclass $\mathbf{D}_0^k$ consisting of points located at distances that do not exceed r ($k = 1, 2, ..., K$). A subclass may consist of one or many natural objects. The training set of class $\mathbf{D}$ includes all $\mathbf{D}_0^k$ ($k = 1, 2, ..., K$). The condition $\mathbf{D}_0 \subseteq \mathbf{D}$ must be modified. Specifically, CLUSTERS uses K subsets $\overline{\mathbf{D}}_0 = \mathbf{D}_0^1 \cup \mathbf{D}_0^2 \cup ... \cup \mathbf{D}_0^K$ instead of $\mathbf{D}_0$; each must have a vector that belongs to $\mathbf{D}$, although some vectors from $\overline{\mathbf{D}}_0$ may belong to $\mathbf{N}$. Note also that the same vector may enter $\overline{\mathbf{D}}_0$ several times.

The learning stage of CLUSTERS differs from that of CORA-3 in the following. First, by definition, a subclass has a trait if it contains a vector with this trait. Some vectors without this trait may belong to this subclass. A trait $\mathbf{A}$ is a characteristic trait of class $\mathbf{D}$ if

$$K^S(\overline{\mathbf{D}}_0, \mathbf{A}) \geq k_1 \quad \text{and} \quad K(\mathbf{N}_0, \mathbf{A}) \leq \overline{k}_1,$$

where $K^S(\overline{\mathbf{D}}_0, \mathbf{A})$ is the number of subclasses that have trait $\mathbf{A}$.

Second, trait $\mathbf{A}$ is a characteristic trait of class $\mathbf{N}$ if $K(\mathbf{N}_0, \mathbf{A}) \geq k_2$ and $K(\overline{\mathbf{D}}_0, \mathbf{A}) \leq \overline{k}_2$. Note that $\overline{\mathbf{D}}_0$ is used here as a set of vectors.

Third, the definition of weaker and equivalent traits for characteristic traits of class $\mathbf{D}$ is different. A characteristic trait $\mathbf{A}_1$ of class $\mathbf{D}$ is weaker than a characteristic trait $\mathbf{A}_2$ of the same class if any subclass that has the trait $\mathbf{A}_1$ also has $\mathbf{A}_2$ and there is at least one subclass, which has trait $\mathbf{A}_2$ but does not have trait $\mathbf{A}_1$. Traits $\mathbf{A}_1$ and $\mathbf{A}_2$ are equivalent if they are found in the same subclasses.

CLUSTERS forms the sets of characteristic traits of classes $\mathbf{D}$ and $\mathbf{N}$ like CORA-3. The stage of voting and classification is the same as in CORA-3. However, Δ is determined from the condition

$$\Delta \leq \min_{1 \leq K} \max_{\mathbf{w}^i \in \mathbf{D}_0^k} \Delta_i, \tag{6.8}$$

where, first, the maximum is taken over the vectors belonging to $\mathbf{D}_0^k$, and, second, the minimum is searched over all K subclasses. The classification of a vector $\mathbf{w}_i \in \mathbf{W}$ is made, as CORA-3. If the number of vectors in $\mathbf{D}$ satisfies (6.2), then the classification may be considered satisfactory.

6.2.2   Evaluating the Reliability of Recognition

The best way of establishing the validity of recognition results is, of course, by comparing them with the locations of earthquake epicenters with $M \geq M_0$ that occurred after relevant publications. Section 6.5 below returns to this point. There are indirect approaches that can help to evaluate the reliability of recognition. They include control tests, statistical analysis of established classifications, the transfer of decision rule for application in similar problems, and other techniques. Here we present a basic set of control tests relevant to the determination of earthquake-prone areas.

"Seismic Future" (SF test). Assume that a long time has elapsed since the classification $\mathbf{W} = \mathbf{D} \cup \mathbf{N}$ was established, so that coming back to pattern recognition, one would find the training sets $\mathbf{D}_0 = \mathbf{D}$ and $\mathbf{N}_0 = \mathbf{N}$. What classification will result in this case? Note that we can try to answer the question right after the classification is obtained. This is the essence of the Seismic Future test; its results are claimed successful if the total change of classification is less than 5% of the total number of recognition objects. We consider a success of a SF test a necessary precondition for claiming that the resultant classification is a solution of the problem.

Note that CORA-3 allows easy repetition of the initial classification if one takes $\bar{k}_1 = \bar{k}_2 = 0$ and sufficiently small k_1 and k_2. Therefore, it is advisable to perform a SF test with nonzero thresholds $\bar{k}_1$ and $\bar{k}_2$ (for example, $\bar{k}_1 = \bar{k}_2 = 1$ or the same as in the initial classification) and large enough k_1 and k_2.

"Testing Stability" (ST tests). An ST seeks a classification $\mathbf{W} = \mathbf{D} \cup \mathbf{N}$ close to the initial one produced by various subsets $\mathbf{D}'_0 \subseteq \mathbf{D}_0$, $\mathbf{N}'_0 \subseteq \mathbf{N}_0$ used for learning. The test is considered successful if the initial classification is stable to substantial changes of the training sets. In pattern recognition of the earthquake-prone areas, the result is usually claimed successful when no more than 10% of vectors change their membership compared to the initial classification.

The choice of $\mathbf{D}'_0$ and $\mathbf{N}'_0$ used as training sets in an ST test can be rather different. For instance, the region at hand can be divided into two parts, and the subsets $\mathbf{D}'_0$ and $\mathbf{N}'_0$ then formed from $\mathbf{D}$ and $\mathbf{N}$ vectors with preimages belong to one part. The other way of selecting $\mathbf{D}'_0$ and $\mathbf{N}'_0$ can be based on voting results in the initial classification: If HAMMING (or HAMMING-1) is used, the vectors $\mathbf{w}_i \in \mathbf{D}$ close to the kernel K can be assigned to $\mathbf{D}'_0$, and those far from it are assigned to $\mathbf{N}'_0$. When CORA-3 (or CLUSTERS) is used, the vectors $\mathbf{w}^i \in \mathbf{D}$ with larger values of Δ_i can be assigned to $\mathbf{D}'_0$, whereas those with small Δ_i form $\mathbf{N}'_0$.

In a special ST test named *Seismic History* (SH test), several last strong earthquakes are excluded from the initial training set. Then, several successive classifications are obtained on adding these earthquakes to training sets one at a time. The test illustrates how these classifications converge to the initial one.

Successful results of different ST tests are appealing indirect arguments favoring the validity of an established classification. At the same time, a success in a single test with an arbitrary choice of $\mathbf{D}'_0$ and $\mathbf{N}'_0$ is by no means a proof of reliability.

"Sliding Control" (SC test). This test is very similar to the well-known "jackknife" method used to decide whether or not the classification of a given vector from the training set changes in response to being excluded from learning. Training sets in an SC test are formed by excluding pairs of vectors, each consisting of one vector $\mathbf{w}^i \in \mathbf{D}_0$ and one vector $\mathbf{w}^k \in \mathbf{N}_0$, $i = 1, 2, ..., n_1$, $k = 1, 2, ..., n_2$, where $n_1 = |\mathbf{D}_0|$ and $n_2 = |\mathbf{N}_0|$. An SC test does not consider all pairs of vectors, but runs through their limited subset; a pair from this subset consists of a vector from a smaller training set (usually $\mathbf{D}_0$) and a vector from the other one (usually $\mathbf{N}_0$). Note that entire subclasses are excluded in turn from $\mathbf{D}_0$ in the case of CLUSTERS.

An SC test is usually claimed successful if less than 20% of excluded vectors are erroneously classified.

"Voting by Equivalent Traits" (VET test). This test is applied only to classifications obtained by CORA-3 (or CLUSTERS). In this case (see Sect. 6.2.1), the result of classification depends on the choice of traits picked up from equivalence groups. The VET test aims at evaluating the classification stability under such a choice.

Denote by u^i_{Dj} and u^i_{Nj} the number of characteristic $\mathbf{D}$ and $\mathbf{N}$ traits of the vector $\mathbf{w}_i$, respectively, equivalent. For each vector $\mathbf{w}_i \in \mathbf{W}$, define the weighted ensemble votes of the equivalent characteristic traits in favor of $\mathbf{D}$ and $\mathbf{N}$, respectively, by formulas

$$u^i_D = \sum_{j=1}^{p_D} \frac{u^i_{Dj}}{p^i_D}\,, \qquad u^i_N = \sum_{j=1}^{p_N} \frac{u^i_{Nj}}{p^i_N}\,, \tag{6.9}$$

where p^i_D and p^i_N are the total numbers of traits equivalent to trait j and p_D and p_N are the total numbers of characteristic traits from the final list that belong to classes $\mathbf{D}$ and $\mathbf{N}$, respectively. The difference $u^i_D - u^i_N$ substitutes in the VET test for the voting $\Delta_i = n^i_D - n^i_N$ in the second stage of the algorithm to determine the classification by the condition, $\mathbf{w}_i \in \mathbf{D}$ if $u^i_D - u^i_n \geq \Delta$ and $\mathbf{w}_i \in \mathbf{N}$, otherwise.

The results of the VET test are claimed successful if it is possible to find Δ such that the total change in classification is less than 5% of the total number of recognition objects. We consider a success of the VET test a necessary precondition for claiming the validity of the resultant classification obtained with CORA-3 or CLUSTERS.

Randomization of data. These tests [GK81] are used to estimate the probability of an erroneous classification and its nonrandomness in the absence of a control sample.

Formal definitions. Let $\mathbf{W} = \{\mathbf{w}^i\}$ be a finite set of pattern recognition objects (vectors). Consider a training set $\mathbf{W}_0 = \{\mathbf{D}_0, \mathbf{N}_0\}$, where $\mathbf{D}_0, \mathbf{N}_0 \subset \mathbf{W}$, and $\mathbf{D}_0 \cap \mathbf{N}_0 = \varnothing$, and a class of pattern recognition algorithms used in the search for a solution $\mathcal{G} = \{g(\mathbf{W}_0) : \mathbf{W} \to \{-1, 1\}\}$. Given a training set $\mathcal{L}$, each algorithm $g \in G$ determines classification $\mathbf{W} = \mathbf{D} \cup \mathbf{N}$, where $\mathbf{D} = g^{-1}(-1)$ and $\mathbf{N} = g^{-1}(1)$ are complete preimages of transform $g(\mathbf{W}_0)$.

Consider a control sample $\mathbf{U} = \{\mathbf{V_D}, \mathbf{V_N}\}$, where $\mathbf{V_D}, \mathbf{V_N} \subset \mathbf{W}$, and assume that setting $\mathbf{V_D} \subset \mathbf{D}$ and $\mathbf{V_N} \subset \mathbf{N}$ does not produce errors. Given $\mathbf{U}$ and $g(\mathbf{W}_0)$, determine the total penalty $r[g(\mathbf{W}_0), \mathbf{U}]$ for misclassification on the control sample. The simplest penalty count is the frequency of misclassification on a control set $\nu[g(\mathbf{W}_0), \mathbf{U}] = |\mathbf{V_D} \cap \mathbf{N} \cup \mathbf{V_N} \cap \mathbf{D}|/|\mathbf{V_D} \cup \mathbf{V_N}|$.

For each training set $\mathcal{L}$, a "shuffled training" variate $\widetilde{\mathcal{L}}$ is combined from "shuffled sets" $\widetilde{\mathbf{D}}_0$ and $\widetilde{\mathbf{N}}_0$ by drawing objects from $\mathbf{W}$ at random; shuffled sets must have the same number of objects as the respective $\mathbf{D}_0$ and $\mathbf{N}_0$.

There are the total of $C_n^{n_1} C_{n-n_1}^{n_2} = n!/n_1!n_2!(n - n_1 - n_1)!$ "shuffled problems," where $n = |\mathbf{W}|$, $n_1 = |\mathbf{D}_0|$, $n_2 = |\mathbf{N}_0|$. A "shuffled control set" $\widetilde{\mathbf{U}}$ is set by analogy.

An estimate of nonrandomness can be based on a comparison of $r[g(\mathbf{W}_0),$ $\mathbf{U}]$ and mean value $\mathbf{M}r(g(\widetilde{\mathcal{L}}), \widetilde{\mathbf{U}})$, for example, on the adapted Student criterion [GK81]. In pattern recognition of earthquake-prone areas, the control set $\mathbf{U}$ is usually very specific: First, $\mathbf{V_N} = \varnothing$ because $\mathbf{W}$ consists of potential earthquake-prone areas. Second, usually $\mathbf{V_D} = \mathbf{D}_0$, due to extremely limited, small samples of places where large earthquakes did happen.

Pattern recognition verification. The reliability of pattern recognition algorithm $g(\mathbf{W}_0)$ is estimated most adequately by the probability of misclassification, $p_{g(\mathbf{W}_0)}$. The frequency of misclassification on a control set $\nu(g(\mathbf{W}_0), \mathbf{U})$, when $\mathbf{V_D} \cap \mathbf{D}_0 \cup \mathbf{V_N} \cap \mathbf{N}_0 = \varnothing$, delivers an unbiased estimate of $p_{g(\mathbf{W}_0)}$. Unfortunately, in many problems (e.g., with small samples) it is not possible to reserve a control set large enough for decent verification. At the same time, the frequency of misclassification of training $\nu[g(\mathbf{W}_0), \mathbf{W}_0]$ is evidently biased. Pinsker [Pin73] suggested the following heuristic assumption:

$$\nu[g(\mathbf{W}_0, \mathbf{U}) - \nu[g(\mathbf{W}_0, \mathbf{W}_0) \leq \mathbf{M}\nu[g(\widetilde{\mathcal{L}}), \widetilde{\mathbf{U}}) - \mathbf{M}\nu[g(\widetilde{\mathcal{L}}), \widetilde{\mathcal{L}}] . \quad (6.10)$$

In other words, on the transition from the control on known to unknown data, losses are generally larger for a random problem. The assumption implies that

$$\nu[g(\mathbf{W}_0, \mathbf{U}) \leq \nu[g(\mathbf{W}_0, \mathbf{W}_0) + \mathbf{M}\nu[g(\widetilde{\mathcal{L}}), \widetilde{\mathbf{U}}) - \mathbf{M}\nu[g(\widetilde{\mathcal{L}}), \widetilde{\mathcal{L}}] . \quad (6.11)$$

Ensemble averaging over control set $\mathbf{U}$ results in $\mathbf{M}\nu(g(\mathbf{W}_0), \mathbf{U}) = p_{g(\mathbf{W}_0)}$ on the left side because the estimate is unbiased. On the right side, $\mathbf{M}\nu(g(\widetilde{\mathcal{L}}),$ $\mathbf{U}) = \mathbf{M}(p_{\widetilde{\mathbf{D}}}|\widetilde{\mathbf{N}}| + p_{\widetilde{\mathbf{N}}}|\widetilde{\mathbf{D}}|)/|\mathbf{W}|$, where $\widetilde{\mathbf{D}} = g^{-1}(\widetilde{\mathcal{L}})(-1)$ and $\widetilde{\mathbf{N}} = g^{-1}(\widetilde{\mathcal{L}})(1)$ are subsets of classification by $g(\widetilde{\mathcal{L}})$ and $p_{\widetilde{\mathbf{D}}}$, $p_{\widetilde{\mathbf{N}}}$ are a priori probabilities of objects from $\widetilde{\mathbf{D}}$ and $\widetilde{\mathbf{N}}$ in control set $\widetilde{\mathbf{U}}$. Now (6.11) takes the form

$$p_{g(\mathbf{W}_0)} \leq \nu[g(\mathbf{W}_0), \mathbf{W}_0] + \mathbf{M}(p_{\widetilde{\mathbf{D}}}|\widetilde{\mathbf{N}}| + p_{\widetilde{\mathbf{N}}}|\widetilde{\mathbf{D}}|)/|\mathbf{W}|$$
$$- \mathbf{M}\nu[g(\widetilde{\mathcal{L}}), \widetilde{\mathcal{L}}] . \quad (6.12)$$

Note that (6.12) does not depend on control set U; this allows using this unbiased upper limit estimate of probability $p_{g(\mathbf{W}_0)}$ in the absence of an unknown control sample, although at the cost of one heuristic assumption. Such a possibility is very important for pattern recognition in earthquake-prone areas.

Naturally, a small value of $p_{g(\mathbf{W}_0)}$ is the argument favoring the validity of classification obtained for the original problem. If the estimation (6.12) results in a large value, it is advisable to return to the original problem. Such a situation may indicate, for instance, an insufficient size of $\mathbf{D}_0$. On the other hand, one should remember that (6.12) estimates the probability of error from above, though its value is usually much less.

Replication tests. These are various tests that attempt to replicate the result obtained by an alternative technique. The application of another pattern recognition algorithm is the simplest example of such a test. For instance, given a classification established by using CORA-3, an attempt can be made to repeat it by applying HAMMING to the same vectors. In the analysis of earthquake-prone areas, such a test is usually considered satisfactory if less than 20% of the total small sample changes class membership.

Note that a classification is favored to be nonrandom and valid when it is obtained from some algorithm and repeated by a simpler algorithm. On the other hand, replication by another algorithm cannot be considered a necessary condition for a valid classification. Nevertheless, we found it helpful to attempt to alter coding or discretization of characteristics or even introducing new characteristics when testing the stability of classification. At the extreme, new natural recognition objects can be introduced [ZRS75, GGK+76, ZK80]. If **D** and **N** areas outlined in such a test on the basis of objects of different type are in agreement, one can consider it a strong argument in favor of the natural validity of the classifications.

"Transfer of Criteria" (TC Test). Such a test, if successful, produces sound arguments for establishing the validity of classification and also a certain universality of the decision rules obtained.

Given two seismic regions and a solution of an earthquake-prone pattern recognition problem in one of them, it is worth testing whether the decision rule obtained applies to predicting locations of strong earthquakes in another region. Note that the definition of strong earthquakes could be different ($M_0 \neq M_0'$) in the two regions. It is enough to determine natural recognition objects of the same kind and measure the same characteristics. Consider $\mathbf{w}^i \in \mathbf{W}$, the vectors from the first region with classification $\mathbf{W} = \mathbf{D} \cup \mathbf{N}$, where $\mathbf{D} = g^{-1}(-1)$ and $\mathbf{N} = g^{-1}(1)$, and $\mathbf{w}'^i \in \mathbf{W}'$, the vectors from the second region; let $\mathbf{D}_0' \subseteq \mathbf{W}'$ be the set of reported earthquakes with $M \geq M_0'$ from the second region. Assume that the same characteristics are discretized (possibly, with different thresholds) and the same coding methods are used in both regions independently, so that vectors from both regions belong to the same binary vector space. Note that $g^{-1}(-1)$ and $g^{-1}(1)$ determine classification $\mathbf{W}' = \mathbf{D}' \cup \mathbf{N}'$ as well. The analysis of this classification finalizes the TC test.

The transfer of criteria can be judged a success or a failure from a priori information on the upper estimate of the number of earthquake-prone objects $\beta'|\mathbf{W}'|$ and the total of misclassified vectors from $\mathbf{D}_0'$. The TC test is claimed successful if $|\mathbf{D}'| \leq \beta'|\mathbf{W}'|$ and all or practically all $\mathbf{w}_i' \in \mathbf{D}_0'$ belong to $\mathbf{D}'$. Note that a failure to transfer criteria does not reject the validity of the initial classification but proves the absence of tectonic or seismic similarities between the regions considered.

Summary. The control tests listed above have not, of course, exhausted all possibilities for evaluating the reliability of classification. However, each of them adds a grain of confidence to the conclusion that a valid classification $\mathbf{W} = \mathbf{D} \cup \mathbf{N}$ is obtained and it reflects the actual division of natural recognition objects into earthquake-prone and other areas.

While checking reliability, it is essential to subject the established classification to other additional information available. For example, the locations of historical earthquakes that might be magnitude $M \geq M_0$ events can confirm or question the classification. Of course, substantial errors can arise in locating historical earthquakes. That is why historical earthquakes are usually not considered in the learning stage.

To verify the results obtained, it is also helpful to analyze and try to explain the decision rule established. Such an analysis is based on reformulating the rule in terms of the initial language of the characteristics used for describing natural recognition objects. This is not a hard task at all for decision rules established by HAMMING, HAMMING-1, CORA-3, or CLUSTERS, due to the above-mentioned relation between the binary vector space and the questionnaire. Section 6.4 avoids binary language and describes the characteristic traits in a natural language.

6.3   Choosing Objects for Recognition

The approach presented here is based on a widely accepted concept that the lithosphere consists of different-scale blocks separated by mobile boundaries (see Chap. 1 and references therein). The division of a territory into a system of hierarchically ordered blocks is rather evident in a present-day topography, which clearly expresses recent tectonics. The joint analysis of geomorphic data and tectonic patterns may suggest an adequate set of objects for recognizing and describing of earthquake-prone areas.

The methodology for recognizing of earthquake-prone areas from Sect. 6.2 is based on the association of large earthquakes with specific structures formed about intersections of fault zones. This section is concerned with the morphostructural zoning method that allows delineating the block-structure of a region and providing the characteristics of recent tectonic motions.

6.3.1   The Basics of Morphostructural Zoning

Morphostructural zoning (MZ) was been initially designed by Gerasimov and Rantsman [GR73] especially for identifying earthquake-prone areas. Gelfand et al. [GGZ⁺73] pioneered applying this method to the Pamirs and Tien Shan. Later, Alexeevskaya et al. [AGG⁺77], Rantsman [Ran79], and Gvishiani et al. [GGR⁺88] developed further enhancements of morphostructural zoning, including a formalization of the method. MZ correlates geomorphic data with available information on geologic structures and tectonic patterns. The analysis also involves specialized interpretation of satellite photographs. *It should be emphasized that MZ is performed without a direct connection with seismic data in a region under study.*

MZ distinguishes three types of morphostructures: (1) territories of different ranks, *blocks*; (2) their boundaries, morphostructural lineaments; (3) sites where lineaments intersect or join, *morphostructural nodes.* In general, the three types reveal different intensities of tectonic motion. Blocks are relatively stable units, whereas major motions take place along lineaments and at nodes. Most epicenters of strong earthquakes nucleate at nodes in all previously studied regions y [GGI$^+$72a, GGZ$^+$73, GGZ$^+$74a, GGZ$^+$74b, ZRS75, GGK$^+$76, GZKK78, ZK80, CKO$^+$80, GK81, Kos83, GS84, GZS84, CGG$^+$85, WGG$^+$86, GNRS87, GGK$^+$87, GZRT91, BCF$^+$92, BRC$^+$94, GKPS00]. Figure 6.1 illustrates the basic scheme of MZ.

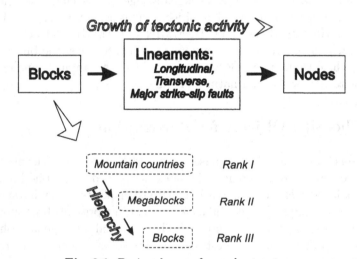

Fig. 6.1. Basic scheme of morphostructures

MZ hierarchically orders blocks and lineaments assigning them ranks from the highest to the lowest. Lower rank blocks are parts of a higher rank block. Usually, three levels of a hierarchy were considered [GGI$^+$72a, GGZ$^+$73, GGZ$^+$74a, GGZ$^+$74b, ZRS75, GGK$^+$76, GZKK78, ZK80, CKO$^+$80, GK81, Kos83, GS84, GZS84, CGG$^+$85, WGG$^+$86, GNRS87, GGK$^+$87, GZRT91, BCF$^+$92, BRC$^+$94, GKPS00]. Blocks of the first rank, *mountain countries*, are divided into blocks of the second rank, *megablocks*. Megablocks are further divided into blocks of the third rank, called *blocks* per se. The hierarchy of blocks could be extended both ways either uniting mountain countries into a larger zero-rank block or subdividing a block into smaller units. In general, the hierarchy covers tectonic units from lithospheric plates to grains of rock (see Chap. 1). A lineament separating two blocks gets their highest rank.

Morphostructural lineaments are linear zones (not lines) of tens, usually hundreds of kilometers long and of several, sometimes tens of kilometers wide. The lineament width depends on the character and intensity of tectonic motion and indicates indirectly the rank of morphostructures involved. With respect to the regional trend of the tectonic structure and topography, one can distinguish the longitudinal and transverse lineaments.

Longitudinal lineaments are approximately parallel to the regional trend of the tectonic structure and that of the topography. Usually, they include long segments of prominent faults. They stretch along the boundaries of large topographic forms, separating relatively uplifted areas from relatively subsided ones.

Transverse lineaments cross the regional trend of tectonic structures. Normally, they appear on the surface as a discontinuous chain of tectonic escarpments, rectilinear parts of river valleys, and partly faults.

An intersection of lineaments covers an area of size greater than their width. An intersection or a union of neighbouring intersections forms a specific morphostructure called a *node*. The boundaries of nodes are not easy to define without some geomorphic fieldwork.

Definitions of large topographic forms. To outline the hierarchy of blocks and lineaments in a given region, MZ starts by reviewing *quantitative indices of large topographic forms* such as mountain ranges and massifs, plateaus and uplands, intermontane and intramontane basins, and longitudinal valleys.

Quantitative index determined from topographic, structural, and geologic maps include the following:

- altitudes of different topographic forms (ranges, massifs, plateaus, highlands, etc.);
- azimuths of axes of different topographic forms (ranges, footlines, basins, intermontane depressions, and longitudinal valleys);
- widths of intermontane basins;
- thicknesses of soft sediments;
- ages of rocks composing a large topographic form.

MZ compares quantitative index of large similar landforms. Alexeevskaya et al. [AGG+77] and Rantsman [Ran79] formalized MZ definitions for cases where conclusive common terms for large landforms were absent.

A *mountain range* is a large elongated elevation clearly recognized by the presence of a highest axial zone, lowest footline, and slopes. A large range (e.g., such as the Main Caucasus Range) that includes large separated topographic forms developed on its slopes (namely, frontal ridges, plateaus and mountain massifs, longitudinal valleys, and intrarmontane depressions) is called a *composite range*. Its axis, the divide of rivers flowing down from its opposite slopes, is called the *main range*. The main range normally stretches continuously along the whole composite range, whereas frontal ridges are much shorter. Altitude quantitative index of the range are determined from the averaged altitudes of peaks, H_{mean}, and the maximum of H_{mean} for several ranges, H_{max}. MZ also includes another quantitative index, the strike of the range axis.

A *mountain massif* is a large isolated elevation without a clear-cut axis. Its quantitative index are the altitude of the summit surface and the age of rocks.

An *intermontane basin* is a wide depression between mountain ranges formed in the process of long-term subsidence. A marginal basin spreads along the boundary of a mountain belt or a platform territory and, as a rule, is not included in a mountain country.

An *intramontane basin* is a structural depression between ranges or along the slope of a composite range. Contrary to intermontane basins, the length of an intramontane basin is shorter than that of the ranges it separates. A *longitudinal valley* is a narrow structural depression between close mountain ranges or along the slope of a composite range usually occupied by a river valley. The quantitative index of inter- and intramontane basins and longitudinal valleys are the mean altitude of the bottom, the width of the basin, the thickness of accumulated sediments, and the azimuth.

Territorial units (blocks) in morphostructural zoning. The uniformity of each morphostructural unit is determined by a certain set of morphometric features and their quantitative index.

A single mountain country is a territory of the same orogenesis and certain appearance of the relief. It can border plain territories and other mountain countries. Each mountain country is outlined by the following geomorphic and tectonic features. Four basic types of mountain countries are distinguished by their orogenesis:

- rejuvenated mountains originated on ancient platforms (epiplatform mountains);
- epogeosynclinal mountains built on the Alpine geosynclines on completion of their folding;
- volcanic mountains resulted from volcanism and/or plutonism (lava plateaus and uplands, volcanic domes);
- continental rifting where orogenesis is due to extension of the crust and where the combination of narrow ranges with deep intramontane basins is in common.

The same orogenesis can result in territories with a different *appearance of relief* (physiography) described by features described below. The territories belong to different mountain countries:

(1) if their levels of maximum altitude H_{max} differ roughly by 2000 meters;
(2) if their combination of large topographic forms are different;

(3) if the dominating strikes of large topographic forms are nearly perpendicular.

Any of these differences allows dividing a territory of the same orogenesis into different mountain countries. A combination of relief elements, like mountain ranges and intermontane basins, mountain ranges and plateaus, piedmont hills and plains, reflects the relationship between the uplift and subsidence that occurred in the course of mountain building and characterizes the uniformity of the mountain country. A uniform mountain country under regional tectonic stress tends to a similar orientation of its morphostructures.

An abrupt change of any of the following quantitative index distinguishes megablocks:

(1) gradual increasing altitude of ranges;
(2) gradual decreasing width of near-parallel basins or a chain of basins;
(3) successive changes in orientation of ranges;
(4) concordant changes in the altitude of near-parallel ranges.

A sharp and considerable change of at least one index distinguishes morphostructural blocks from each other. When changes in index are gradual, the territory is not divided into blocks. Among the index are (1) the difference in altitude of neighbouring summits above one-tenth of the main range's altitude; (2) the bend of a range or of a footline above 20°; and (3) the age of rocks belongs to different geologic periods (Pre-Mesozoic, Mesozoic-Eocene, Oligocene-Neogene, and Quaternary).

Boundaries (lineaments) in MZ. Hobbs [Hob04] was the first to introduce the term "lineament" in 1904. Bogdanovich et al. [BKKM14] used this term in the description of the 1911 earthquake in northern Tien Shan to denote "lines of orographic and tectonic importance." Nowadays, the term is widely attributed to a variety of different linear traces on the surface of Earth or other planets. Geophysical discontinuities are also often called lineaments (magnetic lineament, gravitational lineament etc).

Unlike definitions from [Hob04, BKKM14], a morphostructural lineament is viewed as a boundary zone between territorial units delineated by MZ. There are no morphostructural lineaments in the absence of a recognized territorial unit (a block). MZ distinguishes longitudinal, transverse, and major strike-slip lineaments. Figure 6.2 presents a sample of a transverse lineament.

Longitudinal lineaments follow the boundaries of large topographic forms or slopes of composite ranges. They usually include zones of the prominent faults and, in general, are more evident than transverse lineaments. A lineament zone includes the lower parts of range slopes and marginal parts of the adjacent landforms. Topographic forms along a longitudinal lineament normally change on intersection with a transverse lineament.

Transverse morphostructural lineaments go across the predominant trend of topography and tectonic structures. Normally, they appear discontinuously on the surface. Zones of transverse lineaments are traced along small

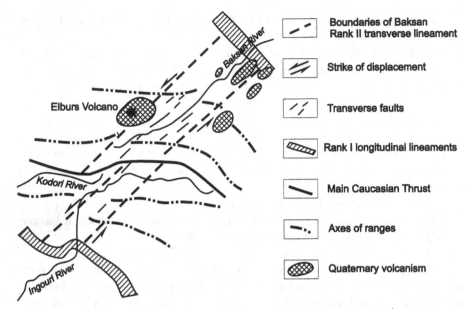

Fig. 6.2. Scheme of the Baksan transverse lineament [GGR$^+$88]

individual ridges and/or chains of small ridges, rectilinear segments of river valleys, elongated erosional landforms, tectonic scarps, faults, flexures and narrow intrusive bodies, and linear contacts in rocks. Although not so evident as the longitudinal, transverse lineaments may cross the entire mountain country and sometimes can be traced across several hundreds of kilometers. Usually, tracing transverse lineaments is a most complicated procedure in MZ. Satellite photos help a lot to identify them.

Major strike-slip lineaments form an individual group of the first rank lineaments along which the regional tectonic units slide across tens of kilometers. MZ denotes as such the Talas-Fergana fault in Central Asia, the San Andreas fault in California, the North Anatolian fault in Asia Minor, etc.

Morphostructural nodes. *Morphostructural nodes* are formed about intersections or junctions of two or several lineaments. A node may include more than one intersection or junction. Lineament zones become wider at nodes. Nodes are characterized by a mosaic combination of various topographic forms and by an increased number of linear topographic forms of various strikes that reveal instability of the area.

River valleys within the nodes are represented by rectilinear segments of various strikes. Knee-like bends are characteristic of the river streams within a node. There are oblique segments of valleys that are discordant with respect to the direction of the maximum gradient. Excessive concentration of water (streams, springs) at nodes has to be noted. It is due to the confluence of several rivers confined to the crossings of lineaments, as well as to the uplift of underground waters, in particular thermal springs. As an example of a node structure, Fig. 6.3 presents the morphostructural scheme of the Shemakha node.

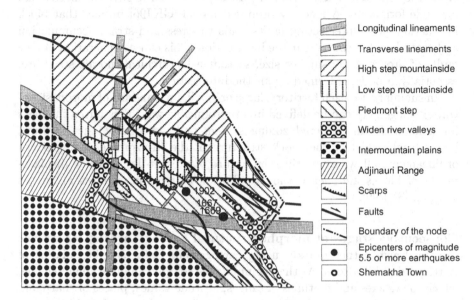

Fig. 6.3. Morphostructural scheme of the Shemakha node [GGR+88]

Naturally, in the course of intensive displacements, intersections and junctions of faults accumulate large amount of dislocated rocks on mountain slopes. Indeed, large rock falls were found concentrated at nodes in the Caucasus [GZ97, GZ98].

The presence of water, the junctions of valleys and, perhaps, attractive landscapes historically facilitated populating the territory of many nodes. At present, many large cities, including megapolises such as Los Angeles, San Francisco, Athens, and Istanbul situated at nodes, are exposed to the highest level of seismic hazard. On the other hand, the impact of intensive tectonic movements along the faults of various strikes masked, sometimes, by manmade changes in topography complicates the delineation of node boundaries. Therefore special fieldwork is needed to outline them reliably. If no field investigations were made, a certain circle could be used as a substitute for a node (for instance, a 25-km radius for seismic regions of moderate activity and $M_0 = 5$–6). Field investigations in the Tien Shan and Caucasus show that nodes are usually asymmetrical with regard to intersections and lineament axes. The sizes of nodes range drastically and depend on the number of intersections or junctions that form a node as well as on the number of lineaments and their ranks.

The fact that earthquakes are nucleated at nodes was first established for the Pamirs and Tien Shan [GGI⁺72a]. Later, the nonrandom nature of the phenomenon was proven statistically for many seismic regions [GS81, KS83, GGKR86]. McKenzie and Morgan [MM69] described a physical mechanism for node formation. A recently proposed model [GKJ96] implies that block interaction along intersecting faults leads to stress and strain accumulation and secondary faulting about the intersection. This causes generation of new faults of progressively smaller size, so that a hierarchical mosaic structure, essentially, a node, is formed about the intersection.

In summary, for each territory, large or small, its standard of morphostructural homogeneity can be defined in a reproducible way. This forms the basis for a hierarchically ordered zoning system and, essentially, the empirical hierarchical model of the block structure of the lithosphere. The pattern of lineaments allows evaluating the nature of present-day tectonic displacements of blocks and hence the distribution of tectonic stresses all across a region [SVP99].

Consecutive stages of morphostructural zoning. A draft MZ scheme can be drawn rather quickly using cartographic data, satellite photos of Earth, and publications. At the beginning, we compile an orographic scheme, which shows axes and footlines of ranges, altitude values, places of sharp and considerable changes in elevation, and orientation of structures along with straightened segments of valleys and other rectilinear landforms.

The first stage of MZ scheme compilation results in outlining mountain countries and first rank morphostructural lineaments. The territories of the same orogenesis are determined from geologic and tectonic maps. The positions of mountain countries and their boundaries locate the morphostructural lineaments of the first rank. In the first approximation, this task is rather simple because the boundaries of such large morphostructural units are well studied and defined.

The second MZ stage is the division of a mountain country into blocks, which are third rank territorial units. Their boundaries complete the network of morphostructural lineaments. These lineaments are compared with the network of reported faults to ascertain whether the territories they separate belong to different blocks. Then, in the third stage, the blocks are united into megablocks promoting some lineaments from the third to the second rank. The experience of compiling MZ schemes at the scale of 1:1,000,000 has proved that such a sequence of analysis is convenient.

The third stage results in defining megablocks and second rank morphostructural lineaments. The blocks determined in the second stage are united if the differences of quantitative index are within the limits required by the definition of a megablock (see above).

The fourth stage locates nodes. A geomorphologist starts from an intersection or a cluster of close intersections of lineaments and, moving along each lineament, marks the place where the dominating linear landforms come back to the strike of the lineament. These marks indicate parts of the node boundary. The presence of lineament zones below the third rank in a node may complicate outlining the entire boundary, even in the course of thorough field mapping.

6.3.2   Objects of Recognition Derived from Morphostructural Zoning

MZ outlines unstable structures of a seismic region and initially restricts the area where strong earthquakes are possible. Specifically, MZ suggests considering nodes as candidates for earthquake-prone areas. Some of these candidates have confirmation in seismic history. For example, there were 16 out of 41 nodes in the Pamirs and Tien Shan, where 22 out of 23 reported magnitude 6.5 or larger earthquakes happened prior to the first pattern recognition analysis [GGI+72a]. When the fourth MZ stage has not been carried out, one may regard intersections of lineaments as candidates that define earthquake-prone areas. Specifically, as already mentioned, circles of a certain radius centered at intersections can be used in place of nodes. The radius depends upon the scale of MZ that corresponds to the level of seismic activity in the region. For example, in the Andes of Southern America with MZ for $M_0 \geq 7.75$ on a scale of 1:7,500,000, the radius equals 75 km, whereas in the Pyrenees with MZ for $M_0 \geq 5.0$ on a scale of 1:1,000,000, the radius was 25 km.

The hypothesis that most strong earthquakes nucleate at nodes (either in the original definition or substituted by circles) has been validated in all seismic regions where MZ and pattern recognition were performed [GGI+72a, GGZ+73, GGZ+74a, GGZ+74b, ZRS75, GGK+76, GZKK78, ZK80, CKO+80, GK81, Kos83, GS84, GZS84, CGG+85, WGG+86, GNRS87, GGK+87, GZRT91, BCF+92, BRC+94, GKPS00].

Each node is assigned an unambiguous description in the form of a common questionnaire for applying the pattern recognition technique. In general, a questionnaire should reflect the characteristics of tectonic movement and geologic environment responsible for the present-day seismic level. Here, we present typical characteristics normally used to classify nodes as **D** or **N**. The list of characteristics was derived from a much larger set in a number studies. Many parameters that can be important for recognition but not available with the same accuracy for the entire region are not included in the list. The requirement of uniform description leads to unavoidable loss of information on some geologic and geomorphic features, as well as some geophysical parameters like heat flow, gravity, and magnetic fields. Of course, other characteristics might be useful for identifying earthquake-prone areas and, hopefully, will be implemented in future studies.

The characteristics listed below belong to three groups describing (A) the contrast and intensity of tectonic movement, (B) the degree of fragmentation, and (C) the heterogeneity at crust depths.

Group A:

1. Maximum altitude, H_{max}.
2. Minimum altitude, H_{min}.
3. Minimum distance l between the points with H_{max} and H_{min}.
4. Relief contrast, $\Delta H = H_{max} - H_{min}$.
5. Measure of slope, $\Delta H/l$.
6. The following combinations of large topographic forms (Yes if a form is present and No if it is absent):

    a mountain range and a piedmont plain (m/p),

    a mountain range, piedmont hills, and a piedmont plain $(m/pd/p)$,

    a mountain range and piedmont hills (m/pd),

    mountain ranges separated by a longitudinal valley (m/m),

    piedmont hills and a piedmont plain (pd/p),

    piedmont plains (p/p).

7. The percentage of Quaternary deposits, Q.

Characteristics 1–6 are measured from topographic maps and characteristic 7 is taken from geologic maps. Schematic profiles in Fig. 6.4 illustrate most common combinations of large topographic forms typical of mountain areas.

Group B:

8. The highest rank of a lineament in the node, R_L.
9. The number of lineaments in the node, n_L.
10. The distance from the node to the nearest intersection, r_{int}.
11. The distance from the node to the nearest first rank lineament, r_1.
12. The distance from the node to the nearest second rank lineament, r_2.
13. The number of faults in the node, N_F.

Characteristics 8–12 are evaluated directly from MZ, and characteristic 13 counts all of the faults reported on a geologic map uniform in detail all across the territory under study.

Group C:

14. Maximum Bouguer anomaly, B_{max}.
15. Minimum Bouguer anomaly, B_{min}.
16. Gravity "relief energy", $\Delta B = B_{max} - B_{min}$.
17. The mean value of the Bouguer anomaly, $B_m = (B_{max} + B_{min})/2$.
18. "Free-air anomaly," $HB = 0.1H_{max} + B_{min}$.
19. The number of isolines (10 mGal multiple) of the Bouguer anomaly in the node, N_B.
20. The number of closed isolines (10 mGal multiple) of the Bouguer anomaly in the node, N_c.
21. The minimum distance between isolines of the Bouguer anomaly (10 mGal multiple), $(\nabla B)^{-1}$.

The gravity characteristics 14-21 are related to Bouguer anomalies. Characteristic 18 is an integral characteristic of Earth's surface relief, density distribution in the crust, and isostatic disbalance (coefficient 0.1 is used when H_{max} is measured in meters and B_{min} in mGal). Gorshkov et al. [GNRS87] demonstrated the efficiency of this characteristic for identifying earthquake-prone areas in the Caucasus.

Fig. 6.4. Several combinations of large topographic forms: A schematic illustration [GGR+88]

6.4 Recognition of Where Strong Earthquakes Can Occur

Out of extensive analyses of seismic regions worldwide [GGI+72a,GGZ+73, GGZ+74a, GGZ+74b, ZRS75, GGK+76, GZKK78, ZK80, CKO+80, GK81, Kos83, GS84, GZS84, CGG+85, WGG+86, GNRS87, GGK+87, GZRT91, BCF+92,BRC+94,GKPS00], we sampled a few studies that are typical applications of the methodology to objects of different kinds. At the same time, the selected studies demonstrate how the method works on different levels of lithospheric and seismic hierarchies. Specifically, they cover at least regional (Caucasus, Alps, and California) and global (Circumpacific and other major seismic belts) structures of the lithosphere and earthquake magnitude ranges from 5 to 8.

6.4.1   The Greater Caucasus

Pattern recognition was applied in this region to two sets of natural recognition objects: morphostructural nodes and intersections of lineaments. The sets were used to distinguish earthquake-prone areas defined by different M_0. We start with pattern recognition of nodes assuming $M_0 = 5.5$ [GGR+88].

Morphostructural zoning. An MZ scheme of the Greater Caucasus compiled on a scale of 1:1,000,000 (Fig. 6.5) resulted from field investigations by E.Ia. Rantsman, A.I. Gorshkov, and M.P. Zhidkov from 1979–1984, which allowed outlining the boundaries of morphostructural nodes. The basics for dividing the Caucasus into territorial units were elaborated by Milanovsky [Mil68], who considered the Greater Caucasus a neotectonic structure of the highest rank. According to MZ [GGKR86, GGR+88], the Greater Caucasus is a first rank territorial unit, specifically, a mountain country of epigeosynclinal orogenesis, which was most active in the Pliocene - Pleistocene, extended to the Alpine through rigid massifs to the south and north of it (Precaucasus and Transcaucasus).

The orogenic movements resulted in a folded block-mountain structure in the Greater Caucasus. The combination of the unified Main Range (west-northwest strike) with front ridges, mountain massifs, intramontane depressions, and longitudinal valleys that complicate its flanks characterizes the relief pattern of the Greater Caucasus. Khain [Kha82] distinguished three large Alpine structures of the preorogenic stage on his scheme of the Greater Caucasus: (1) the Precaucasus (Scythian) plate that forms the Main Range in the central part of the Caucasus and most of its northern flank; (2) geosynclinal trough south of the plate, where sediments compose the southern flank of the Greater Caucasus; (3) the crystalline Transcaucasus massif involved in the present-day uplift of the central part of the southern flank.

The first rank lineament 1–58 separates the Greater Caucasus from the Precaucasus plate in the north [GGR+88]. The southern boundary, lineament 2–65, runs along the Alpine trough deposits of the Greater Caucasus and cuts across the Transcaucasus massif at its central segment. Transverse lineaments limit the Greater Caucasus on the west (the Anapa 1–2 and the Yashma 64–65 Lineaments, respectively, Fig. 6.5).

The Greater Caucasus megablocks are recognized either from the changes in altitude and the strike of the Main Range or from a conjunction of relief elements that complicates the Greater Caucasus flanks. These regular topographic patterns supported with other morphostructural characteristics divide the mountain country into 10 megablocks. Their morphostructural homogeneity is evident in the patterns of large river valleys that differ from one megablock to the other (Fig. 6.6).

The work [GGR+88] presents a detailed description of all 10 megablocks, a few blocks and two sample nodes of the Greater Caucasus, as well as their boundaries, i.e., morphostructural lineaments of the second and third ranks.

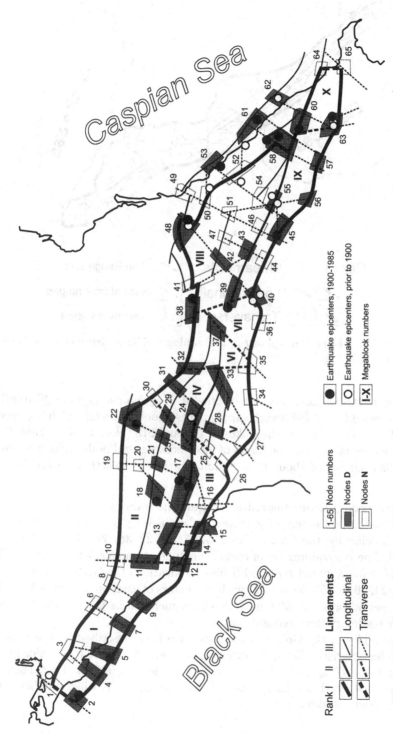

Fig. 6.5. The Greater Caucasus: MZ map and pattern recognition of earthquake-prone nodes ($M \geq 5.5$) [GGR[+]88]

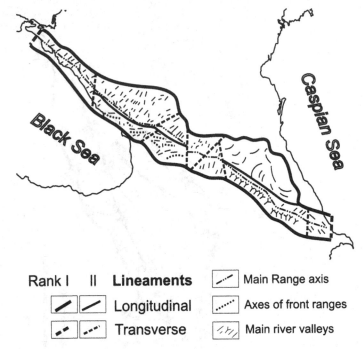

Rank I   II   **Lineaments**   Main Range axis

Longitudinal   Axes of front ranges

Transverse   Main river valleys

Fig. 6.6. The scheme of morphostructural zoning and basic elements of the Greater Caucasus [GGR+88]

Strong earthquakes. The list of crustal $M \geq 5.5$ earthquakes (Table 6.1) was compiled from [KS77] with supplements [Kon80,Kon82]. With the exception of the two historic earthquakes in 743 and 918 (Nos. 4 and 5, Table 6.1), all epicenters fall within the limits of morphostructural nodes (Fig. 6.5). Note that the accuracy of about 1° is common for ancient earthquakes [KS77].

Pattern recognition: Learning stage. The training set $\mathbf{D}_0$ consists of 13 nodes hosting epicenters of earthquakes with $M \geq 5.5$ that occurred after 1900. Specifically, these are nodes 2, 5, 17, 18, 22, 38, 39, 45, 48, 53, 60, 61, and 63. The $\mathbf{N}_0$ training set of class $\mathbf{N}$ includes 42 nodes, where earthquakes with $M \geq 5.5$ were not reported before publication of [GGKR86]. Nine nodes (1, 15, 24, 40, 50, 52, 55, 59, 62) with epicenters of earthquakes with $M \geq 5.5$, in the period prior to 1900, have not been included in $\mathbf{D}_0$ and $\mathbf{N}_0$ but were left for testing pattern recognition.

The nodes of the Greater Caucasus have been described by the uniform set of characteristics listed in Table 6.2. Among them, the most informative appear to be $H_{\min}$, ΔH, Q, n_{L}, r_{int}, and r_2, defined in Sect. 6.2.2. For each of these characteristics R_{k} (see Sect. 6.2.1) exceeds 17%, which implies satisfactory discretization.

Table 6.1. Earthquakes with $M \geq 5.5$ in the Greater Caucasus

#	Date	Coordinates		Magnitude	Intensity	Depth	Node Number
		$\varphi°$N	$\lambda°$E	M	I_0	km	(Fig. 6.5)
1	50	42.9	41.0	5.5	VIII	3–30	15
2	400	42.9	41.0	5.5	VIII	3–30	15
3	650	42.6	47.7	6.1	VIII	7–60	50
4	743	42.1	48.2	5.5	VII	7–60	outside nodes
5	918	42.1	48.2	5.5	VIII	7–60	outside nodes
6	957	41.5	49.0	5.5	VII	7–60	62
7	1250	41.6	47.2	5.7	VII–VIII	5–50	55
8	1350	43.0	43.0	6.5	VIII–IX	10–40	24
9	1530	42.0	45.4	5.7	VIII	5–50	40
10	8.06.1652	42.1	47.7	5.8	VIII–IX	3–30	52
11	17.12.1667	41.7	47.3	6.5	VIII	20–45	55
12	1667	40.9	48.2	7.0	IX–X	6–24	63
13	14.01.1669	40.6	48.6	6.0	IX	5–20	63
14	24.07.1742	42.1	45.2	6.2	VII	5–50	40
15	9.08.1828	40.7	48.4	5.7	VIII	5–20	63
16	9.03.1830	43.0	47.0	6.3	VIII–IX	11–24	48
17	11.06.1859	40.7	48.5	5.9	VIII–IX	7–15	63
18	28.01.1872	40.6	48.7	5.7	VIII–IX	3–14	63
19	4.05.1878	41.6	48.1	5.7	VII	7–60	58
20	9.10.1879	45.1	37.8	5.7	VII	22	1
21	26.06.1889	42.5	48.0	5.9	VI	28–63	53
22	13.02.1902	40.7	48.6	6.9	VIII–IX	15	63
23	5.07.1903	41.8	48.7	5.5	VI	40	61
24	21.10.1905	43.3	41.7	6.4	VII	35	17
25	21.10.1905	43.6	41.2	5.6	VI	32	18
26	20.02.1906	41.5	48.4	5.9	VI	25	58
27	30.10.1909	42.4	48.0	5.8	VI	40	53
28	25.03.1913	41.8	48.3	5.7	VII	15	57
29	29.06.1921	43.9	42.8	5.6	VII	22	22
30	15.08.1947	42.5	45.0	5.5	VII	25	39
31	29.06.1948	41.6	46.4	6.1	VII	48	45
32	16.07.1963	43.25	41.58	6.4	IX	5	17
33	12.07.1966	44.7	37.3	5.8	VII	55	2
34	14.05.1970	43.0	47.09	6.6	VIII–IX	13	48
35	20.12.1971	41.23	48.38	5.5	VII	5	60
36	28.06.1976	43.1	45.5	6.4	–	–	38
37	3.09.1978	44.38	38.03	5.5	VI	25	5

Table 6.2. Characteristics of nodes of the Greater Caucasus

Characteristics	First threshold	Second threshold
Maximum altitude H_{max}, m	1401	2814
Minimum altitude H_{min}, m	200	600
$\Delta H = H_{max} - H_{min}$, m	1200	2100
The percentage of Quaternary deposits, $Q\%$	5	20
The highest rank of the lineament R_L	1	2
The number of lineaments n_L	2	3
The distance to the nearest intersection r_{int}, km	27	–
The distance to the nearest first rank lineament r_1, km	12.5	30.5
The distance to the nearest second rank lineament r_2, km	29	62.5
The number of faults in the node N_F	3	5
"Free-air anomaly", $HB = 0.1 H_{max} + B_{min}$, mGal	750	1050
Combinations of Large Topographic Forms		
Mountain ranges separated by a longitudinal valley (m/m)		
Continental slope (CS)		
Piedmont plains (p/p)		
A mountain range and a piedmont plain (m/p)		
A mountain range and piedmont hills (m/pd)		
Piedmont hills and a piedmont plain (pd/p)		

Pattern recognition: Earthquake-prone nodes ($M \geq 5.5$). The main case recognition of **D** and **N** nodes in the Greater Caucasus was obtained through CORA-3 with $k_1 = 5$, $\bar{k}_1 = 6$, $k_2 = 16$, $\bar{k}_2 = 1$ (see Sect. 6.2.1). Table 6.3 lists the characteristic traits of **D** and **N** nodes obtained (**D** and **N** traits, respectively). Class **D** nodes are those with $\Delta^i = n_D^i - n_N^i \geq \Delta = -2$, where n_D^i and n_N^i are the numbers of **D** and **N** traits of the node i. Table 6.4 presents the results of classification. Out of 64, 38 nodes of the Greater Caucasus are recognized as earthquake-prone for $M \geq 5.5$, whereas 26 nodes are not. Figure 6.5 shows the positions of **D** and **N** nodes.

Pattern recognition: Control tests. The SF test (see Sect. 6.2.2) with $\bar{k}_1 = \bar{k}_2 = 0$ and $\bar{k}_1 = \bar{k}_2 = 1$ gives the same classification as in the main case (Table 6.4). When $\bar{k}_1 = \bar{k}_2 = 2$, the classification changes for just one object (node 54 is then recognized as **D** node). In the "jackknife" test, only 15% of D_0 (two nodes, 2 and 22) and about 10% of N_0 (nodes 25, 27, 49 and 54) fail to confirm their classifications. In a "seismic history" (SH) test [GGK⁺76], nodes with epicenters of earthquakes prior to 1971 formed a new training set D'_0, whereas the rest of the original D_0 (two nodes, 5 and 38) added to N_0 formed a new training set N'_0. The values of $\bar{k}_1$ and $\bar{k}_2$ were the same as those in the main case. The threshold Δ was chosen so that the number of **D** nodes did not exceed 38. The test replicated the classification of the main case.

Table 6.3. Characteristic traits selected by algorithm CORA-3 for recognition of earthquake-prone nodes of the Greater Caucasus ($M \geq 5.5$)

#	H_{min}, m	ΔH, m	CS	m/p	m/pd	pd/p	Q, %	r_{int}, km	N_F	HB, mGal
				Topographic form combinations						
						D traits				
1							> 5	≤ 27		
2	≤ 600	> 2100			No					
3		≤ 2100					> 20			
4					No		> 5			> 5
5	≤ 600					No				> 5
6				No			> 20			≤ 5
7			No				> 20			
						N traits				
1		≤ 2100						> 27	≤ 5	
2		≤ 2100			No			> 27		
3					No			> 27	≤ 5	
4			No	No				> 27		
5		≤ 2100				No				≤ 1050
6		≤ 2100					≤ 20			
7						No			≤ 5	≤ 1050
8							≤ 20			≤ 1050
9							≤ 20		≤ 5	

Table 6.4. Classification of earthquake-prone nodes of the Greater Caucasus ($M \geq 5.5$)

Node number	Class	Node number	Class	Node number	Class	Node number	Class
Training set D_0		6	N	28	D	51	N
2	D	7	D	29	D	54	N
5	D	8	N	30	N	56	D
17	D	9	D	31	D	57	D
18	D	10	N	32	D	64	N
22	D	11	D	33	D	65	N
38	D	12	N	34	N	*Testing*	
39	D	13	D	35	N	1	N
45	D	14	D	36	N	15	D
48	D	16	N	37	D	24	D
53	D	19	N	41	N	40	D
60	D	20	N	42	D	50	N
61	D	21	N	43	D	52	N
63	D	23	D	44	N	55	D
Training set N_0		25	N	46	N	58	D
3	N	26	N	47	N	62	D
4	D	27	N	49	N		

The following test has been carried out as well. Each pair of nodes belonging to D_0 was removed in turn from learning. The pattern recognition algorithm used the remaining 11 D_0 objects and 42 N_0 objects as the training set for classification, the excluded pair in particular. Thresholds k_1, $\overline{k}_1$, k_2, and $\overline{k}_2$ took on the same values as in the main case, and the value of Δ was set to keep $|D| \leq 38$. The vectors of excluded pairs were not recognized as N 28 times out of 156 cases, i.e., in 18% of the total.

Tests with random data produce the following results. One hundred training sets $\overline{D}_0$ and $\overline{N}_0$ were generated at random with the same number of nodes as in the original D_0 and N_0, in the main case. The results of solving these 100 problems satisfied the condition $\overline{D}_0 \subseteq D$ (Sect. 6.2.2) only in 10 cases. Therefore the nonrandomness for the main case was estimated as 0.1.

An attempt was made to repeat the results of the main case by using a simpler HAMMING-1 (see Sect. 6.2.1). The relevant vector space was reduced to characteristics that were selected in the main case by CORA-3. The weights for the Hamming distance were set proportional to the number of times the given component entered the traits selected by CORA-3. The resulting classification differs from the main case on just two objects. Thus, a different recognition algorithm practically replicates the classification of the main case and allows simplifying the interpretation of characteristic traits.

The Greater Caucasus, $M \geq 5.0$. Here, we show pattern recognition of earthquake-prone areas determined by the analysis of intersections of morphostructural lineaments and $M_0 = 5.0$ [GGR+88].

Earthquakes with $M \geq 5.0$ and a learning stage. Identification of earthquake-prone areas for magnitude threshold $M_0 = 5.0$ in the Greater Caucasus followed the analysis of nodes aimed at larger magnitude earthquakes [GGK+76]. The location of an earthquake of this size is often outside the nodes. Still they tend to appear next to the intersections of morphostructural lineaments (i.e., within 25 km). This allows us to regard these intersections as objects of recognition. Table 6.5 lists earthquakes with $M \geq 5.0$ that occurred in the Greater Caucasus after 1900.

D_0 and N_0 were formed in the following way. Each epicenter from Table 6.5 became a center of a circle 25 km in radius. If there existed one and only one intersection in this circle, it was put in D_0, whereas in the case of ambiguity the intersections were reserved for control testing. N_0 included all intersections at distances of 50 km or more from the nearest epicenter. As a result, 19 intersections constituted D_0 (numbers 2, 6, 25, 29, 36, 38, 47, 55, 60, 62, 72, 76, 78, 83, 85, 88, 93, 98, and 100 from Fig. 6.7). N_0 consisted of 23 intersections (numbers 7, 8, 10, 11, 12, 13, 15, 16, 17, 18, 26, 27, 41, 46, 50, 51, 52, 53, 54, 75, 81, 101, and 102 from Fig. 6.7). The remaining 59 intersections were not included in the training sets.

Table 6.5. Earthquakes with $M \geq 5.0$ in the Greater Caucasus since 1900

#	Date	Coordinates		Magnitude	Intersections at a distance r	
		$\varphi°N$	$\lambda°E$	M	$r \leq 25$ km	$r \leq 30$ km
1	13.02.1902	40.7	48.6	6.9	100	
2	21.02.1902	41.8	48.6	5.6	87, 88	99
3	18.06.1902	43.0	42.0	5.2	36	29
4	05.07.1903	41.8	48.7	5.5	87, 88, 91	
5	02.11.1903	41.1	47.1	5.0	93	
6	28.04.1904	40.7	48.5	5.1	100	
7	05.07.1904	40.8	48.7	5.1		100
8	04.10.1905	44.7	37.4	5.1	2	
9	21.10.1905	43.3	41.7	6.4	29	24
10	21.10.1905	43.6	41.2	5.6	23, 25	
11	20.02.1906	41.5	48.4	5.9	92, 98	89, 96
12	25.09.1906	43.4	42.8	5.0	31, 45	
13	21.08.1907	42.7	48.3	5.3	83	
14	30.10.1909	42.4	48.0	5.8	83, 84	
15	25.03.1913	41.8	48.3	5.7	89, 91	87
16	14.01.1915	42.8	44.7	5.4	55	56, 57
17	29.06.1921	43.9	42.8	5.6		30
18	09.10.1924	40.9	48.6	5.2	100	
19	10.02.1929	43.1	43.9	5.3	47	48
20	05.12.1930	41.4	48.7	5.2	97, 98	
21	17.05.1931	40.5	48.5	5.0		100
22	24.10.1933	42.9	45.9	5.2	63, 64	
23	02.09.1936	41.5	46.6	5.3	78	
24	15.08.1947	42.5	45.0	5.5	60	55
25	29.06.1948	41.6	46.4	6.1	77, 78	
26	30.08.1948	41.9	48.0	5.4	85	89
27	02.11.1951	42.3	45.3	5.3	59, 60	
28	30.04.1953	41.0	48.1	5.0	95	
29	21.03.1956	40.92	48.39	5.3	100	
30	26.01.1957	42.58	42.3	5.3		38
31	29.01.1957	42.47	42.45	5.1	38	
32	06.05.1958	43.15	47.77	5.5		73
33	16.07.1963	43.25	41.58	6.4	24, 29	
34	20.04.1966	41.79	48.15	5.4	89, 91	
35	12.07.1966	44.7	37.3	5.8	2	5
36	17.06.1969	43.27	45.19	5.1	62	61
37	14.05.1970	43.0	47.09	6.6	72	
38	17.05.1970	42.98	46.92	5.2	71, 72	
39	04.12.1970	43.84	39.34	5.1		9
40	20.12.1971	41.23	48.38	5.5	92, 96	
41	04.08.1974	42.2	45.9	5.1		68
42	23.12.1974	42.95	46.82	5.0	71, 72	
43	09.01.1975	43.09	47.1	5.2	72	
44	28.07.1976	43.17	45.60	6.2	62, 63	

Table 6.5. (Continued)

#	Date	Coordinates $\varphi°N$	$\lambda°E$	Magnitude M	Intersections at a distance r $r \leq 25$ km	$r \leq 30$ km
45	26.05.1978	41.91	46.48	5.3	76	
46	03.09.1978	44.38	38.03	5.5	6	
47	01.03.1983	43.2	45.30	5.4	62	

The characteristics used for recognition are given in Table 6.6. Most of them were measured within a circle of radius 25 km surrounding the intersection of lineaments.

The objective discretization of parameters was used for coding (see Sect. 6.2.1). The number of parts was three for H_{max}, H_{min}, ΔH, $\Delta H/l$, Q, r_1, r_2,

Table 6.6. Characteristics of intersections of morphostructural lineaments in the Greater Caucasus

Characteristics	First threshold	Second threshold
Maximum altitude H_{max}, m	2274	3562
Minimum altitude H_{min}, m	90	460
$\Delta H = H_{max} - H_{min}$, m	2294	3237
Altitude gradient $\Delta H/l$, m km^{-1}; l is the distance between the points where H_{max} and H_{min} were measured	57.25	83
The percentage of Quaternary deposits, $Q\%$	6.5	32.5
The highest rank of the lineament R_L	2	–
The number of lineaments, n_L	2	–
The number of lineaments in the circle of radius 25 km, N_L	2	–
The distance to the nearest intersection r_{int}, km	26.5	–
The distance to the nearest first rank lineament r_1, km	6.5	29.5
The distance to the nearest second rank lineament r_2, km	5.5	51.5
The number of faults on the geological map, N_F	5	12
Maximum value of the Bouguer anomaly B_{max}, mGal	−51	−17.5
Minimum value of the Bouguer anomaly B_{min}, mGal	−72.5	−101
$\Delta B = B_{max} - B_{min}$, mGal	47.5	-
Minimum distance between two Bouguer anomaly isolines (10 mGal multiple), $(\nabla B)^{-1}$, km	2.4	3.85
Combinations of Large Topographic Forms *(Yes or No)*		
Mountain ranges separated by a longitudinal valley (m/m)		
Continental slope (CS)		
A mountain range and a piedmont plane (m/p)		
A mountain range and piedmont hills and a piedmont plane $(m/pd/p)$		
A mountain range and piedmont hills (m/pd)		
Piedmont hills and a piedmont plain (pd/p)		

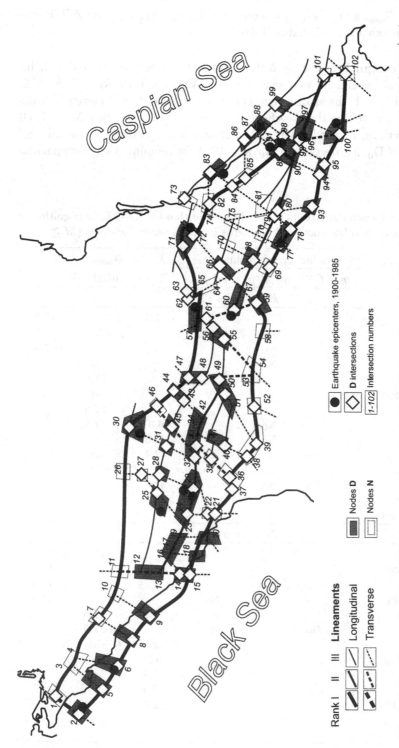

Fig. 6.7. The Greater Caucasus: MZ map and pattern recognition of intersections of earthquake-prone lineaments ($M \geq 5.0$) [GGR[+]88]

N_F, B_{max}, B_{min}, and $(\nabla B)^{-1}$, whereas for R_L, n_L, N_L, r_{int}, and ΔB, it was two. The thresholds are listed in Table 6.6.

Pattern recognition ($M \geq 5.0$). Similar to nodes (see Sect. 6.3.1), intersections of lineaments were classified by CORA-3. With $k_1 = 7$, $\overline{k}_1 = 2$, $k_2 = 4$, and $\overline{k}_2 = 1$, the algorithm selected eleven **D** traits and eleven **N** traits (Table 6.7), controlling the classification given in Table 6.8, when $\Delta = -1$. Of 102 intersections, 72 and 30 were put in **D** and N, respectively. **D** includes all 20 objects of $\mathbf{D}_0$, 9 objects from $\mathbf{N}_0$, and 43 of the remaining set. Comparison

Table 6.7. Characteristic traits selected by algorithm CORA-3 for recognition of intersections of morphostructural lineaments in the Greater Caucasus ($M \geq 5.0$)

#	H_{min}, m	Topographic form combinations				Q, %	B_{max}, mGal	ΔB,
		m/m	CS	$m/pd/p$	pd/p			
				D traits				
1						> 32.5		> 47.5
2				No			> −51	> 47.5
3		No		No				> 47.5
4				No			> −51	
							≤ −17.5	
5	≤ 460						> −51	
							≤ −17.5	
6				No		> 6.5	≤ −17.5	
7				No		> 32.5	> −51	
8			No			> 32.5	> −51	
9	≤ 460				Yes		> −51	
10			No	No		> 32.5		
11	> 90 ≤ 460					> 6.5		
				N traits				
1					No		> −17.5	≤ 47.5
2	≤ 460						> −17.5	≤ 47.5
3				Yes				≤ 47.5
4	> 460				No			≤ 47.5
5						≤ 32.5	> −17.5	
6			No		No		> −17.5	
7						≤ 32.5	≤ −51	
8	> 90					> 6.5 ≤ 32.5		
9		No			No	≤ 32.5		
10	≤ 460				No	≤ 32.5		
11	> 90 ≤ 460					≤ 6.5		

of Tables 6.8 and 6.5 shows that each epicenter of $M \geq 5.0$ recorded in the Greater Caucasus has at least one intersection in its narrow vicinity assigned to class **D**.

Five characteristics that compose the decision rule (Table 6.7) are essential for recognition. The classification alters for more then 10% if any of them is excluded from the binary vector spaces. The minimum change (12 objects) occurs with the exclusion of a "combination of large topographic forms," whereas the maximum change (17 objects) follows from excluding either Q or B_{max}.

Table 6.8. Classification of intersections of the Greater Caucasus ($M \geq 5.0$)

Intersection number	Class	Intersection number	Class	Intersection number	Class	Intersection number	Class
Training set D_0		15	D	22	D	66	D
2	D	16	N	23	N	67	D
6	D	17	N	24	D	68	D
25	D	18	N	28	D	69	D
29	D	26	N	30	D	70	D
36	D	27	D	31	D	71	D
38	D	41	N	32	D	73	D
47	D	46	D	33	D	74	D
55	D	50	D	34	N	77	D
60	D	51	N	35	D	79	D
62	D	52	N	37	D	80	D
72	D	53	N	39	D	82	D
76	D	54	N	40	D	84	D
78	D	75	N	42	N	86	D
83	D	81	N	43	D	87	D
85	D	101	D	44	D	89	D
88	D	102	D	45	D	90	D
93	D	$W \setminus (D_0 \cup N_0)$		48	D	91	D
98	D	1	D	49	D	92	D
100	D	3	N	56	D	94	D
Training set N_0		4	N	57	N	95	D
7	D	5	D	58	N	96	D
8	D	9	D	59	D	97	D
10	N	14	D	61	D	99	D
11	N	19	N	63	D		
12	N	20	N	64	N		
13	N	21	D	65	N		

Control tests. *The Seismic Future test* gives the same classification as the main case when $\bar{k}_1 = \bar{k}_2 = 0$, $k_1 = k_2 = 3$, and $\Delta = 1$. When $\bar{k}_1 = \bar{k}_2 = 1$, the classification differs in two objects (object 25 passes to class **N** and object 97 goes to class $\mathbf{D}$). The parameters $\bar{k}_1 = 2$, $\bar{k}_2 = 1$ (or even less restrictive $\bar{k}_1 = \bar{k}_2 = 2$) yield the classification that differs from the main case in the same two objects. This is less than 2% of the total number of recognition objects.

Three tests to validate the classification stability were done. In the first test, the training sets included all objects, so that $\mathbf{D}'_0 = \mathbf{D}$ and $\mathbf{N}'_0 = \mathbf{N}$ of the main case classification. After new learning, the algorithm determined the classification that differed from the main case on three objects only (object 7 goes to $\mathbf{N}$, objects nos. 26 and 97 pass to $\mathbf{D}$), which is less than 3% of their total number.

In the second test, objects 11–20, 31–40, 51–60, 71–80, and 91–100 were removed from the training sets of the first test. The classification obtained from the reduced learning differed from that of the main case in five objects (objects 18, 20, 51, 74, and 97 passed to $\mathbf{D}$), which is less than 5% of the total number of recognition objects.

In the third test, the algorithm used the reduced learning sets complementary (in $\mathbf{D}$ and $\mathbf{N}$) to those of the second test. The resultant classification differs from the main case in objects 26 and 90 (both go from $\mathbf{N}$ to $\mathbf{D}$) which is less than 2% of the total number of objects.

The "jack-knife" test changes the classification of three objects when they are excluded from the training sets. Object 25 passes from $\mathbf{D}$ to $\mathbf{N}$; objects 26 and 51 go from $\mathbf{N}$ to $\mathbf{D}$.

The classification obtained in the VSEF experiment (see Sect. 6.2.2) coincides exactly with that of the main case, confirming that it does not depend on the choice of traits from the equivalence groups.

Discussion of results. In the Greater Caucasus, the characteristic traits of nodes of high seismic potential prone to $M \geq 5.5$ earthquakes (Table 6.3) include alliances of the following seven characteristics: minimum altitude $H_{\min}$, relief energy ΔH, combinations of topographic forms, percentage of Quaternary deposits Q, distance between the nearest intersections in neighboring nodes r_{int}, the number of faults in the geologic map N_F, and the gravity characteristic H_B. They indicate intensive neotectonic movements evident from small values of $H_{\min}$, large values of ΔH, a relatively large percentage of Quaternary deposits ($Q > 5\%$ or $Q > 20\%$), and the absence of a noncontrast combination of topographic forms (mountain range/piedmont hills and piedmont hills/plain). In addition, the traits of nodes $\mathbf{D}$ point to increased fragmentation of the crust characterized by small distances $r_{\text{int}} \leq 27$ km and by a large number of faults, $N_F > 5$.

Nodes $\mathbf{N}$ are characterized by complementary values of the same characteristics. Three traits of nodes $\mathbf{N}$ include small values of HB ($HB \leq 1050$) indicating, although indirectly, a more homogeneous crust in a typical node $\mathbf{N}$ compared to a node $\mathbf{D}$.

In addition to D_0 nodes hosting earthquakes with $M \geq 5.5$ that occurred after 1900, the other 25 nodes have been assigned to class **D**. Six of them (15, 24, 40, 55, 58, and 62) host historic earthquakes. Three nodes (1, 50, and 52) with historic epicenters have been recognized as **N**, which is probably due to the poor accuracy of historic records of about 1° and 0.5 of magnitude unit [KS77].

Earthquakes with $M \geq 5.5$ have not occurred so far at 19 out of 39 nodes **D**. Six of them (9, 29, 32, 37, 43, and 56) host earthquakes with $5.0 \leq M \leq 5.4$ that occurred after 1900.

There are 20 nodes **D** situated along the boundaries of the mountain country (on first rank lineaments) and 19 nodes in the Greater Caucasus interior (see Fig. 6.5). Three groups, each of the same type of **D** traits, can be recognized among nodes **D**. The first group (see Fig. 6.5) is at the boundary between the Greater Caucasus and the Black Sea basin (nodes 2, 4, 5, 7, 9, 12, 14, 15). The second group sits in the junction zone of the southern flank of the Greater Caucasus in the Agrichai-Alazany basin (nodes 40, 45, 56, 57). The third group includes nodes from the three locations: in the central segment of the first rank northern lineament (nos. 22, 31, 32, 38), within the zone of the Main Caucasus Thrust (11, 13, 17, 28), and on the Pshekish-Tyrnyauz lineament (18, 21, 23, 29).

D nodes of the first group located along the Black Sea coast are characterized by intensive neotectonic movements evident from high relief energy ($\Delta H > 2100$ m), a large percentage of Quaternary deposits ($Q > 20\%$), and lowered areas ($H_{\min} \leq 200$ m). Nodes of the second group possess a more complicated set of **D** traits that combine indicators of increased fragmentation of the crust ($r_{\text{int}} \leq 27$ km and $N_F > 5$) and intensity of neotectonic movements ($Q > 20\%$ and $H_{\min} \leq 600$ m). Indicators of increased fragmentation dominate in the third group.

It's worth mentioning that the Shemakha (63) and Sulak (48) nodes, where the strongest earthquakes were observed in the Greater Caucasus after 1900, possess the same **D** traits combining indicators of increased crust fragmentation ($N_F > 5$) and intensive neotectonic movements ($H_{\min} \leq 600$ m, $\Delta H > 2100$ m). The same traits characterize the Sochi node (12), where strong earthquakes are so far unknown.

Twenty six nodes of the Greater Caucasus were assigned to class **N**. Similar to **D** nodes, they cluster in certain regions of the mountain country (see Fig. 6.5). **N** nodes 1, 3, 6, 8, 10, and 19 sit along the western segment of the first rank northern lineament. **N** nodes 16, 26, 27, 34, 35, and 36 are located in a compact area of the central section of the first rank southern lineament, nodes 50, 56, 64, and 65 are in the northwestern and eastern boundaries of the mountain country, and nodes 47, 46, 51, 54 are in Daghestan. The first rank lineament zones with these nodes include the lower part of the mountain flank and piedmont hills which indicates relatively weak intensity of movements in the Quaternary there.

Comparison of the two classifications. It is natural that both classifications of nodes and of intersections are very similar with respect to spatial distribution of earthquake-prone areas as well as in parametric description.

We postulate in this comparison that each **D** node capable of $M \geq 5.5$ earthquakes must include at least one **D** intersection prone to $M \geq 5.0$. The comparison of both classifications (Tables 6.4, 6.8, Figs. 6.5 and 6.7) shows that there is disagreement only in four **D** nodes: 11, 13, 14, and 15 containing **N** intersections 12, 13, 16, 17, 18, 19, and 20. All of these objects are closely located in the Sukhumi-Sochi region comprising a relatively small area of the Greater Caucasus (see Fig. 6.7). Both classifications are even in better agreement across the rest of the territory.

The comparison of the **D** and **N** traits of nodes and intersections shows their consistency. At the same time they indicate a certain difference. The **D** traits of nodes (see Table 6.3) include characteristics from all three groups mentioned above in Sect. 6.2.2, whereas the **D** traits of intersections include those only from groups A and C.

Thus, either classification suggests a reliable solution in distinguishing the earthquake-prone areas of the Greater Caucasus. The earthquakes that happened after publishing the results confirm this conclusion (see Sect. 6.5).

6.4.2   The Western Alps

Unlike other searches for earthquake-prone areas, the Western Alps is the region where MZ was applied to part of the mountain country, i.e., the Alps [GR82]. It was also the first time that earthquake-prone areas were recognized for such a relatively small magnitude, i.e., $M_0 = 5.0$ [CGG$^+$85].

Morphostructural zoning of the Western Alps. We define the Western Alps as a part of the Alps west of the line connecting Boden and Como lakes. The morphostructural map of the Western Alps (Fig. 6.8) was compiled on a scale of 1:1,000,000 through joint analysis of geomorphic, tectonic, geologic and satellite data [GR82, CGG$^+$85].

Lineaments of the first rank correspond to prominent regional faults separating the folded mountain structure of the Western Alps from the adjacent platform areas and the deep Ligurian Sea basin. Lineaments of the second rank divide this mountain country into eight megablocks that have different relief heights, orientations of mountain ranges, and geologic structures. The megablocks have different geometrical shapes and linear dimensions. They extend from 50 to 150 km along the mountain chain. Their width varies from 40 – 50 and 75 – 100 km.

Second rank longitudinal lineaments correspond to major tectonic faults. The zone of lineament 13–42 (Brianconnais fault) divides the External Alps (megablocks I, III, and VII) (Fig. 6.8) and the Internal Alps (megablocks II, IV, VI, and VIII) composed of Helvetic and Pennine rock complexes,

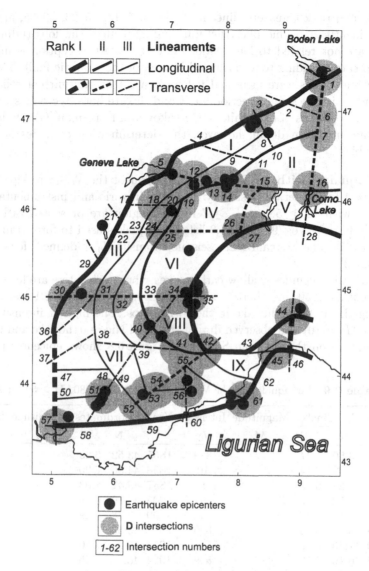

Fig. 6.8. The Western Alps: MZ map and pattern recognition of earthquake-prone intersections of lineaments ($M \geq 5.0$) [CGG$^+$85]

respectively. The Internal Alps are overthrusting the External Alps along the fault; the process caused by pressure exerted by the Adria plate; this has produced a deep-seated thrust dipping 35–45° in the crystalline basement [Bea81]. Lineament 7–14 corresponds to the extension of the Brianconnais fault into the Swiss Alps. A seismic geotraverse [Ryb79] showed that the fault plane is subvertical here. Lineament 15–16 represents the western part of the Insubric fault. The fault plane has a low-angle northward dip [Ryb79].

Second rank transverse lineaments 1–16, 5–14, 15–26, 30–35, and 42–52 were identified by morphostructural zoning. Unlike the longitudinal ones, they are not related to long major fault zones. An exception is lineament 15–26 corresponding to a southward branch of the Insubric fault. The transverse lineaments were identified following a set of intermittent relief forms extending in the same direction: fault and flexure lines, tectonic scarps, and straight segments of river valleys. Gorshkov and Rantsman [GR82] described the lineaments in detail; however, the determination of nodes is not yet available.

Earthquakes with $M \geq 5.0$. The region of the Western Alps is characterized by moderate seismicity. Data on historic and instrumental earthquakes with $M \geq 5.0$ or $I_0 \geq 7$ in this region are presented rather completely [Cap81, Vog79]. Table 6.9 lists earthquakes used to form training sets [CGG+85]. The Mercalli macroseismic intensity I_0 is defined, for example, in [Bol93]

Table 6.9 includes shallow earthquakes whose epicenters are located with an accuracy of 50 km or better. The largest magnitude, $M = 6.1$, was recorded during the 1946 earthquake in the Swiss Alps. The highest intensity in the region ($I_0 = 10$) was observed during the 1564 Nice earthquake and the 1887 earthquake on the Ligurian Sea coast. Figure 6.8 shows epicenter positions

Table 6.9. Earthquakes in the Western Alps: 1500–1980, $M \geq 5.0$ or $I_0 \geq 7$

Year	Coordinates		Magnitude	Intensity	Year	Coordinates		Magnitude	Intensity
	$\varphi°$N	$\lambda°$E	M	I_0		$\varphi°$N	$\lambda°$E	M	I_0
1509	43.50	5.47		8–9	1855	43.52	6.26		8
1564	44.02	7.14		10	1858	44.50	7.20		8
1584	46.23	6.56		8	1887	43.42	8.03		10
1601	46.58	8.22		9	1905	43.59	6.56		8
1644	43.53	7.18		9	1909	43.48	5.19		9
1708	43.48	5.46		8–9	1913	43.53	5.51		7–8
1720	47.45	9.60		7–8	1914	45.05	7.20	5.2	7
1755	46.19	7.59		8–9	1924	46.15	7.55	5.1	7
1759	45.00	7.10		9	1935	44.35	6.38	5.0	7–8
1767	45.15	7.30		8	1945	44.81	9.12	5.0	7
1773	44.22	4.48		7–8	1946	46.19	7.30	6.1	8
1774	46.63	8.38		8	1947	44.45	7.16	5.5	7–8
1796	47.21	8.40		8	1954	46.15	7.17	5.1	7
1808	44.52	7.14		8–9	1955	44.32	7.18	5.0	7
1812	43.45	5.42		8	1959	44.32	6.47	5.5	7–8
1822	45.49	5.49		8	1962	45.01	5.32	4.8	7–8
1831	43.51	7.51		9	1964	46.55	8.16	5.3	7
1854	43.45	7.50		9	1980	45.01	7.24	5.3	7
1855	46.18	7.53		9					

from Table 6.9 on a MZ map. Practically all epicenters are nucleated about lineament intersections at a distance not exceeding 25 km. An exception is the epicenter of the 1767 earthquake.

Pattern recognition: Learning stage. Intersections of morphostructural lineaments [CGG+85] were natural objects of pattern recognition. All intersections were divided a priori into three sets: training sets D_0 and N_0 and the remaining objects. D_0 includes 14 intersections located near epicenters of earthquakes with $M \geq 5.0$, 1900–1980. If an epicenter was at a distance less than 25 km from two intersections, both of them were included in D_0. As a result, 14 intersections (3, 12, 13, 14, 20, 30, 31, 35, 40, 41, 42, 44, 51, and 57) constituted D_0. Intersections 1, 5, 6, 8, 53, 55, 56, 58, 60, and 61 hosting historic earthquakes with $I_0 \geq 7$ were excluded from N_0, as well as intersections 18 and 19. The latter are near the 1905 epicenter represented in D_0 by the nearest intersection 20. The remaining 36 intersections composed the training set N_0.

Table 6.10 lists the characteristics and the discretization thresholds used for recognition. Except for the combination of topographic forms, their binary coding was S type (see Sect. 6.2).

Table 6.10. Characteristics of intersections in the Western Alps

Characteristics	First threshold	Second threshold
Maximum altitude H_{max}, m	2686	4807
Minimum altitude H_{min}, m	325	–
$\Delta H = H_{max} - H_{min}$, m	2500	–
The percentage of Quaternary deposits, Q %	10	–
The highest rank of the lineament R_L	1	2
The number of lineaments n_L forming an intersection	2	–
The number of lineaments in a circle of 25 km radius, N_L	3	4
The distance to the nearest intersection r_{int}, km	20	31
The distance to the nearest first rank lineament r_1, km	0	32
The distance to the nearest second rank lineament r_2, km	0	40
The maximum value of the Bouguer anomaly B_{max}, mGal	−82	−8
The minimum value of the Bouguer anomaly B_{min}, mGal	−145	−85
$\Delta B = B_{max} - B_{min}$, mGal	45	65
Minimum distance between two Bouguer anomaly isolines spaced at 10 mGal $(\nabla B)^{-1}$, km	2	3
Combinations of Large Topographic Forms	(Yes or No)	
Mountain ranges separated by a longitudinal valley (m/m)		
A mountain range and a piedmont plain (m/p)		
A mountain range and piedmont hills (m/pd)		
Piedmont hills and a piedmont plain (pd/p)		

Table 6.11. Characteristic traits selected by algorithm CORA-3 for the Western Alps

#	Q, %	n_L	N_L	r_1, km	r_2, km	$(\nabla B)^{-1}$, mGal	ΔB, km
					D *traits*		
1				≤ 32		≤ 2	≤ 65
2				> 0		≤ 2	≤ 65
3			> 3	≤ 32	0		≤ 65
4			> 4		0		≤ 65
5						≤ 3	> 45
6					$\geq 0;\ \leq 40$		> 45
7		2		> 32			45
8		2		> 32		≤ 3	
9		> 2	≤ 3			≤ 2	
10	> 10		> 3		≤ 40		
11	> 10	> 2		≤ 32			
					N *traits*		
1						> 2	≤ 45
2					> 0		≤ 45
3		2					≤ 45
4					> 40		≤ 45
5					> 40	> 2	
6		2			> 40		
7		2	≤ 3		> 0		
8		2		0			

Table 6.12. Classification of intersections in the Western Alps

Number	Class	Number	Class	Number	Class	Number	Class
Training set D_0		*Training set* D_0		29	N	59	N
3	D	2	N	32	D	62	N
12	D	4	N	33	N	$W \setminus (D_0 \cup N_0)$	
13	D	9	N	34	D	1	D
14	D	10	N	36	N	5	D
20	D	11	N	37	N	6	D
30	D	15	D	38	N	8	D
31	D	16	D	39	N	18	D
35	D	17	N	43	N	19	D
40	D	21	N	45	D	53	D
41	D	22	N	46	N	55	D
42	D	23	N	47	N	56	D
44	D	24	N	48	N	58	N
51	D	25	D	49	N	60	N
57	D	26	D	50	N	61	D
		27	D	52	D		
		28	N	54	D		

The main case of classifying the 62 binary vectors was obtained through CORA-3 with $k_1 = 3$, $\bar{k}_1 = 2$, $k_2 = 11$, $\bar{k}_2 = 1$, $= 1$, and $\Delta = 0$. The main case resulted in the eleven **D** traits and eight **N** traits listed in Table 6.11; 34 vectors were assigned to **D**, and the remaining 28 to **N** (Table 6.12). **D** included all $\mathbf{D}_0$, 11 objects from $\mathbf{N}_0$, and 9 objects from outside the training sets.

Control tests. *The SF test* run with $\bar{k}_1 = \bar{k}_2 = 0$, $k_1 = 12$, and $k_2 = 10$ replicates the main case. When $\bar{k}_1 = \bar{k}_2 = 1$, the classification differed from the main case only in three objects (33, 43, and 59 passed to **D**). When the parameters of the main case $\bar{k}_=2$ and $\bar{k}_2 = 1$ were used in the SF test, the classification obtained differed again from the main case in three objects (29, 33, and 43 passed to **D**).

Two *stability validation tests* were executed. Training sets of the first test included vectors with 1–10, 21–30, and 41–50. $\mathbf{D}'_0$ and $\mathbf{N}'_0$ were combined from these vectors assigned to **D** and **N** in the main case, respectively. The classification obtained from $\mathbf{D}'_0$ and $\mathbf{N}'_0$ differed from the main case in six objects: 33, 39, and 59 passed to **D**, whereas 18, 51, and 56 moved to **N**. Sets $\mathbf{D}'_0$ and $\mathbf{N}'_0$ in the second test were formed in a similar way from vectors 11–20, 31–40, and 51–62. The result differed from the main case in four objects: 29 moved to **D**, whereas 1, 3, and 30 passed to **N**.

The classification obtained in the *VSEF test* (see Sect. 6.2.2) completely agrees with the main case.

For tests with random data, 100 pairs of training sets $\overline{\mathbf{D}}_0$ and $\overline{\mathbf{N}}_0$ with the same number of vectors as in the main case were generated at random. In recognizing these 100 random problems, $\overline{\mathbf{D}}_0 \subseteq \mathbf{D}$ was obtained in 17 cases. The measure of nonrandomness of the main case is, therefore, 0.17.

Discussion of results. All intersections of morphostructural lineaments in the Western Alps were assigned to **D** and **N** classes using their characteristic traits.

It is seen from Fig. 6.8 that **D** intersections form compact clusters along second rank, mostly transverse lineaments. Larger clusters relate to second rank transverse lineaments 1–16, 5–15, and 42–52. Twelve **D** intersections are on first rank lineaments. Most **D** intersections (22 of 34) belong to second rank lineaments. Only three **D** intersections (18, 51, and 56) are formed by third rank lineaments. This indicates indirectly that active tectonic processes concentrate along higher ranking morphostructural boundaries, mostly within the inner part of the Western Alps. **D** intersections 1, 3, 5, 26, 27, 30, 34, and 42 on first rank lineaments are located in the areas of topographic contrasts where mountain flanks steeply join piedmont plains. For the most part, **D** traits indicate increased fragmentation of the crust (see Table 6.11) characterized by large values of n_L and N_L, closeness of higher rank lineaments ($r_2 \leq 32$ km or $r_1 = 0$ km), large values of ΔB, and small values of

$(\nabla B)^{-1}$. **D** intersections possess various combinations of **D** traits. Those on first rank lineaments are characterized by large values of Q accompanied by large n_L and N_L. Those inside the mountain country are characterized by indicators of increased crust fragmentation and by closeness to second rank lineaments.

N traits point to weaker crust fragmentation ($n_L = 2$; $N_L \leq 3$) and remoteness from second rank lineaments ($r_2 > 40$ km). Note that the distance between **N** intersections and second rank lineaments is very important for identification. Five out of eight **N** traits include the condition $r_2 > 40$ km.

Control tests justified a high reliability level of the solution obtained in distinguishing earthquake-prone areas of the Western Alps. The earthquakes that happened after the first publication [CGG+85] confirm this conclusion (see Sect. 6.5).

6.4.3 Pattern Recognition Applied to Earthquakes in California

This section presents patterns distinguishing places in California and adjacent territories of Nevada where epicenters of large earthquakes occurred in the past and can occur in the future. Pattern recognition problems solved in 1976 [GGK+76] treated two different sets of natural recognition objects: regularly spaced points along major strike–slip faults and intersections of morphostructural lineaments.

Earthquake-prone segments of major strike–slip faults of California. Strong strike–slip earthquakes with $M \geq 6.5$ listed in Table 6.13 were

Table 6.13. Strike–slip earthquakes with $M \geq 6.0$ in California

#	Date	$\varphi°$N	$\lambda°$W	Magnitude M	#	Date	$\varphi°$N	$\lambda°$W	Magnitude M
1	1836	37.50	121.90	> 7	13	08.06.1934	36.00	120.50	6.0
2	1836	37.70	122.10	> 7	14	30.12.1934	32.25	115.50	6.5
3	1836	37.80	122.30	> 7	15	31.12.1934	32.00	114.75	7.0
4	1857	34.70	116.80	> 7	16	25.03.1937	33.50	116.50	6.0
5	18.04.1906	38.25	122.95	8.25	17	19.05.1940	32.70	115.50	6.7
6	01.07.1911	37.25	121.75	6.6	18	21.10.1942	33.00	116.00	6.5
7	21.04.1918	33.75	117.00	6.8	19	04.12.1948	33.90	116.40	6.5
8	10.03.1922	35.75	120.25	6.5	20	08.10.1951	40.25	124.50	6.0
9	22.01.1923	40.50	124.50	7.2	21	22.11.1952	35.80	121.20	6.0
10	23.07.1923	34.00	117.25	6.25	22	19.03.1954	33.38	116.18	6.2
11	22.10.1926	35.75	122.00	6.1	23	09.04.1968	33.40	116.20	6.9
12	11.03/1933	33.60	118.00	6.25					

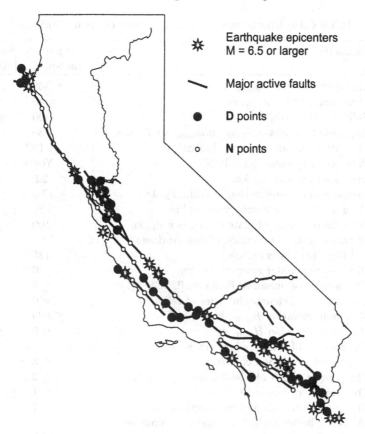

Earthquake epicenters
M = 6.5 or larger

— **Major active faults**

● **D points**

○ **N points**

Fig. 6.9. Pattern recognition of earthquake-prone segments of major faults in California ($M \geq 6.5$) [GGK$^+$76]

selected for training sets [GGK$^+$76]. It was assumed that epicenters associate with major strike–slip faults (referred to as faults in what follows). [USGS62] with a few additions is the principal source of information on the system of faults (Fig. 6.9). The problem was to find where on these faults earthquakes can occur.

Natural recognition objects were selected as follows. Epicenters of earthquakes with $M \geq 6.5$ (see Table 6.13) were projected on the nearest faults. Additional points on faults were spaced about 50 km apart. Additional points neighboring the projections of epicenters (called their associates) were at a distance about 25 km from respective projections. The projections and the additional points were combined into the training set.

Pattern recognition: Learning stage. Training sets were defined as follows: D_0 included the projections and their associates that corresponded to earthquakes with $M \geq 6.5$ (Table 6.13). A projection together with its associates formed a cluster. Two clusters were excluded because the corre-

Table 6.14. Characteristics of points along faults in California[a]

#	Characteristics	First threshold	Second threshold
1	Maximum altitude h_{max}, m	≤ 500	≤ 1250
2	Minimum altitude h_{min}, m	≤ 50	≤ 500
3	Relief contrast Δh, m	≤ 500	≤ 100
4	$\Delta h/l$, where points of h_{max} and h_{min} are l km apart	≤ 30	≤ 70
5	Percentage of Quaternary sediments, $Q\%$	≤ 10	≤ 50
6	Presence of igneous rocks, IGN	Yes or No	
7	Distance to a fault r_1, km	≤ 12.5	≤ 37.5
8	Distance to an intersection of faults r_2, km	≤ 12.5	≤ 37.5
9	Distance to a geothermal zone r_3, km	≤ 25	≤ 75
10	Distance to a zone of plate divergence r_4, km	≤ 100	≤ 200
11	Distance to the intersection of San Andreas and Big Pine faults r_5, km	≤ 125	≤ 375
12	Distance to a water reservoir r_6, km	≤ 0	≤ 25
13	The number of unnamed faults on [USGS62], n_1	≤ 3	≤ 5
14	The number of changes in types of relief, n_2	$= 0$	
15	Maximum elevation H_{max}, m	≤ 500	≤ 1500
16	Minimum elevation H_{min}, m	≤ 0	≤ 200
17	The number of contacts between rocks of different age [USGS65], n_3	≤ 8	
18	The number of parallel faults, n_4	≤ 2	≤ 3
19	The number of faults, n_5	≤ 1	≤ 3
20	The number of ends and intersections, n_6	≤ 1	≤ 2
21	The angle between the fault and the dominant structural trend in the region α, degrees	≤ 10	≤ 20

[a]Characteristics 1–6, 13, and 14 were measured inside circles of radius 12.5 km; characteristics 15–17 and 19 were measured inside circles of radius 25 km

sponding epicenters occur too far from the nearest fault. The remaining 13 clusters were used as $\mathbf{D_0}$. $\mathbf{N_0}$ contained the total of 34 points. These were at distances more than 25 km from all projections, and none of them could be associated with earthquakes of $6.0 \leq M < 6.5$.

Table 6.14 lists the characteristics used to describe points along the faults. Most of them characterize the degree of intensity and the contrast of tectonic movement expressed in the topography as well as in the fragmentation of the crust. Some characteristics reflect various hypotheses on conditions favorable for the occurrence of strong earthquakes.

Pattern recognition: Earthquake-prone segments of major strike–slip faults. Table 6.15 displays the characteristic traits found by CLUSTERS. Each of the eight $\mathbf{D}$ traits and eleven $\mathbf{N}$ traits combines either two or three characteristics.

Table 6.15. Characteristic traits selected by algorithm CLUSTERS for recognition of points on major faults in California

#	h_{max}	$\Delta h/l$	Q	IGN	r_2	r_3	r_6	H_{min}	n_3	n_5	α
					D *traits*						
1					≤ 37.5	≤ 75					≤ 10
2				No	≤ 37.5						≤ 10
3				No	≤ 37.5					≤ 3	
4		> 10			≤ 37.5						> 1
5					≤ 37.5			≤ 200	≤ 8		
6		> 10			≤ 37.5			≤ 0			
7				No	≤ 37.5	> 25					
8			.50		≤ 37.5	> 25					
					N *traits*						
1							> 20		> 8	≤ 3	
2					> 37.5			≤ 0		≤ 3	
3				Yes		> 25				≤ 3	
4	> 500			Yes						≤ 3	
5					> 12.5		> 20		> 8		
6					> 12.5	> 75			> 8		
7						> 75			> 8		
8				Yes	> 37.5	> 25		≤ 200			
9					> 37.5			≤ 200			
10	> 500	> 10			> 37.5			≤ 200			
11	> 500	> 10			> 12.5						

The traits were used to recognize all 100 selected points. The recognition succeeded in finding many characteristic traits of **D** and **N** classes and in a distinctive division of points into two groups. The results of classification are shown in Fig. 6.9.

Pattern recognition: Control tests. The stability *SH test* appears successful since the results were based, on the whole, on a partial catalog. In the Seismic History test, at least one point in each cluster was recognized as **D** before the corresponding earthquake occurred, and information about it was used in learning. Moreover, at least one point in each cluster was recognized as **D** at all steps since 1911 with three exceptions; the clusters corresponding to the earthquakes of 1923, 1940, and 1942 were not recognized with learning during periods terminating at 1918 and 1923. However, at least one object in these clusters was recognized as **D** before the actual earthquake for all other steps in the test.

  The SF test illustrates the limiting stability of classification: The identification of only seven points changes. Four **D** points are recognized as **N**, and vice versa three **N** points are classified as **D**.

Thus, one may conclude from the SH and SF tests that earthquake-prone areas neither wander nor proliferate across all Californian faults, but strong earthquakes are predicted for the limited number of **D** points identified.

Inverse SH test. In this ST test, earthquakes were excluded from being used in learning in the order of their occurrence, beginning with 1836. Recognition remains successful in predicting the area of the excluded epicenter until the 1906 earthquake. Points near its epicenter are not recognized as **D**. On removing the 1911 epicenter, a **D** point is recognized in each cluster. When learning is based on seven earthquakes from 1918–1948, the sites of all earthquakes from 1836–1911, except for 1906, were recognized as **D**.

In other ST tests, several areas were selected. The epicenters in these areas were assumed to be unknown, and the corresponding clusters were eliminated from training sets. The results suggested that recognition was reasonably stable under such variations of training sets. They also indicated the degree of homogeneity in spreading the training set across the **D** area necessary for successful classification.

Variations of vector space were also performed to evaluate the informational power of the characteristics. In this test, each characteristic (one at a time) was removed from the data, and then learning and classification were redone. As a result, the distance to a fault end or an intersection of faults, r_2, it was found, was a key characteristic. In its absence, it was necessary to set significantly weaker limitations for characteristic traits (see Sect. 6.4.1) to get a decent number of them.

Variation of the algorithm had also shown certain stability. In this test, D_0 consisted of the epicenter projections, N_0 remained unchanged, and CORA-3 substituted CLUSTERS. The classification was very similar to the main case.

Recognition in a random vector space that simulated the problem was also carried out. In this test, binary vectors describing points were selected at random. CLUSTERS used such descriptions along with the initial training sets and thresholds required for the characteristic traits. In this case, 32 **D** traits and no **N** traits were found. The result suggests that the **N** traits selected in the main case are not random. When the numbers of characteristic traits were fitted to those in the main case, the separation of vectors by voting appeared to be rather vague and was not delivering a satisfactory classification without errors.

Discussion of results. The characteristics that make up **D** and **N** traits help to understand what distinguishes earthquake-prone areas from other areas in California. Although the 21 characteristics indicated in Table 6.15 were considered at the learning stage, only eleven were incorporated in the characteristic traits and the decision rule.

The distance to the closest intersection of faults or to the closest end of a fault, r_2, it was found, was a dominant characteristic; it was included in all eight **D** traits and in seven out of eleven **N** traits. The fact that r_2 is apparently small for **D** and large for **N** supports the hypothesis [GGI$^+$72a] that epicenters of strong earthquakes tend to originate near the intersections of major faults and less evident morphostructural lineaments.

Geologic maps of the region suggest that most other characteristics in Table 6.15 favor an indirect indication of low elevation in earthquake-prone areas. These characteristics include the absence of igneous rocks, the proximity to a large area of recent sediments, and a small number of contacts on the geologic map (n_3). The last characteristic behaves contrary to the common expectation that large n_3 should associate with intensive fragmentation of the crust, which is evident for **D** [GGI$^+$72a, GGZ$^+$73, GGZ$^+$74a, GGZ$^+$74b]. However, according to these works, intensive fracturing should be related to neotectonic processes, whereas n_3 represents tectonic history integrated across a longer period. A survey of places with large and small n_3 indicates that small values of n_3 tend to go with subsidence and low elevations.

A small distance r_6 to a large water reservoir is another characteristic of **D** areas. This indirectly implies relatively low elevation and implies ideas on the role of water in triggering earthquakes. However, small r_6 indicates the proximity of many points to the ocean, making the interpretation rather loose.

Small values of α measuring the deviation of a fault from the dominant strike of the San Andreas Fault characterize **D**, thus suggesting that strike–slip earthquakes in the San Andreas Fault system tend to occur away from its bends, whereas earthquakes with substantial dip–slip components are more frequent at the bends.

A large distance from the intersection of the San Andreas and the Big Pine faults is a characteristic of **D**, perhaps for the reason that dip–slip earthquakes tend to occur in Transverse Ranges adjacent to the Big Bend of San Andreas Fault. Other characteristics, such as the distance to a geothermal zone (r_3) and the number of parallel faults (n_4), are also important in **D** or **N** traits.

Some characteristics appearing with the same values in both **D** and **N** traits (for example $n_4 \leq 3$), are somewhat puzzling. They either entered the traits by chance, due to a random interplay of events, or signify a characteristic subdivision of points. It is hard to distinguish the two cases and attribute geologic meaning to such a characteristic without combining it with others that are different in **D** and **N** traits.

Gelfand et al. [GGK$^+$76] drew a qualitative conclusion that **D** areas are characterized by proximity to the end or to an intersection of major faults in association with low relief and often with some kind of downward neotectonic movement expressed in topography and geology. The fact that minimum and maximum elevations are simultaneously small for many **D** points suggests that they are often characterized by relatively low relief or subsidence on the background of a weak uplift. Higher elevations or an uplift with lesser contrasts in relief characterize **N** points near intersections or ends of major faults.

Earthquake-prone intersections of morphostructural lineaments in California and Nevada. *Morphostructural Zoning.* The morphostructural map of California and adjacent territories of Nevada shown in Fig. 6.10 was compiled in 1976 on a scale of 1:2,500,000 [GGK+76]. The map is based on the synthesis of published data available at that time, among them [Atw40, Ric58, USGS62, USGS65, Kha71]. Most longitudinal lineaments correspond to prominent faults. Although transverse lineaments are not that evident, their location is justified and supported by satellite photographs of the region [Ran79].

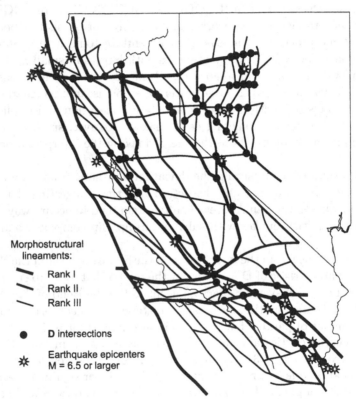

Fig. 6.10. California and adjacent territories of Nevada: MZ map and pattern recognition of earthquake-prone intersections of lineaments ($M \geq 6.5$) [GGK+76]

Pattern recognition: Learning stage. Objects of recognition were selected as 165 intersections of lineaments. Pattern recognition was first made for the land part of California bounded by the latitude of the Mendocino fault on the north, the Mexican border on the south, and the axis of the Great Valley on the east.

The characteristics of the main case recognition are listed in Table 6.16. Characteristic H_1 is determined directly at the point of intersection. The characteristics marked by asterisks in Table 6.16 are determined within 12.5 km

Table 6.16. Characteristics of intersections of morphostructural lineaments in California and adjacent regions of Nevada

Characteristics	First threshold	Second threshold
Absolute altitude H_1, m	≤ 200	≤ 1200
*Maximum altitude H_{max}, m	≤ 1500	
*Minimum altitude H_{min}, m	≤ 100	
$\Delta H/l$	≤ 40	
Type of intersection (crossing or joint)	Yes or No	
How pronounced is the lineament	Strong	Feeble
Distance to the closest lineament r_1, km	0	≤ 25
Distance to the second closest lineament r_2, km	≤ 25	≤ 37.5
Number of lineaments, n_L		
Distance to the nearest intersection r_{int}, km	0	≤ 37.5
Distance to the second nearest of rank 2, R_{22}, km	≤ 50	≤ 100
**The number of intersections, n_2	≤ 4	≤ 7
**The weighted rank of the lineament, $\mathcal{M}$	≤ 15	≤ 17
**Combination of large topographic forms	Yes or No	
Mountain ranges separated by a longitudinal valley, m/m		
A moumtain range and a piedmont plain, m/p		

from the respective intersections. Those marked by double asterisks are taken from a larger circle of radius 62.5 km. The weighted rank $\mathcal{M}$ of a lineament takes the form

$$\mathcal{M} = \sum_{j=1}^{n} \frac{62.5 m_i}{62.5 - R_{ij}} \, ,$$

where i is the rank of the lineament, m_i is the weight ($m_1 = 5$, $m_2 = 4$, $m_3 = 3$), n is the number of lineaments in the circle of radius 62.5 km about the intersection, and R_{ij} is the distance to the jth lineament of rank i.

Pattern recognition: Earthquake-prone intersections of lineaments. Algorithm CLUSTERS was used to classify the intersections of lineaments. The main case characteristic traits are given in Table 6.17 [GGK$^+$76]. Figure 6.10 shows the location of **D** areas.

The recognition is claimed successful [GGK$^+$76] because most intersections are assigned by voting to two rather distinctive groups; only a small number of intersections is given a "neutral" vote. The results are also encouraging for the following reasons.

(1) A noninstrumental catalog for the nineteenth century reports four earthquakes with estimated magnitude 7 or more. Their epicenters are close to **D** intersections.

(2) The intersections closest to epicenters from Table 6.13 are recognized as **D** with only two exceptions: the 1906 San Francisco and the 1971 San Fernando earthquakes. In both cases, certain intersections slightly farther away than the closest were recognized as **D**.

Table 6.17. Characteristic traits selected by CLUSTERS for earthquake-prone intersections of morphostructural lineaments in California ($M \geq 6.5$)

#	H_1, m	m/m	m/p	r_1	r_2	R_{22}	r_{int}, km	H_{max}, m	H_{min}, m	$\Delta H/l$	n_2	$\mathcal{M}$
							D *traits*					
1		Yes				> 50						> 17
2	≤ 200	No										> 17
3								≤ 100				> 15
4							≤ 12.5					> 15
5	> 200						≤ 12.5					> 15
6			Yes		> 25							> 15
7		Yes	≠ 0									> 15
8					> 25			≤ 100			> 4	
9					> 25			≤ 100		> 40		
10				≤ 25				≤ 100		> 40		
11		No			≤ 50			≤ 100				
12				≤ 25				≤ 100				
13	≤ 200			≤ 25	> 25			≤ 1500				
							N *traits*					
1								> 1500				≤ 15
2						> 50					≤ 4	≤ 17
3								> 1500	> 100			≤ 17
4						> 50		> 1500				≤ 17
5		No					≤ 37.5					≤ 17
6		No			≤ 100							≤ 17
7				≤ 25			≤ 37.5					≤ 15
8	≤ 1200	No										≤ 15
9								> 1500				≤ 4
10							≤ 37.5					≤ 4
11					≤ 100					≤ 40		≤ 4
12	≤ 1200	No										≤ 4

Pattern recognition: Control tests. The SF test shows the stability of the classification; only four intersections were shifted to **D**, and two **D** intersections were transferred to **N**. In an *SH test* starting from 1940, the algorithm recognizes at least one **D** intersection in the cluster about the epicenter of each subsequent strong earthquake. The characteristics and their coding were the same at each step of the SH test. Therefore, the results illustrate the stability of recognition under variations of training sets.

To test the sensitivity of results to a particular algorithm used, CORA-3 was applied to the same data. $\mathbf{D_0}$ contained intersections closest to the epicenters, and $\mathbf{N_0}$ included the rest. The results obtained confirmed the main case classification obtained by CLUSTERS.

To check whether an epicenter has some special meaning or it is a random point where stress first reaches the critical strength, the epicenters were replaced by intersections on the entire fault break of the 1906 San Francisco earthquake. Clusters were formed about each of these intersections. The results of recognition appeared to be poor. No intersections were recognized as **D** near 9 out of 12 epicenters.

Two of the earthquakes, the 1952 Kern County and 1971 San Fernando used at the learning stage in the main case are of dip–slip variety. The clusters determined by these earthquakes were excluded from $\mathbf{D}_0$. Characteristic traits did not change noticeably in this test. However, the number of **D** intersections decreases significantly, from 25 in the main case to 17. Only three intersections outside clusters from the learning set are recognized as **D**, whereas a group of **D** intersections of the main case from the Transverse Ranges is eliminated, which is not surprising from a geologic viewpoint.

Discussion of results. One can see good agreement between **D** areas from two classifications: the first obtained for intersections of morphostructural lineaments and the second found for points on major faults mentioned above (see Figs. 6.9 and 6.10). The same five groups of earthquake-prone areas show up in both cases. Slight differences are due to the fact that the intersections of interest cover a larger territory.

Characteristic traits (Table 6.17) show that **D** and **N** intersections are often characterized by low and high elevations (H_1, $H_{\min}$, and $H_{\max}$), respectively. This supports the idea derived from recognition of points that **D** intersections often associate with neotectonic subsidence on top of a background weak uplift.

Additional evidence corroborating the importance of vertical movements at **D** intersections is the presence of large gradients ($\Delta H/l$) and mountain-plain contacts occurring in many **D** traits along with their alternatives holding for **N** traits.

The fragmentation of the crust is more pronounced near **D** than near **N** intersections. This is evident from the presence of high rank lineaments in **D** traits and their absence in **N** traits. Many **N** traits also combine small number of intersections within 62.5 km, n_2, with other characteristics.

The results presented in this section were published in 1976 [GGK+76]. The subsequent earthquakes with $M \geq 6.5$ confirmed the validity of earthquake-prone areas and their description determined at that time (see Sect. 6.5).

6.4.4 Pattern Recognition of the Great ($M \geq 8.2$) Earthquake-Prone Segments of Major Seismic Belts

It is general knowledge that most of the largest earthquakes occur along the Circumpacific rim framing Earth's largest Pacific Plate. The remaining quarter of recorded great earthquakes split nearly-equally between the Alpine-Himalayan seismic belt and intraplate regions (mostly in eastern Asia). Unraveling global morphological structures all over Earth remains a challenging problem. In the absence of a detailed global network of morphostructural lineaments, one can try associating locations of great earthquake epicenters with segments of major seismic belts per se. This association became the basis for a pattern recognition study, which eventually gave a simple characterization of places where great earthquakes can occur.

Major seismic belts of Earth are by no means homogeneous in the character and history of faults that form them. Therefore, there was no a priori assurance that global characteristic traits of the areas prone to the largest earthquakes exist. However the chances were not negligible. The largest earthquakes are large-scale events of several hundred km in linear size and cut the entire or nearly all of the lithosphere.

Gutenberg and Richter [GR54] distinguished as the largest, "class a," earthquakes those of magnitude 7.75 or greater. To reduce the possible influence of errors in magnitude, the analysis started with locations of great $M \geq 8.2$ events; sites of earthquakes with $M = 7.9$–8.1 were not considered at the learning stage [GZKK78, GZKK80, Kos80, Kos84]. Gvishiani and Kossobokov [GK81] used randomization tests extensively and demonstrated that pattern recognition yields better results [GZKK80] for earthquakes of magnitude $M \geq 8.0$. Further analysis of the decision rules obtained suggested a significant simplification [GK84], by combining only the volcanism type and the topographic differential.

Pattern recognition: Learning stage. The analysis started with the Circumpacific seismic belt extended to the adjacent Java Trench and the Caribbean. The natural recognition objects were selected as points set along the belt approximated by a line. The line combines axes of deep-sea trenches and the feet of the continental slopes (Fig. 6.11). A total of 226 points was selected; they were either projections of earthquakes with $M \geq 8.2$ from 1904–1975 on the line or points spaced between them nearly uniformly with a step of about 300 km. Each point represents an adjacent segment.

Training sets were parts of $\mathbf{D_0}$ that included the projections of earthquakes with $M \geq 8.2$ and $\mathbf{N_0}$ that contained 133 points representing segments of the seismic belt with no magnitude $M \geq 7.9$ events. The remaining 41 points did not take part in the learning. In the main case of recognition, training sets were taken from the eastern part of the Circumpacific, whereas the western part was reserved as a control sample. Having in mind the heterogeneity of the Circumpacific belt, satisfactory classification was defined by two restrictions: (1) there must be no more than 5 errors of the total of 52 $\mathbf{D_0}$ vectors and (2) there must be no more than 133 vectors in $\mathbf{D}$.

The majority of the characteristics used in the analysis are elevations and their gradients that naturally reflect the intensity and contrast of neotectonic movement. Two more parameters were evaluated in circles of radius 300 km centered at each point: the presence of a transverse structure, (TS), and of

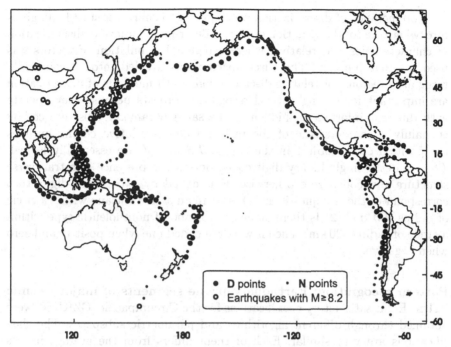

Fig. 6.11. Earthquake-prone segments of major seismic belts ($M \geq 8.2$): Composite recognition in the Circumpacific by different algorithms [GZKK80]

a volcanism type V. Specifically, the following characteristics were used to describe the points along major seismic belts.

1. Maximum elevation measured within a circle of radius 100 km, h_{max}.
2. Minimum elevation measured within a circle of radius 100 km, h_{min}
3. Elevation contrast, $\Delta h = h_{max} - h_{min}$.
4. Gradient, $\Delta h/l$, where l is the distance between the points of h_{max} and h_{min}).
5. Maximum elevation measured within a circle of radius 300 km, H_{max}.
6. Minimum elevation measured within a circle of radius 300 km, H_{min}.
7. Elevation contrast, $\Delta H = H_{max} - H_{min}$.
8. Gradient, $\Delta H/L$, where L is the distance between the points of H_{max} and H_{min}.
9. Elevation at the point, h.
10. Gradient, $|h/\rho|$, where ρ is the distance to the shoreline.
11. The presence of a transverse structure within a circle of radius 300 km (a deep-sea trench, a foot of continental slope, a major fault, a ridge, an oceanic rise, a chain of volcanoes), TS.
12. The presence of volcanism within a circle of radius 300 km (extinct volcanoes, active volcanoes of tholeiitic type, active volcanoes of nontholeiitic type), V.

Two methods of discretization were applied to characteristics 1–10: absolute, where the total population of vectors determined objective discretization of characteristics and relative, where a regional population of vectors was used for that purpose. The Circumpacific was divided into 13 commonly recognized regions for relative discretization. Both methods of discretization are important for coding natural recognition objects because they operate with different information hidden in the same parameter: absolute coding is mainly a characteristic of the region, whereas relative coding outlines the position of the object in the region. TS and V are essentially logical. TS requires a single binary digit coded one in the presence of a transverse structure and coded zero otherwise. V is coded by two binary digits that correspond to the two questions: (1) Is there an active volcano within a circle of radius 300 km? (2) Is there an active volcano of nontholeiitic type within a circle of radius 300 km? The answers are coded one when positive and zero when negative.

Pattern recognition: Earthquake-prone segments of major seismic belts. Eight satisfactory classifications for the Circumpacific [GZKK80] were obtained through different algorithms and parametric subspaces. The classifications are very similar. Each of them differs from the average in less than 10% of the points considered. Therefore, it is natural to conclude that any of the eight satisfactory classifications delivers a solution of the pattern recognition problem with about the same accuracy (nearly 10%).

The decision rules [GZKK80] showed that the characteristics measured within circles of radius 100 km were less informative than similar ones from larger circles. Therefore, these characteristics were not included in either of the eight decision rules. Moreover, only H_{max} represented relative coding. Other characteristics were used in absolute coding indicating, perhaps, the global character of a catastrophe associated with a great earthquake. The condition $H_{min} \leq 5$ km favored the presence of deep-sea trenches. High values of elevation contrast ΔH and gradient $|h/\rho|$ characterized contrasting neotectonic movements. Contrary to previous experience (e.g., see Sect. 6.4.3), the presence of transverse structures appeared less informative compared with other characteristics. This may result from a crude determination of parameters in the absence of a global geomorphological reconstruction. The presence of active volcanism is another indication of the largest scale phenomena associated with great earthquakes. The condition on the nontholeiitic type of volcanism might indicate indirectly large depths involved in neotectonic movements. According to [Yod76], tholeiitic magmas originate at shallower depths than nontholeiitic ones.

Pattern recognition: Control tests. In the main case recognition, as well as in the six out of eight satisfactory cases, a significant part of the total $\mathbf{D_0}$ was reserved for control testing, which allowed evaluating the nonrandomness of classifications beyond any doubt (above 99.98%). Moreover, the transfer of criteria to the Alpine seismic belt did confirm this conclusion and extended the limit of their universality [Kos80]; six out of eight projections of great earthquakes were determined correctly, whereas about a quarter of the total was recognized as **D**. It was noted that the fractions of earthquake-prone areas (both instrumentally reported and obtained by pattern recognition) in the Circumpacific and the Alpine seismic belts were about the same [Kos80], thus favoring the common understanding that the Circumpacific is the main source of great earthquakes. It also allowed us to suggest that about 200 years of global instrumental data would be enough to outline the locations of nearly all great earthquake-prone areas.

Pattern recognition: A hyperplane solution. This approach aims at determining a hyperplane in a relevant space that separates two given point sets in the best possible way. Originally, the application of a traditional linear discriminator [DH73] did not provide a satisfactory classification [GZKK80]. Perhaps, this situation is common due to the objective difficulty of finding the global maximum of a step function (naturally, the number of errors is such a function). Later, Gurvich and Kossobokov [GK84] modified the algorithm and used the gradient method to find the global maximum of an error function in a 16-dimensional space of characteristics. The coefficients of the hyperplane were such that its slightest rotation delivered a satisfactory classification determined by a single inequality without loss of efficiency:

$$\Delta H + 2V \geq 11 \; ;$$

where ΔH is given in km and $V = 0$, 1, or 2 stands for no active volcano and the presence of an active volcano of tholeiitic or nontholeiitic type, respectively. Figure 6.12 displays the distribution of vectors in the two-dimensional space $\Delta H \times V$. The line (the hyperplane) divides the total of 226 points into 124 and 102 belonging to **D** and **N**, respectively. The efficiency achieved is the same as that of the main case classification by CORA-3 [GZKK80].

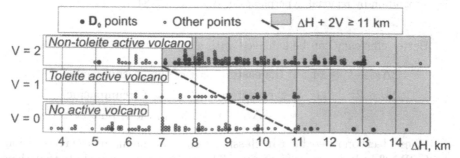

Fig. 6.12. Earthquake-prone segments of major seismic belts ($M \geq 8.2$): Distribution of vectors in $\Delta H \times V$. The domain of the **D** criterion is shaded gray

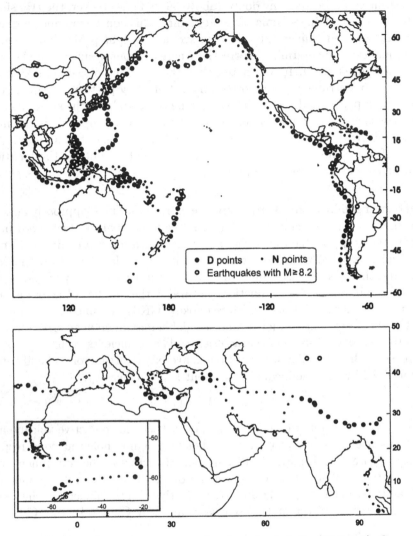

Fig. 6.13. Earthquake-prone segments of major seismic belts ($M \geq 8.2$): Composite recognition by regional hyperplanes in $\Delta H \times V$ [GK84]

Better efficiency can be achieved with a set of regional inequalities:

$$3\Delta H + V \geq 24 \quad \text{in the eastern part of the Circumpacific,}$$
$$\Delta H + 2V \geq 11 \quad \text{in the western part of the Circumpacific,}$$
$$2\Delta N + V \geq 13 \quad \text{in the Alpine belt.}$$

These decision rules differ in classification from the main case recognition by CORA-3 in less than 10% of sites. Therefore, we may conclude that they determine great earthquake-prone areas with the same accuracy as any of the eight satisfactory rules from [GZKK80]. At the same time, such a simple description is a naturally preferred alternative. Figure 6.13 displays the classification of points along major seismic belts worldwide determined by regional inequalities.

The rest of the reported great earthquakes are located in the Transasian seismic belt outside its Alpine zone (Fig. 6.14). This is a much more complicated network of deep-laid faults whose crude approximation was used for selecting natural recognition objects. As with the Circumpacific and Alpine belts, we have selected 142 points along the axes of deep-laid faults [Kos84]. Two parameters were assigned to each point, H_{max} and H_{min} measured within a circle of radius 300 km. Figure 6.15 displays the distribution of vectors in this two-dimensional space.

Two populations of points were distinguished. The first combines points within a distance of 1000 km from the Himalayas where the reduced decision rule for the Alpine belt worked, i.e., $\Delta H \geq 6.5$ km. The second included 89 points outside the Himalayas and was rather different: $\Delta H \geq 6.5$ km was satisfied only for four points. Although two of them are projections of $M \geq 8.2$ earthquakes, which by themself have a low chance (< 0.001) of happening at random, the inequality is by no means a reasonable classification of the 89 vectors. As evident from Fig. 6.15 (the lower panel), $H_{max} + 4H_{min} \geq 6.5$ km holds for 20 out of 89 vectors. All six projections of great earthquakes ($M \geq 8.2$, 1904–1975) are among them. The probability of getting such

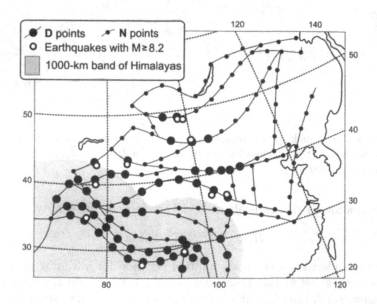

Fig. 6.14. Earthquake-prone segments in the Transasian seismic belt outside its Alpine zone ($M \geq 8.2$): Composite recognition by hyperplanes in $H_{max} \times H_{min}$ [Kos84]. The reduced decision rule for the Alpine belt works within 1000-km distances from Himalayas (shaded grey) [GK84], i.e. $\Delta H \geq 6, 5$ km there

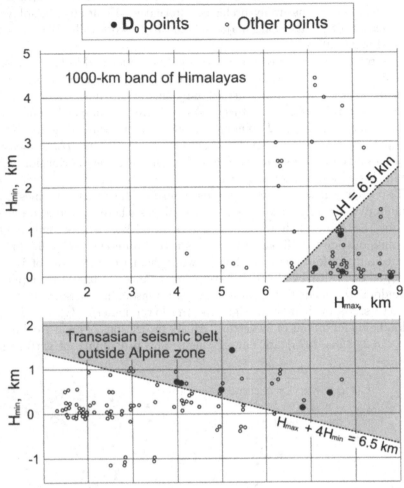

Fig. 6.15. Earthquake-prone segments in the Transasian seismic belt outside its Alpine zone ($M \geq 8.2$): Distribution of vectors in $H_{\max} \times H_{\min}$. Domains of **D** criteria are shaded gray.

a classification by chance is about 1%. Therefore, the inequality,

$$H_{\max} + 4H_{\min} \geq 6.5\,\text{km} ,$$

can be considered a first approximation criterion for outlining great earthquake-prone areas in the Transasian seismic belt outside the Alpine zone. Of course, small sample statistics do not permit accurate estimation of the reliability of the criterion.

The simplifications suggest that great earthquake-prone areas are characterized either by high elevation contrast combined with active deep-laid volcanism within circles of radius 300 km at the contact of tectonic plates or by present-day intensive large-scale uplift of lithospheric blocks (of about 100,000 km^2) in front of the collision zone of the Indian continent and Asia. Both criteria indicate a high degree of instability and favor the hypothesis that sites of extreme tectonic deformations can be identified by the amplitude of the solid Earth surface and the depth of magma chambers.

The case history of great earthquake-prone areas points to the necessity of systematic analysis of pattern recognition results. A solution of the pattern recognition problem does not ensure the absence of other, perhaps, better classifications. Some characteristics can be essentially irrelevant although supported by theoretical hypothesizing.

6.5   Conclusion: Confirmation of Pattern Recognition Results by Subsequent Large Earthquakes

In the past three decades, pattern recognition of earthquake-prone areas based on morphostructural zoning was performed in eleven regions listed in Tables 6.18 and 6.19. All MZ schemes were developed through the same methodology (see Sect. 6.3), although on different scales and for different magnitudes of strong earthquakes. The scale was 1:7,500,000 for the Andes of South America and for Kamchatka. It was 1:1,000,000 for the Western Alps, the Pyrenees, the Greater and Minor Caucasus. The scale 1:2,500,000 was used for zoning in other regions. The list of characteristics used for recognition varied, as well as the definition of natural objects. The bulk of the characteristics was usually formed from morphometric measures of relief and the parameters of MZ schemes. However, gravitational parameters were used in the Caucasus, Alps, and Pyrenees.

Guberman and Rotwain [GR86] were the first to test classifications of earthquake-prone areas in four regions for newly occurring earthquakes. Since then, the natural control set experienced a multiple increase, which allows more adequate and reliable conclusions now.

We used global catalogs [GHDB89, PDE] as the sources of uniform data on earthquakes in the eleven regions. The test consists of comparing locations of earthquake-prone areas with epicenters of strong earthquakes in the region that occurred after publishing pattern recognition results. The starting date of the test differs from region to region; the ending date is 1 July, 2000. The earthquake was considered strong if any of the four magnitudes reported in [GHDB89, PDE] was equal to or larger than M_0 defined in the region previously. Table 6.18 lists all earthquakes used for testing (including 1991/10/19 that occurred in the Himalayas after the manuscript was submitted, although before its publication). Table 6.19 presents the eleven regions and the respective values of M_0. Pattern recognition was performed in some regions several times for different magnitudes M_0 (up to 3 in the Greater Caucasus). Table 6.19 includes the Minor Caucasus although no magnitude $M \geq 5.5$ earthquakes occurred there since publication.

Table 6.18. Earthquakes that tested pattern recognition results

Date	Latitude	Longitude	Depth, km	M	Comment
Tien Shan and Pamirs ($M \geq 6.5$)					
1974/08/11 01:13	39.45°N	73.83°E	9	7.3	D
1978/03/24 21:05	42.83°N	78.60°E	33	7.1	D
1978/11/01 19:48	39.34°N	72.61°E	40	6.8	Outside nodes
1985/08/23 12:41	39.43°N	75.22°E	7	7.5	D
1992/08/19 02:04	42.14°N	73.57°E	27	7.5	Outside nodes
1998/05/30 06:22	37.11°N	70.11°E	33	7.0	D*
Balkans, Asia Minor, Transcaucasia ($M \geq 6.5$)					
1975/03/27 05:15	40.41°N	26.13°E	5	6.7	D
1975/09/06 09:20	38.47°N	40.72°E	26	6.7	D
1976/05/11 16:59	37.56°N	20.35°E	33	6.7	D
1976/11/24 12:22	39.12°N	44.02°E	36	7.3	D
1978/06/20 20:03	40.73°N	23.22°E	3	6.6	D
1979/04/15 06:19	42.09°N	19.20°E	10	7.3	D
1981/02/24 20:53	38.22°N	22.93°E	33	6.8	D*
1982/01/18 19:27	40.00°N	24.31°E	10	7.0	D
1983/01/17 12:41	38.02°N	20.22°E	14	7.2	D
1983/08/06 15:43	40.14°N	24.76°E	2	7.3	D*
1983/10/30 04:12	40.33°N	42.18°E	12	6.9	D
1988/12/07 07:41	40.99°N	44.19°E	5	7.0	D*
1992/03/13 17:18	39.71°N	39.60°E	27	6.9	D
1995/05/13 08:47	40.15°N	21.69°E	14	6.8	D*
1995/06/15 00:15	38.40°N	22.28°E	14	6.5	D
1997/10/13 13:39	36.38°N	22.07°E	24	6.7	D*
1997/11/18 13:07	37.57°N	20.66°E	33	6.6	D
1998/06/27 13:55	36.88°N	35.31°E	33	6.6	N
1999/08/17 00:01	40.74°N	29.86°E	17	7.8	D
1999/11/12 16:57	40.75°N	31.16°E	10	7.5	D
California and Nevada ($M \geq 6.5$)					
1979/10/15 23:16	32.63°N	115.33°W	12	7.0	D
1980/05/25 16:33	37.60°N	118.84°W	5	6.5	D*
1980/05/25 19:44	37.57°N	118.82°W	15	6.7	D*
1983/05/02 23:42	36.22°N	120.32°W	10	6.7	Outside nodes
1986/07/21 14:42	37.54°N	118.45°W	9	6.5	D*
1987/11/24 01:54	33.08°N	115.78°W	4	6.5	D
1987/11/24 13:15	33.01°N	115.84°W	2	6.7	D
1989/10/18 00:04	37.04°N	121.88°W	18	7.1	D
1992/04/25 18:06	40.37°N	124.32°W	15	7.1	D
1992/04/26 07:41	40.42°N	124.60°W	20	6.6	D
1992/04/26 11:18	40.38°N	124.57°W	22	6.6	D
1992/06/28 11:57	34.20°N	116.44°W	1	7.6	D
1992/06/28 15:05	34.20°N	116.83°W	5	6.7	D*

* Marks strong earthquakes that occurred at sites **D** where no earthquakes that large were reported before performing pattern recognition

Table 6.18. (Continued)

Date	Latitude	Longitude	Depth, km	M	Comment
1994/01/17 12:30	34.21°N	118.54°W	18	6.8	**D***
1999/10/16 09:46	34.59°N	116.27°W	0	7.4	Outside nodes
Italy ($M \geq 6.0$)					
1980/11/23 18:34	40.91°N	15.36°E	10	7.2	**D**
1984/04/29 05:03	43.26°N	12.55°E	12	6.1	Outside nodes
1984/05/07 17:49	41.76°N	13.89°E	10	6.0	**D***
1997/09/26 09:40	43.08°N	12.81°E	10	6.4	Outside nodes
1998/04/12 10:55	46.25°N	13.65°E	10	6.0	**D**
Andes of South America ($M \geq 7.75$)					
1985/03/03 22:47	33.13°S	71.87°W	33	7.8	**D**
1995/07/30 05:11	23.34°S	70.29°W	45	8.0	**D***
Kamchatka ($M \geq 7.75$)					
1997/12/05 11:26	54.84°N	162.04°E	33	7.8	**D**
Western Alps ($M \geq 5.0$)					
1985/10/20 08:43	44.81°N	6.63°E	10	5.0	**D**
1991/02/11 15:43	44.87°N	6.70°E	14	5.1	**D**
1991/11/20 01:54	46.78°N	9.52°E	10	5.1	**D***
1992/05/08 06:44	47.27°N	9.50°E	5	5.0	**D**
1994/12/14 08:55	46.01°N	6.35°E	10	5.1	**N**
Greater Caucasus ($M \geq 5.0; 5.5; 6.5$)					
1981/02/23 04:06	41.79°N	45.97°E	33	5.1	**D**
1981/05/12 17:43	42.83°N	45.76°E	33	5.1	**D**
1981/10/18 05:22	43.25°N	45.30°E	33	5.7	**D**
1981/11/19 14:10	40.73°N	49.19°E	33	5.0	Outside nodes
1981/11/29 23:37	40.83°N	48.07°E	45	5.1	**D**
1984/03/04 19:24	43.07°N	45.61°E	33	5.3	**D**
1985/07/04 05:08	42.13°N	45.86°E	33	5.2	Outside nodes
1988/05/03 09:15	42.47°N	47.66°E	22	5.1	**D***
1991/04/29 09:12	42.45°N	43.67°E	17	7.3	**D***
1991/06/15 00:59	42.46°N	44.01°E	9	6.5	**N**
1992/10/23 23:19	42.59°N	45.10°E	16	6.8	**D**
1993/02/22 04:24	42.56°N	43.86°E	10	5.0	**D***
1994/04/17 08:02	41.95°N	46.32°E	33	5.0	**D**
1999/01/31 05:07	43.16°N	46.84°E	33	5.8	**D**
Pyrenees ($M \geq 5.0$)					
1996/02/18 01:45	42.83°N	2.53°E	10	5.0	**N**
1999/10/04 18:14	42.90°N	0.60°E	10	5.1	**D**
Himalayas ($M \geq 6.5; 7.0$)					
1991/10/19 21:23	30.78°N	78.77°E	10	7.0	**D***
1999/03/28 19:05	30.51°N	79.40°E	15	6.6	**D**

* Marks strong earthquakes that occurred at sites **D** where no earthquakes that large were reported before performing pattern recognition

Table 6.19. Summary of test results

Region	M_0	Node size, R, km	Publication year	Total number of strong earthquakes	Number of earthquakes in $\mathbf{D}(\mathbf{D}^*)^a$	Number of earthquakes in $\mathbf{N}$
Tien Shan and Pamirs	6.5		1972	6	4 (1)	
Balkans, Asia Minor, Transcaucasia	6.5	62,5	1974	20	19(5)	1
California and Nevada	6.5	25	1976	15	13(5)	
Italy	6.0	35	1979	5	3(1)	
Andes of South America	7.75	75	1982	2	2(1)	
Kamchatka	7.75	75	1984	1	1	
Western Alps	5.0	25	1985	5	4(1)	1
Greater	6.5	25	1986	3	2(1)	1
Caucasus[b]	5.5	25	1987	5	4(1)	1
	5.0	25	1988	14	11(3)	1
Pyrenees	5.0	25	1987	2	1	1
Minor Caucasus	5.5	25	1991	0		
Himalayas[b]	7.0	50	1994	1	1(1)	
	6.5	50	1992	2	2(1)	
Total				**72**	**60(19)**	**4**

[a]$\mathbf{D}^*$ marks nodes $\mathbf{D}$ where no strong earthquakes were reported before the publication of pattern recognition results
[b]For the Greater Caucasus and Himalayas, the statistics at lower magnitude cutoff include those for higher magnitude ranges; therefore statistics for only the lowest magnitude cutoff was included in the total

As a result, we found a total of 72 strong earthquakes in the eleven regions out of which 60 (83%) occurred in $\mathbf{D}$, four (6%) occurred in $\mathbf{N}$, and 8 (11%) could not be associated with a natural recognition object.

First of all, let us note that 64 strong earthquakes did occur within nodes or near intersections of morphostructural lineaments. This fact confirms the initial hypothesis of association between large earthquakes and morphostructural nodes. The 11% of earthquakes without this association do not refute the hypothesis. A similar percentage of strong earthquakes was not associated with nodes at the problem setting stage. Perhaps, this indicates that a small portion of strong earthquakes can happen at nodes formed by lower rank lineaments than those determined in the existing MZ schemes. This is indirectly confirmed by the fact that seven out of eight epicenters not associated with nodes fall onto reported MZ lineaments. Most of these cases appear in regions where MZ was done on the scale of 1:2,500,000. There is just one such case related to MZ on the scale of 1:1,000,000 (the Greater Caucasus). It is premature to conclude that MZ determined on larger scales improves the association of strong earthquakes and morphostructural nodes, although this appears very likely.

It is seen that 18 out of 60 strong earthquakes in D, i.e., 30%, occurred at sites where no earthquakes that large had been reported before pattern recognition was performed. Such a high percentage of "surprises" highlights the necessity of using pattern recognition results along with historical seismic data in estimating seismic risk, e.g., in seismic zoning. The other 42 strong earthquakes occurred in nodes where such events were already reported, most of them in recent decades. This, perhaps, evidences the existence of long-term clustering of large earthquakes.

The probability of classification error can be estimated by the ratio of the number of strong earthquakes in N to the total number of strong earthquakes associated with nodes. This ratio equals $4/64 = 6.25\%$ for our control sample, confirming the high reliability of classifications that define earthquake-prone areas.

Taking into account the size of the control sample set, which is not small in total, we may suggest that the proportion of errors will not change much with time.

The test confirms that the pattern recognition approach to the determination of earthquake-prone areas described in this chapter is most sensitive to the following factors:

- the completeness of the MZ scheme;
- an adequate characterization of nodes by parameters naturally related to seismogenic processes;
- sufficient size of the training set D_0.

It is hard to estimate the completeness and quality of an MZ scheme. Although the methodology is basically formalized, it is not free from some subjectivism. A certain arbitrariness appears mainly in ranking blocks (starting from the second rank) and locating transverse lineaments. Earthquakes remain the only objective estimate of MZ compiled for earthquake-prone area studies. From this viewpoint, the quality of MZ in the eleven regions should be claimed satisfactory.

The selection of natural recognition objects and their characterization is the most important stage of any pattern recognition study. The list of parameters describing nodes is restricted by the condition of uniformity across the entire territory considered. Many parameters of instrumental measurements, which are potentially of great importance, cannot be used in the description for that reason.

The size of a training sample was usually about a dozen or two, which a priori was not enough for reliable justification of the classification quality. However, positive results of multiple control tests [GGI$^+$72a, GGK$^+$76] seemed to favor a conclusion that the classifications obtained were rather reliable in advance testing by newly occurring earthquakes.

In the framework of the methodology presented in this chapter, the quality of the classification of nodes into earthquake-prone areas and others can be improved by designing automated computer analysis of the initial geologic and morphological data [GKSS98], by using additional characteristics that became available recently, and by perfecting pattern recognition techniques aimed at a priori achievement of better stability and reliability.

References

[AE91]   D. C. Agnew and W. L. Ellsworth. Earthquake prediction and long-term hazard assessment. *Rev. Geophys. Suppl.*, 29: 877–889, 1991.

[AEH+76]   C. R. Allen (Chaiman), W. Edwards, W. J. Hall, L. Knopoff, C. B. Raleigh, C. H. Savit, M. N. Toksoz, and R. H. Turner. Predicting earthquakes: A scientific and technical evaluation – with implications for society. Panel on Earthquake Prediction of the Committee on Seismology, Assembly of Mathematical and Physical Sciences, National Research Council, U.S. National Academy of Sciences, Washington, D. C., 1976.

[AGG+77]   M. A Alekseevskaya, A. M. Gabrielov, A. D. Gvishiani, I. M. Gelfand, and E. Ya. Ranzman. Formal morphostructural zoning of mountain territories. *J. Geophys.*, 43: 227–233, 1977.

[AJ87]   H. J. Anderson and J. Jackson. Active tectonics of the Adriatic region. *Geophys. J. R. Astron. Soc.*, 91: 937–938.

[AK89]   B. Akasheh and V. G. Kossobokov. Premonitory clustering before strong earthquakes in Iran-Afgan region. *Bollettino di Geofisica Teorica ed Applicata*, XXXI(122): 159–162, 1989.

[AKB+84]   C. Allen, V. I. Keilis-Borok, I. V. Kuznetsov, and I. M. Rotwain. Long-term earthquake prediction and self-similarity of seismic precursors. In *Achievements and Problems of Modern Geophysics*, pp. 152–165, Nauka, Moscow, 1984 (in Russian).

[Aki81]   K. Aki. A probabilistic synthesis of precursory phenomena. In D. V. Simpson and P. G. Richards, editors, *Earthquake Prediction. An International Review*, Manrice Ewing Ser., Vol.4, pp. 566–574, AGU, Washington, D. C., 1981.

[Aki96]   K. Aki. Scale dependence in earthquake phenomena and its relevance to earthquake prediction.*Proc. Natl. Acad. Sci. USA*, 93: 3740–3747, 1996.

[ALM82]   C. J. Allegre, J. L. Le Mouël, and A. Provost. Scaling rules in rock fracture and possible implications for earthquake prediction. *Nature* 297: 47–49, 1982.

[AMCN95]   C. J. Allegre, J.-L. Le Mouel, H. D. Chau, and C. Narteau. Scaling organization of fracture tectonics (SOFT) and earthquake mechanism. *Phys. Earth Planet. Int.*, 92: 215–233, 1995.

[Ari74]   St. Arinei. *The Romanian Teritory and Plate Tectonics*, Technical Publishing House, Bucharest, 1974 (in Rumanian).

[ARS+92]   E. Arieh, I. Rotwain, J. Steinberg, I. Vorobieva, and F. Abramovici. Diagnosis of time of increased probability of strong earthquakes in the Jordan-Dead Sea rift zone. *Tectonophysics*, 202: 351–359, 1992.

[Atw40]   W. Atwood. *The Physiographic Provinces of North America*, Ginn, Boston, Mass., 1940.

[Bar93]   G. I. Barenblatt. Micromechanics of fracture. In E. R. Bodner, J. Singer, A. Solan and Z. Hashin, editors, *Theoretical and Applied Mechanics*, pp. 25–52, Elsevier, Amsterdam, 1993.

[Bar96]   G. Barenblatt. *Scaling, Self-Similarity, and Intermediate Asymptotics*, Cambridge University Press, Cambridge, 1996.

[Bat65]   M. Bath. Lateral inhomogeneties of the upper mantle. *Tectonophysics*, 2: 483–514, 1965.

[BCF+92]   S. C. Bhatia, T. R. K. Chetty, M. Filimonov, A. Gorshkov, E. Rantsman, and M. N. Rao. Identification of potential areas for the occurrence of strong earthquakes in Himalayan arc region. *Proc. Indian Acad. Sci. (Earth Planet. Sci.)*, 101(4): 369–385, 1992.

[BCT92]   P. Bak, K. Chen, and C. Tang. A forest-fire model and some thoughts on turbulence. *Phys. Lett. A*, 147: 297–300, 1992.

[Bea81]   A. Beach. Thrust tectonics and crustal shattering in the external French Alps based on a seismic cross section. *Tectonophysics*, 7 (1/2): T1–T6, 1981.

[Ben51]   H. Benioff. Global strain accumulation and release as related by great earthquakes. *Bull. Geol. Soc. Am.*, 62: 331–338, 1951.

[Ben00]   Y. Ben-Zion. Fault interaction, seismicity, and criticality. *Extended Abstract for the 17th Course of International School of Geophysics*, Erice, June 17–23, 2000.

[Ben01]   Y. Ben-Zion. Dynamic ruptures in recent models of earthquake faults. *J. Mech. Phys. Solids*, 49(9): 2209–2244, 2001.

[Bil65]   P. Billingsley. *Ergodic Theory and Information*, Wiley, New York, 1965.

[Bir98]   P. Bird. Testing hypotheses on plate-driving mechanisms with global lithosphere models including topography, thermal structure, and faults. *J. Geophys. Res.*, 103(B5): 10115–10129, 1998.

[BK67]   R. Burridge and L. Knopoff. Model and theoretical seismicity. *Bull. Seismol. Soc. Am.*, 57: 341–360, 1967.

[BKKM14]   K. I. Bogdanovich, I. M. Kark, B. Ya. Korol'kov, and D. I. Mushketov. Earthquakes in Northern chains of Tien-Shan. In *Proc. Geol. Counc.*, 89, St. Petersburg, 1914 (in Russian).

[BKM83]   G. M. Barenblatt, V. I. Keilis-Borok, and A. S. Monin. Filtration model of earthquake sequence. *Trans. (Doklady) Acad. Sci. SSSR*, 269: 831–834, 1983.

[BL85]   W. H. Bakun and A. G. Lindh. The Parkfield, California, earthquake prediction experiment. *Science*, 229: 619–624, 1985.

[Bol93]   B. A. Bolt. *Earthquakes - Newly Revised and Expanded, Appendix C.* 331 pp. Freeman, New York, 1993.

[Bon67]   M. M. Bongard. *The Problem of Recognition*, Nauka, Moscow, 1967 (in Russian).

[BOS+98]   D. D. Bowman, G. Ouillon, G. G. Sammis, A. Sornette, and D. Sornette. An observational test of the critical earthquake concept. *J. Geophys. Res.*, 103: 24359–24372, 1998.

[BRC+94]   S. G. Bhatia, M. N. Rao, T. R. K. Chetty, E. Ya. Rantsman, A. I. Gorshkov, M. B. Filimonov, and N. V. Shtock, Recognition of earthquake-prone areas. XIX. The Himalaya, $M \geq 7.0$. In V. I. Keilis-Borok and

G. M. Molchan, editors, *Theoretical Problems of Geodynamics and Seismology, Comput. Seismol.*, *27*, pp. 280–287, Nauka, Moscow, 1994 (in Russian).

[BS97]   E. M. Blanter and M. G. Shnirman. Simple hierarchical systems: Stability, SOC and catastrophic behavior. *Phys. Rev. E*, 55: 6397–6403, 1997.

[BS99]   E. M. Blanter and M. G. Shnirman. Mixed hierarchical model of seismicity: Scaling and prediction. *Phys. Earth Planet. Inter.*, 111: 295–303, 1999.

[BSL98]   E. M. Blanter, M. G. Shnirman, and J.-L. Le Mouël. Hierarchical model of seismicity: Scaling and predictability. *Phys. Earth Planet. Int.*, 103: 135–150, 1998.

[BSL99]   E. M. Blanter, M. G. Shnirman, and J.-L. Le Mouël. Temporal variation of predictability in a hierarchical model of dynamical self-organized criticality. *Phys. Earth Planet. Int.*, 111: 317–327, 1999.

[BSLA97]   E. M. Blanter, M. G. Shnirman, J.-L. Le Mouël, and C. J. Allégre. Scaling laws in blocks dynamics and dynamic self-organized criticality. *Phys. Earth. Planet. Int.*, 99: 295–307, 1997.

[BT89]   P. Bak and C. Tang. Earthquakes as a self- organized critical phenomenon *J. Geophys. Res.*, 94: 15635–15637, 1989.

[BT94]   B. Barriere and D. L. Turcotte. Seismicity and self-organized criticality. *Phys. Rev. E*, 49(2): 1151–1160, 1994.

[BV93]   C. G. Bufe and D. J. Varnes. Predictive modeling of the seismic cycle of the greater San Francisco Bay region. *J. Geophys. Res.*, 98: 9871–9883, 1993.

[BVG+66]   M. M. Bongard, M. I. Vaintsveig, Sh. A. Guberman, M. L. Izvekova, and M. S. Smirnov. 1966. The use of self- learning programs in the detection of oil containing layers. *Geol. Geofiz.*, 6: 96–105, 1966 (in Russian).

[Cap81]   M. Caputo. Critical study of ENEL catalogue of Italian earthquakes from the year 1000 through 1975. *Rassegna Lavori Pubblici*, 2: 3–16, 1981.

[CCG+83]   M. Caputo, R. Console, A. M. Gabrielov, V. I. Keilis-Borok, and T. V. Sidorenko. Long-term premonitory seismicity patterns in Italy. *Geophys. J. R. Astron. Soc.*, 75: 71–75, 1983.

[CE84]   L. Constantinescu and D. Enescu. A tentative approach to possibly explaining the occurrence of the Vrancea earthquakes. *Rev. Roum. Geol. Geophys. Geogr.*, 28: 19–32, 1984.

[CFPS86]   J. P. Crutchfield, J. D. Farmer, N. H. Packard and R. S. Shaw. Chaos *Sci. Am.*, 255; 46–57, 1986.

[CGG+85]   A. Cisternas, P. Godefroy, A. Gvishiani, A. I. Gorshkov, V. Kosobokov, M. Lambert, E. Ranzman, J. Sallantin, H. Saldano, A. Soloviev, and C. Weber. A dual approach to recognition of earthquake prone areas in the western Alps. *Annales Geophysicae*, 3(2): 249–270, 1985.

[CGK+77]   M. Caputo, P. Gasperini, V. I. Keilis-Borok, L. Marcelli, and I. Rotwain. Earthquake's swarms as forerunners of strong earthquakes in Italy. *Annali di Geofisica*, XXX(3/4): 269–283, 1977.

[CKO+80]   M. Caputo, V. I. Keilis-Borok, E. Oficerova, E. Ranzman, I. Rotwain, and A. Solovjeff. Pattern recognition of earthquake-prone areas in Italy. *Phys. Earth Planet. Inter.*, 21: 305–320, 1980.

[CL82]   J. M. Carlson and J. S. Langer. Mechanical model of an earthquake fault. *Phys. Rev. A*, 40: 6470–6484, 1982.

[CMTC94]   *CMT, Centroid Moment Tensor Catalogue, 1977-1994*, Harvard University, Department of Earth and Planetary Sciences, Cambridge, 1994.

[DF91]   S. D. Davis and C. Frohlich. Single link cluster analysis of the earthquake aftershocks: decay laws and regional variations. *J. Geophys. Res.*, 96: 6335–6350, 1991.

[DeM80]   J. De Maré. Optimal prediction of catastrophes with applications to Gaussian processes. *Ann. Probab.*, 8: 841–850, 1980.

[DH73]   R. O. Duda and P. E. Hart. *Pattern Classification and Scene Analysis*, Willey, New York, 1973.

[DP84]   A. M. Dziewonski and A. G. Prozorov. Self-similar definition of clustering of earthquakes. In V. I. Keilis-Borok, editor, *Mathematical Modelling and Interpretation of Geophysical Data, Comput. Seismol. 16*, pp. 10–21, Nauka, Moscow, 1984 (in Russian).

[Dud65]   S. J. Duda. Secular seismic energy release in the circum Pacific belt. *Tectonophysics*, 2(5): 409–452, 1965.

[Ell85]   S. P. Ellis. An optimal statistical decision rule for calling earthquake alerts. *Earthquake Prediction Res.*, 3: 1–10, 1985.

[FBB+79]   K. Fuchs, K.-P. Bonjer, G. Bock, I. Cornea, C. Radu, D. Enescu, D. Jianu, A. Nourescu, G. Merkler, T. Moldoveanu, and G. Tudorache. The Rumanian earthquake of March 4, 1977. II. Aftershocks and migration of seismic activity. *Tectonophysics*, 53: 225–247, 1979.

[Fri95]   U. Frisch. *Turbulence: The legacy of A. N. Kolmogorov*. Cambridge University Press, Cambridge, 1995.

[FS87]   J. D. Farmer and J. Sidorowich. Predicting chaotic time series. *Phys. Rev. Lett.*, 59: 845, 1987.

[FSB+77]   S. A. Fedotov, G. A. Sobolev, S. A. Boldyrev, et al. Long- and short-term earthquake prediction in Kamchatka. *Tectonophysics*, 37: 305–321, 1977.

[GBB97]   J. Gomberg, M. L. Blanpied, and N. M. Beeler. Transient triggering of near and distant earthquakes. *BSSA*, 87(2): 294–309, 1997.

[GDK+86]   A. Gabrielov, O. E. Dmitrieva, V. I. Keilis-Borok, V. G. Kossobokov, I. V. Kuznetsov, T. A. Levshina, K. M. Mirzoev, G. M. Molchan, S. Kh. Negmatullaev, V. F. Pisarenko, A. G. Prozoroff, W. Rinehart, I. M. Rotwain, P. N. Shebalin, M. G. Shnirman, and S. Yu. Shreider. Algorithm of long-term earthquakes' prediction. Centro Regional de Sismologia para America del Sur, Lima, Peru, 1986.

[Gel94]   M. Gell-Mann. *The Quark and the Jaguar: Adventures in the Simple and the Complex*. Freeman, New York, 1994.

[Gel97]   R. J. Geller. Earthquake prediction: A critical review. *Geophys. J. Int.* 131: 425–450, 1997.

[GF98]   R. Girbacea, and W. Frisch. Slab in the wrong place: Lower lithospheric mantle delamination in the last stage of the eastern Carpathian subduction retreat. *Geology*, 26: 611–614, 1998.

[GG90]   A. E. Gripp and R. G. Gordon. Current plate velocities relative to the hotspots incorporating the NUVEL-1 global plate motion model. *Geoph. Res. Lett.*, 17(8): 1109–1112, 1990.

[GGI⁺72a] I. M. Gelfand, Sh. A. Guberman, M. L. Izvekova, V. I. Keilis-Borok, and E. Ya. Rantzman. On criteria of high seismicity. *Trans. (Doklady) Acad. Sci. SSSR*, 202: 1317–1320, 1972.

[GGI⁺72b] I. M. Gelfand, Sh. A. Guberman, M. L. Izvekova, V. I. Keilis-Borok, and E. Ya. Ranzman. On criteria of high seismicity. *Trans. (Doklady) Acad. Sci. SSSR*, 202(6): 1317–1320, 1972.

[GGI⁺72c] I. M. Gelfand, Sh. A. Guberman, M.L. Izvekova, V. I. Keilis-Borok, and E. Ya. Ranzman. Criteria of high seismicity determined by pattern recognition. *Tectonophysics*, 13(1/4): 415–422, 1972.

[GGK⁺76] I. M. Gelfand, Sh. A. Guberman, V. I. Keilis-Borok, L. Knopoff, F. Press, E. Ya. Ranzman, I. M. Rotwain, and A. M. Sadovsky. Pattern recognition applied to earthquake epicenters in California. *Phys. Earth Planet. Int.*, 11: 227–283, 1976.

[GGKR86] A. Gorshkov, A. Gvishiani, V. Kossobokov, and E. Rantsman. Morphostructures and places of earthquakes in the Greater Caucasus. *Izvestia Acad. Sci. SSSR,: Phys. Earth*, 9: 24–35 (in Russian).

[GGK⁺87] A. Gvishiani, A. Gorshkov, V. Kosobokov, A. Cisternas, H. Philip, and C. Weber. Identification of seismically dangerous zones in the Pyrenees. *Annales Geophysicae*, 5B(6): 681+, 1987.

[GGR⁺88] A. Gvishiani, A. Gorshkov, E. Rantsman, A. Cisternas, and A Soloviev. *Identification of Earthquake-prone-areas in the regions of Moderate Seismicity*. Nauka, Moscow, 1988 (in Russian).

[GGZ⁺73] I. M. Gelfand, Sh. A. Guberman, M. P. Zhidkov, M. S. Kaletzkaya, V. I. Keilis-Borok, and E. Ia. Ranzman. Experience of high seismicity criteria transfer from the Central Asia onto the Anatolia and adjacent regions. *Trans. (Doklady) Acad. Sci. SSSR*, 210(2): 327+, 1973.

[GGZ⁺74a] I. M. Gelfand, Sh. A. Guberman, M. P. Zhidkov, M. S. Kaletzkaya, V. I. Keilis-Borok, E. Ia. Ranzman, and I. M. Rotwain. Recognition of places where strong earthquakes may occur. II. Four regions in Asia Minor and South-Eastern Europe. In V. I. Keilis-Borok, editor, *Computer Analysis of Digital Seismic Data, Comput. Seismol.*, 7, pp. 3+, Nauka, Moscow, 1974 (in Russian).

[GGZ⁺74b] I. M. Gelfand, Sh. A. Guberman, M. P. Zhidkov, V. I. Keilis-Borok, E. Ia. Ranzman, and I. M. Rotwain. Recognition of places where strong earthquakes may occur. III. The case when the boundaries of disjunctive knots are unknown. In V. I. Keilis-Borok, editor, *Computer Analysis of Digital Seismic Data, Comput. Seismol.*, 7, pp. 41+, Nauka, Moscow, 1974 (in Russian).

[GHDB89] *Global Hypocenters Data Base CD-ROM*, NEIC/USGS, Denver, CO, 1989.

[GHDB94] *Global Hypocenters Data Base CD-ROM, version III*, NEIC/USGS, Denver CO., 1994.

[Ghi94] M. Ghil. Cryothermodynamics: The chaotic dynamics of paleoclimate. *Physica D*, 77: 130–159, 1994.

[GJK97] R. J. Geller, D. D. Jackson, Y. Y. Kagan, and F. Mulargia. Earthquakes cannot be predicted *Science*, 275: 1616–1617, 1997.

[GK74] J. Gardner and L. Knopoff. Is the sequence of earthquakes in S. California with aftershocks removed Poissonian? *Bull. Seismol. Soc. Am.*, 64(5), 1363–1367, 1974.

[GK81]   A. D. Gvishiani and V. G. Kosobokov. On grounds of results of predic-
           tion of strong earthquake prone-areas obtained by pattern recognition
           methods. *Izv. Acad. Sci. SSSR, Phys. Earth*, 2: 21–36, 1981 (in Rus-
           sian).

[GK83]   A. M. Gabrielov and V. I. Keilis-Borok. 1983. Patterns of stress corro-
           sion: Geometry of the principal stresses. *Pure Appl. Geophys.*, 121(3):
           477–494, 1983.

[GK84]   V. A. Gourvitch and V. G. Kossobokov. On volcanism and elevations'
           amplitude connections with the strongest earthquakes' epicenters. In
           V. I. Keilis-Borok and A. L. Levshin, editors, *Mathematical Modeling
           and Interpretation of Geophysical Data, Comput. Seismol., 17*, pp. 88–
           93, Nauka, Moscow, 1984 (in Russian).

[GKJ96]   A. M. Gabrielov, V. I. Keilis-Borok, and D. D. Jackson. Geometric in-
           compatibility in a fault system *Proc. Natl. Acad. Sci. USA*, 93: 3838–
           3842, 1996.

[GKPS00]   A. I. Gorshkov, I. V. Kuznetsov, G. F. Panza, and A. A. Soloviev. Identi-
           fication of future earthquake sources in the Carpatho-Balkan orogenic
           belt using morphostructural criteria. *Pure Appl. Geophys.*, 157(1/2):
           79–95, 2000.

[GKR+97]   A. Gorshkov, V. I. Keilis-Borok, I. Rotwain, A. Soloviev, and I. Voro-
           bieva. On dynamics of seismicity simulated by the models of blocks-
           and-faults systems. *Annali di Geofisica*, XL(5): 1217–1232, 1997.

[GKSS98]   A. I. Gorshkov, I. N. Kandoba, E. L. Safronovich, and I. V. Sladkov.
           Automated image processing of geologicgeomorphologic information
           in morphostructural zoning. In V. I. Keilis-Borok and G. M. Molchan,
           editors, *Problems in Geodynamics and Seismology, Comput. Sesmol.,
           30*, pp. 336–347, Geos, Moscow, 1998 (in Russian).

[GKZN00]   A. M. Gabrielov, V. I. Keilis-Borok, I. V. Zaliapin, and W. I. Newman.
           Critical transitions in colliding cascades. *Phys. Rev. E*, 62: 237–249,
           2000.

[GLR90]   A. M. Gabrielov, T. A. Levshina, and I. M. Rotwain. Block model of
           earthquake sequence. *Phys. Earth Planet. Int.*, 61: 18–28, 1990.

[GN94]   A. Gabrielov and W. I. Newman. Seismicity modelling and earthquake
           prediction: A review. In W. I. Newman, A. Gabrielov, and D. L. Tur-
           cotte, editors, *Nonlinear Dynamics and Predictability of Geophysical
           Phenomena*, pp. 7–13, Am. Geophys. Union, Int. Union of Geodesy
           and Geophys., 1994, (Geophysical Monograph 83, IUGG Vol. 18).

[GNRS87]   A. I. Gorshkov, G. A. Niauri, E. Ya. Ranzman, and A. M. Sadovsky.
           Use of gravimetric data for recognition of places of possible occurance
           of strong earthquakes in the Great Caucasus. In V. I. Keilis-Borok and
           A. L. Levshin, editors, *Theory and Analysis of Seismological Informa-
           tion, Comput. Seismol., 18*, pp. 117+, Allerton Press, New York, 1987.

[GR44]   B. Gutenberg and C. F. Richter. Frequency of earthquakes in California.
           Bull. Seismol. Soc. Am., 34: 185–188, 1944.

[GR54]   B. Gutenberg and C. F. Richter: *Seismicity of the Earth*, 2nd ed.,
           Princeton University Press, Princeton, N.J., 1954.

[GR56]   B. Gutenberg and C. F. Richter. Earthquake magnitude, intensity, en-
           ergy and acceleration. *Bull. Seismol. Soc. Am.*, 46: 105–145, 1956.

[GR73]   I. P. Gerasimov and E. Ya. Ranzman. Morphostructure of orogens and
           their seismicity. *Geomorphology*, 1: 3–13, 1973 (in Russian).

[GR82]   A. I. Gorshkov and E. Ia. Ranzman. Morphostructural lineaments of the Western Alps. *Geomorphology*, 4: 64+, 1982 (in Russian).

[GR86]   Sh. A. Guberman and I. M. Rotwain. Checking of the predictions of earthquake prone areas (1974–1984). *Izv. Acad. Sci. SSSR, Phys. of the Earth*, 12: 72–84, 1986. (in Russian).

[GR98]   T. V. Garianova and I. M. Rotwain. The properties of seismicity for the simplest type tectonic movement. The block model and reality. In V. I. Keilis-Borok and G. M. Molchan, editors, *Problems in Geodynamics and Seismology, Comput. Seismol., 30*, pp. 289–299, GEOS, Moscow, 1998.

[GS81]   A. D. Gvishiani and A. A. Soloviev. Association of the epicenters of strong earthquakes with the intersections of morphostructural lines in South America. In V. I. Keilis-Borok and A. L. Levshin, editors, *Interpretation of Seismic Data – Methods and Algorithms, Comput. Seismol., 13*, pp. 42+, Allerton Press, New York, 1981.

[GS84]   A. D. Gvishiani and A. A. Soloviev. Recognition of places on the Pacific coast of the South America where strong earthquakes may occur. *Earthquake Prediction Res.*, 2: 237+, 1984.

[GS96]   A. Gorshkov and A. Soloviev. The Western Alps: Numerical modelling of block structure and seismicity. In *Abstracts XXV General Assembly of ESC*, September 9-14, 1996, p. 66, Reykjavik, Iceland, 1996.

[Gus76]   A. A. Gusev. Indicator earthquakes and prediction. In *Seismicity and Deep Structure of Siberia and Far East*, pp. 241–247, Nauka, Novosibirsk, 1976 (in Russian).

[GZ97]   A. I. Gorshkov and M. P. Zhidkov. Recognition of large-scale mass displacements (Lesser Caucasus). *Trans. (Doklady) Russ. Acad. Sci.*, 356(6): 789–791, 1997.

[GZ98]   A. I. Gorshkov and M. P. Zhidkov. Recognition of large collapse-landslide deformations: Implications for seismological risk assessment. *Phys. Earth*, 3: 92–95, 1998 (in Russian).

[GZKK78]   A. D. Gvishiani, A. V. Zelevinsky, V. I. Keilis-Borok, and V. G. Kosobokov. Study of the violent earthquake occurrences in the Pacific Ocean Belt with the help of recognition algorithms. *Izv. Acad. Sci. SSSR, Phys. Earth*, 9: 31–42, 1978 (in Russian).

[GZKK80]   A. D. Gvishiani, A. V. Zelevinsky, V. I. Keilis-Borok, and V. G. Kosobokov. In V. I. Keilis-Borok and A. L. Levshin, editors, *Methods and Algorithms for Interpretation Seismological Data, Comput. Seismol., 13*, pp. 28–38, Allerton Press, New York, 1980.

[GZNK00]   A. M. Gabrielov, I. V. Zaliapin, W. I. Newman, and V. I. Keilis-Borok. Colliding cascade model for earthquake prediction. *Geophys. J. Int.*, 143(2): 427–437, 2000.

[GZRT91]   A. Gorshkov, M. Zhidkov, E. Rantsman, and A. Tumarkin. Morphostructures of the Lesser Caucaus and places of earthquakes, $M \geq 5.5$. *Izv. Acad. Sci. SSSR, Phys. Earth*, 6: 30-38, 1991 (in Russian).

[GZS84]   A. D. Gvishiani, M. P. Zhidkov, and A. A. Soloviev. Transfer of the high-seismicity criteria of the Andes mountain belt onto Kamchatka. *Izv. Acad. Sci. SSSR, Phys. Earth*, 1: 20+, 1984 (in Russian).

[Har98]   R. A. Harris. Forecasts of the Loma Prieta, California, earthquake. *Bull. Seismol. Soc. Am.*, 88: 898–916, 1998.

[Haw71]   A. G. Hawkes. Spectra of some self-exciting and mutually exciting point processes. *Biometrica*, 58: 83–90, 1971.

[Hab82]   R. E. Habermann. Consistency of teleseismic reporting since 1963. *Bull. Seismol. Soc. Am.*, 72: 93–111, 1982.

[Hab87]   R. E. Habermann. Man-made changes in seismicity rates *Bull. Seismol. Soc. Am.*, 77: 141–159, 1987.

[Hab91]   R.E. Habermann. Seismicity rate variations and systematic changes in magnitudes in teleseismic catalogues. *Tectonophysics*, 193: 277–289, 1991.

[Hat74]   S. Hattori. Regional distribution of b-value in the world. *Bull. Int. Inst. Seismol. Earth Eng.*, 12: 39–58, 1974.

[HC90]   R. E. Haberman and F. H. Creamer. Prediction of large aftershocks on the basis of quiescence. *The 7th US - Japan Seminar on Earthquake Prediction*, Vol.1, pp. 93–96, 1990.

[HC94]   R. E. Habermann and F. Creamer. Catalog errors and the M8 earthquake prediction algorithm. *Bull. Seismol. Soc. Am.*, 84: 1551–1559, 1994.

[HKD92]   J. H. Healy, V. G. Kossobokov, and J. W. Dewey. A test to evaluate the earthquake prediction algorithm, M8. *U.S. Geol. Surv. Open-File Report 92-401*, 1992.

[Hob04]   W. N. Hobbs. Lineaments of the Atlantic border region. *Bull. Geol. Soc. Am.*, 15: 483+, 1904.

[Hol95]   J. H. Holland. *Hidden Order: How Adaptation Builds Complexity*. Addison-Wesley, Reading, Mass., 1995.

[How71]   R. A. Howard. *Dynamic Probabilistic Systems*. Wiley, New York, 1971.

[HRM⁺93]   D. P. Hill, P. A. Reasenberg, A. Michael, W. J. Arabasz, G. Beroza, D. Brumbaugh, J. N. Brune, R. Castro, S. Davis, D. de Polo, W. L. Ellsworth, J. Gomberg, S. Harmsen, L. House, S. M. Jackson, M. Johnston, L. Jones, R. Keller, S. Malone, L. Munguia, S. Nava, J. C. Pechmann, A. Sanford, R. W. Simpson, R. S. Smith, M. Stark, M. Stickney, A. Vidal, S. Walter, V. Wong, and J. Zollweg. Seismicity remotely triggered by the magnitude 7.3 Landers, California, earthquake. *Science*, 260: 1617–1623, 1993.

[HSSS98]   Y. Huang, H. Saleur, C. Sammis, and D. Sornette. Precursors, aftershocks, criticality and self-organized criticality. *Europhys. Lett.*, 41: 43–48, 1998.

[IKS99]   A. T. Ismail-Zadeh, V. I. Keilis-Borok, and A. A. Soloviev. Numerical modelling of earthquake flow in the south-eastern Carpathians (Vrancea): Effect of a sinking slab. *Phys. Earth Planet. Int.*, 111: 267–274, 1999.

[IPGP01]   *Bulletins Sismiques des Observatoires des Antilles, 1979–2001*. Departement des Observatoires Volcanologiques, Institut de Physique du Globe de Paris, 2001.

[IPN00]   A. T. Ismail-Zadeh, G. F. Panza, and B. M. Naimark. Stress in the descending relic slab beneath the Vrancea region, Romania. *Pure Appl. Geophys.*, 157: 111–130, 2000.

[Ito92]   K. Ito. Towards a new view of earthquake phenomena. *Pure Appl. Geophys.*, 138: 531–548, 1992.

[Jen77]   C. W. Jennings. *Geological Map of California*. California Division of Mines and Geology, Sacramento, 1:750 000 scale, 1977.

[Jen94]   C. W. Jennings. *Fault Activity Map of California and Adjacent Areas.*
            Division of Mines and Geology, Sacramento, CA, 1:750:000 scale, 1994.

[JK99]   D. D. Jackson and Y. Y. Kagan. Testable earthquake forecasts for 1999.
            Seismol. Res. Lett., 70(4): 393–403, 1999.

[JS99]   S. C. Jaumé and L. R.Sykes. Evolving towards a critical point: A review
            of accelerating seismic moment/energy release prior to large and great
            earthquakes. *Pure Appl. Geophys.*, 155: 279–306, 1999.

[Kag73]   Y. Kagan. On a probabilistic description of the seismic regime. *Fizika
            Zemli*, 4: 110–123, 1973.

[Kag97]   Y. Y. Kagan. Are earthquakes predictable? *Geophys. J. Int.*, 131: 505–
            525, 1997.

[Kad76]   L. P. Kadanoff. Scaling, universality and operator algebras. In C. Domb
            and M. S. Green, editors, *Phase Transitions and Critical Phenomena*,
            Vol. 5a, Academic Press, London, 1976.

[KC95]   V. G. Kossobokov and J. M. Carlson. Active zone size vs. activity:
            A study of different seismicity patterns in the context of the prediction
            algorithm M8. *J. Geophys. Res.*, 100: 6431–6441, 1995.

[Kei64]   V. I. Keilis-Borok. Seismology and logics. *Research in Geophysics*,
            Vol.2, M.I.T. Press, Cambridge, pp. 1–79, 1964.

[Kei82]   V. I. Keilis-Borok. A worldwide test of three long-term premonitory
            seismicity patterns: A review. *Tectonophysics*, 85: 47–60, 1982.

[Kei90a]   V. I. Keilis-Borok. The lithosphere of the Earth as a nonlinear system
            with implications for earthquake prediction. *Rev. Geophys.*, 28: 19–34,
            1990.

[Kei90b]   V. I. Keilis-Borok, editor. Intermediate-term earthquake prediction:
            Models, algorithms, worldwide tests. *Phys. Earth Planet. Int.*, 61, spe-
            cial issue: 1–2, 1990.

[Kei92]   V. I. Keilis-Borok. Chaos in solid Earth. In *Proceedings of Conference
            on Chaos*, UN University, Tokyo, Japan, 1992.

[Kei94]   V. I. Keilis-Borok. Symptoms of instability in a system of earthquake-
            prone faults. *Physica D*, 77: 193–199, 1994.

[Kei96a]   V. I. Keilis-Borok. Intermediate-term earthquake prediction. *Proc. Nat.
            Acad. Sci. USA*, 93: 3748–3755, 1996.

[Kei96b]   V. I. Keilis-Borok. Non-seismological fields in earthquake prediction re-
            search. In J. Lighthill, editor, *A Critical Review of VAN*, pp. 357–372,
            World Scientific, Singapore-New Jersey-London-Hong Kong, 1996.

[Kha71]   V. E. Khain. 1971. *Regional Geotectonics*, Vol.1, Moscow University,
            Moscow, 1971 (in Russian).

[Kha82]   V. E. Khain. Correlation between fixist and mobilist models of tectonic
            evolution in the Greater Caucasus. *Geotectonics*, 4: 3+, 1982 (in Rus-
            sian).

[KHD97]   V. G. Kossobokov, J. H. Healy, and J. W. Dewey. Testing an earthquake
            prediction algorithm. *Pure Appl. Geophys.*, 149: 219–232, 1997.

[KHD+96]   V. G. Kossobokov, J. H. Healy, J. W. Dewey, P. N. Shebalin, and
            I. N. Tikhonov. A real-time intermediate-term prediction of the Octo-
            ber 4, 1994 and December 3, 1995 Southern-Kuril Islands earthquakes,
            Computational Seismology, 28: 46–55, Nauka, Moscow, 1996.

[Kin83]   G. King. The accommodation of large strains in the upper lithosphere of
            the earth and other solids by self-similar fault systems: The geometrical
            origin of *b*-value. *Pure Appl. Geophys.*, 121: 761–815, 1983.

320   References

[Kin86]   G. C. P. King. 1986. Speculations on the geometry of the initiation an termination processes of earthquake rupture and its relation to morphology and geological structure. *Pure Appl. Geophys.*, 124: 567–583, 1986.

[KJ91]   Y. Y. Kagan and D. D. Jackson. Seismic gap hypothesis: Ten years after. *J. Geophys. Res.*, 96: 21419–21431, 1991.

[KJ00]   Y. Kagan and D. Jackson. Probabilistic forecasting of earthquake. *Geophys. J. Int.*, 143: 438–453, 2000

[KK78]   Y. Kagan and L. Knopoff. Statistical study of the occurrence of shallow earthquakes. *Geophys. J. R. Astron. Soc.*, 55: 67-86, 1978.

[KK84]   V. I. Keilis-Borok and V. G. Kossobokov. A complex of long-term precursors for the strongest earthquakes of the world. In *Proceedings 27th Geological Congress, 61*, pp. 56–66, Nauka, Moscow, 1984.

[KK86]   V. I. Keilis-Borok, and V. G. Kossobokov. Periods of high probability of occurrence of the world's strongest earthquakes. In V. I. Keilis-Borok and A. L. Levshin, editors, *Mathematical Methods in Seismology and Geodynamics, Comput. Seismol., 19*, pp. 48–57, Allerton Press, New York, 1986.

[KK90]   V. I. Keilis-Borok and V. G. Kossobokov. Premonitory activation of earthquake flow: Algorithm M8. *Phys. Earth Planet. Int.*, 61: 73–83, 1990.

[KK91]   L. V. Kantorovich and V. I. Keilis-Borok. Earthquake prediction and decision-making: social, economic and civil protection aspects. In *International Conference on Earthquake Prediction: State-of-the-Art*, pp. 586–593, Scientific-Technical Contributions, CSEM-EMSC, Strasbourg, France, 1991.

[KK92]   A. A. Khokhlov and V. G. Kossobokov. Seismic flux and major earthquakes in the Northwestern Pacific. *Trans. (Doklady) Russ. Acad. Sci.*, 325(1): 60–63, 1992 (in Russian).

[KK94]   A. V. Khokhlov and V. G. Kossobokov. Seismic flux and major earthquakes in the Northwestern Pacific. In V. I. Keilis-Borok and G. M. Molchan, editors, *Geodynamics and Earthquake Prediction, Comput. Seismol., 26*, pp. 3–8, Nauka, Moscow, 1994 (in Russian).

[KK97]   I. V. Kuznetsov and V. I. Keilis-Borok. The interrelation of earthquakes of the Pacific seismic belt. *Trans. (Doklady) Russ. Acad. Sci., Earth Sci. Sect.*, 355A(6), 869–873, 1997.

[KKC00]   V. G. Kossobokov, V. I. Keilis-Borok, and B. Cheng. Similarities of multiple fracturing on a neutron star and on the Earth. *Phys. Rev. E*, 61(4): 3529–3533, 2000.

[KKM74]   L. V. Kantorovich, V. I. Keilis-Borok, and G. M. Molchan. Seismic risk and principles of seismic zoning. In *Seismic design decision analysis*. Department of Civil Engineering, MIT. Internal Study Report, 43, 1974.

[KKMV70]   L. V. Kantorovich, V. I. Keilis-Borok, G. M. Molchan, and E. V. Vil'kovich. Statistical model of seismicity and estimation of basic seismic effects. *Izv. Acad. Nauk SSSR. Fizika Zemli*, 5: 85–101, 1970 (in Russian).

[KKR80]   V. I. Keilis-Borok, L. Knopoff, and I. M. Rotwain. Bursts of aftershocks, long-term precursors of strong earthquakes. *Nature*, 283: 258–263, 1980.

[KKRA88] V. I. Keilis-Borok, L. Knopoff, I. M. Rotwain, and C. R. Allen. Intermediate-term prediction of occurrence times of strong earthquakes. *Nature*, 335(6192): 690–694, 1988.

[KKS90] V. G. Kossobokov, V. I. Keilis-Borok, and S. W. Smith. Localization of intermediate-term earthquake prediction. *J. Geophys. Res.*, 95(B12): 19763–19772, 1990.

[KKTM00] V. G. Kossobokov, V. I. Keilis-Borok, D. L. Turcotte, and B. D. Malamud. Implications of a statistical physics approach for earthquake hazard assessment and forecasting. *Pure Appl. Geophys.*, 157: 2323–2349, 2000.

[KL94] V. E. Khain and L. I. Lobkovsky. Conditions of existence of the residual mantle seismicity of the Alpine belt in Eurasia. *Geotectonics*, 3: 12–20, 1994.

[KLJM82] V. I. Keilis-Borok, R. Lamoreau, C. Johnson, and B. Minster. Swarms of main shocks in Southern California. In T. Rikitake, editor, *Earthquake Prediction Research*, Elsevier, Amsterdam, 1982

[KLKM96] L. Knopoff, T. Levshina, V. I. Keilis-Borok, and C. Mattori. Increased long-range intermediate-term earthquake activity prior to strong earthquakes in California. *J. Geophys. Res.*, 101(B3): 5779–5796, 1996.

[KM64] V. I. Keilis-Borok and L. N. Malinovskaya. One regularity in the occurrence of strong earthquakes. *J. Geophys. Res.*, 69: 3019–3024, 1964.

[KM94] V. G. Kossobokov and S. A. Mazhkenov. On similarity in the spatial distribution of seismicity. In *Computational Seismology and Geodynamics* Vol. 1, pp. 6–15, AGU, Washington, D. C., 1994.

[KMU99] V. G. Kossobokov, K. Maeda, and S. Uyeda. Precursory activation of seismicity in advance of Kobe, 1995, M = 7.2 earthquake. *Pure Appl. Geophys.*, 155: 409–423, 1999.

[Kon80] N. V. Kondorskaya, editor. *USSR Earthquakes in 1976*, Nauka, Moscow, 1980 (in Russian).

[Kon82] N. V. Kondorskaya, editor. *USSR Earthquakes in 1978*, Nauka, Moscow, 1982 (in Russian).

[Kos80] V. G. Kossobokov. An experiment of transferring the criteria of high seismicity ($M \geq 8.2$) from the Pacific to the Alpine belt. In V. I. Keilis-Borok and A. L. Levshin, editors, *Methods and Algorithms for Interpretation of Seismological Data, Comput. Seismol., 13*, pp. 44–46. Nauka, Moscow, 1980 (in Russian).

[Kos83] V. G. Kosobokov. Recognition of the sites of strong earthquakes in East Central Asia and Anatolia by Hamming's method. In V. I. Keilis-Borok and A. L. Levshin, editors, *Mathematical Models of the Structure of the Earth and Earthquake prediction, Comput. Seismol., 14*, pp. 78–82, Allerton Press, New York, 1983.

[Kos84] V. G. Kossobokov. General features of the strongest (with $M \geq 8.2$) earthquake-prone areas in the non-Alpine zone of the Transasian seismic belt. In V. I. Keilis-Borok and A. L. Levshin, editors, *Logical and Computational Methods in Seismology, Comput. Seismol., 17*. pp. 69–72, Nauka, Moscow, 1984

[Kos86] V. G. Kossobokov. The test of algorithm M8. In M. A. Sadovsky, editor, *Algorithms of Long-Term Earthquake Prediction*, pp. 42–52, CERESIS, Lima, Peru, 1986.

322   References

[Kos94]   V. G. Kossobokov. Intermediate-term changes of seismicity in advance of the Guam earthquake on August 8, 1993. *EOS Trans.*, 75(25); *AGU 1994 Western Pacific Geophysics Meeting*, Additional Abstracts, SE22A-10, 1994.

[Kos97]   V. G. Kossobokov. User Manual for M8. In J. H. Healy, V. I. Keilis-Borok, and W. H. K. Lee, editors, *Algorithms for Earthquake Statistics and Prediction*. IASPEI Software Library, Vol. 6. Seismol. Soc. Am., El Cerrito, CA, 1997.

[KP80]   V. I. Keilis-Borok and F. Press. On seismological applications of pattern recognition. In C. J. Allegre, editor, *Source Mechanism and Earthquake Prediction Applications, Editions du Centre national de la recherche scientifique*, pp. 51–60, Paris, 1980.

[KR90]   V. I. Keilis-Borok and I. M. Rotwain. Diagnosis of time of increased probability of strong earthquakes in different regions of the world: algorithm CN. *Phys. Earth Planet. Int.*, 61: 57-72, 1990.

[Kra93]   Yu. A. Kravtsov, editor, *Limits of Predictability*, Springer-Verlag, Berlin-Heidelberg, 1993.

[KRKH99]   V. G. Kossobokov, L. L. Romashkova, V. I. Keilis-Borok, and J. H. Healy. Testing earthquake prediction algorithms: Statistically significant advance prediction of the largest earthquakes in the Circum-Pacific 1992–1997. *Phys. Earth Planet. Int.*, 111: 187–196, 1999.

[Kro84]   T. L. Kronrod. Seismicity parameters for the main high-seismicity regions of the world. In V. I. Keilis-Borok and A. L. Levshin, editors, *Mathematical Modeling and Interpretation of Geophysical Data, Comput. Seismol., 17*, pp. 35–55, Allerton Press, New York, 1985.

[KRS94]   V. I. Keilis-Borok, I. M. Rotwain, and P. N. Shebalin. The spatial redistribution of low-level seismicity in focal zones of large earthquakes. In *Seismicity and Related Processes in the Environment. Global Changes of Environment and Climate*. Federal Research Program of Russia, Vol. 1, pp. 9–13, 1994.

[KRS97]   V. I. Keilis-Borok, I. M. Rotwain, and A. A. Soloviev. Numerical modelling of block structure dynamics: dependence of a synthetic earthquake flow on the structure separateness and boundary movements. *J. Seismol.*, 1(2): 151–160, 1997.

[KS77]   N. V. Kondorskaya and N. V. Shebalin, editors, *New Catalog of Large Earthquakes in the USSR from Antiquity to 1975*, Nauka, Moscow, 1977 (in Russian).

[KS83]   V. G. Kosobokov and A. A. Soloviev. Disposition of epicenters of earthquakes with $M \geq 5.5$ relative to the intersection of morphostructural lineaments in the East Central Asia. In V. I. Keilis-Borok and A. L. Levshin, editors, *Mathematical Models of the Structure of the Earth and Earthquake Prediction, Comput. Seismol., 14*, pp. 75+. Allerton Press, New York, 1983.

[KS99]   V. I. Keilis-Borok and P. N. Shebalin, editors. Dynamics of lithosphere and earthquake prediction. *Phys. Earth Planet. Int., Special Issue*, III: 179–330, 1999.

[KT96]   V. G. Kossobokov and D. L. Turcotte. A systematic global assessment of the seismic risk. *EOS Trans., AGU*, 77, S12E-03, 1996.

[Leh60]   E. L. Lehman. *Testing Statistical Hypothesis*, Wiley, New York, 1960.

[Lin85]   G. Lindgren. Optimal prediction of level crossing in Gaussian processes and sequences. *Ann. Probab.*, 13: 804–824, 1985.

[Lin96]   H.-G. Linzer. Kinematics of retreating subduction along the Carpathian arc, Romania. *Geology*, 24: 167–170, 1996.

[Lom91]   J. Lomnitz-Adler. Model for steady state friction. *J. Geophys. Res.*, 96: 6121–6131, 1991.

[Lor63]   E. N. Lorenz. Deterministic nonperiodic flow. *J. Atmos. Sci.*, 20: 130–141, 1963.

[LV92]   T. Levshina and I. Vorobieva. Application of algorithm for prediction of a strong repeated earthquake to Joshua Tree and Landers. In *Fall Meeting AGU, Abstracts*, p. 382, 1992.

[Man83]   B. Mandelbrot. *The Fractal Geometry of Nature*. Freeman, New York, 1983.

[MaS76]   Ma Shang-Keng. *Modern Theory of Critical Phenomena*, W. A. Benjamin, Reading, Mass., 1976.

[Mat86]   R. S. Matsu'ura. Precursory quiescence and recovery of aftershock activities before some large aftershocks. *Bull. Earth Res. Inst. Univ. Tokyo*, 61: 1–65, 1986.

[McK70]   D. P. McKenzie. Plate tectonics of the Mediterranean region. *Nature*, 226: 239–243, 1970.

[McK72]   D. P. McKenzie. Active tectonics of the Mediterranean region. *Geophys. J. R. Astron. Soc.*, 30: 109–185, 1972.

[MD90]   G. Molchan and O. Dmitrieva. Dynamics of the magnitude-frequency relation for foreshocks. *Phys. Earth Planet. Int.*, 61: 99–112, 1990.

[MD92]   G. M. Molchan and O. E. Dmitrieva. Aftershock identification: Methods and new approaches. *Geophys. J. Int.*, 109(3): 501–516, 1992.

[MDRD90]   G. M. Molchan, O. E. Dmitrieva, I. M. Rotwain, and J. Dewey. Statistical analysis of the results of earthquake prediction, based on burst of aftershocks. *Phys. Earth Planet. Int.*, 61: 128–139, 1990.

[Men79]   G. Menard. Relations entre structures profondes et structures superficielles dans le sud-est de la France. Essai d'ulitisation des donnees geophysiques. These de doctorat de specialite. Universite S.M. Grenoble, 1979.

[MFZ+90]   Z. Ma, Z. Fu, Y. Zhang, C. Wang, G. Zhang, and D. Liu. *Earthquake Prediction: Nine Major Earthquakes in China*, Springer-Verlag, New York, 1990.

[MGS86]   K. C. McNally, J. R. Gonzalez-Ruizand, and C. Stolte. Seismogenesis of the 1985 great ($M_s = 8.1$) Michoacan, Mexico earthquake. *Geophys. Res. Lett.*, 13: 585–588, 1986.

[Mil68]   E. E. Milanovsky. *Recent Tectonics of the Caucasus*, Nedra, Moscow, 1968 (in Russian).

[MJCB84]   J. B. Minster and T. H. Jordan. In J. K. Crouch and S. B.Bachman, editors, *Tectonics and Sedimentation Along the California Margin: Pacific Section*, Vol.38, pp. 1–16, Academic, San Diego, 1984.

[MK92]   G. M. Molchan and Y. Y. Kagan. Earthquake prediction and its optimization. *J. Geophys. Res.*, 97: 4823–4838, 1992.

[MM69]   D. P. McKenzie and W. J. Morgan. The evolution of triple junctions. *Nature*, 224: 125–133, 1969.

[Moc93]   V. I. Mocanu. Final report "Go West" Programme. Proposal No: 4609. Contract No: CIPA3510PL924609. Subject: Methods for Investigation of Different Kinds of Lithospheric Plates in Europe. Period September–December 1993.

[Mog62]   K. Mogi. Magnitude-frequency relation for elastic shocks accompanying fractures of various materials and some related problems in earthquakes. *Bull. Earthquake Inst. Tokyo Univ.*, 40: 831–853, 1962.

[Mog68]   K. Mogi. Migration of seismic activity. *Bull. Earth Res. Inst. Univ. Tokyo*, 46(1): 53–74, 1968.

[Mog74]   K. Mogi. Active periods in the world chief seismic belts. *Tectonophysics*, 22: 265–282, 1974.

[Mog85]   K. Mogi. *Earthquake Prediction*, Academic Press, Tokyo, 1985.

[Mol90]   G. M. Molchan. Strategies in strong earthquake prediction. *Phys. Earth Planet. Int.*, 61: 84–98, 1990.

[Mol91]   G. M. Molchan. Structure of optimal strategies of earthquake prediction. *Tectonophysics*, 193: 267–276, 1991.

[Mol92]   G. M. Molchan. Models for optimization of earthquake prediction (in Russian). In V. I. Keilis-Borok and A. L. Levshin, editors, *Computational Seismology*, Vol. 25, pp. 7–28, Nauka, Moscow, 1992.

[Mol94]   G. M. Molchan. Models for optimization of earthquake prediction. In D. K. Chowdhury, editor, *Computational Seismology and Geodynamics*, Vol. 2, pp. 1–10, Am. Geophys. Union, Washington, D. C., 1994.

[Mol97]   G. M. Molchan. Earthquake prediction as a decision-making problem. *Pure Appl. Geophys.*, 149: 233-247, 1997.

[MP67]   D. P. McKenzie and R. L. Parker. The North Pacific: An example of tectonics on a sphere. *Nature*, 216(5122): 1276–1280, 1967.

[MS99]   V. I. Maksimov and A. A Soloviev. Clustering of earthquakes in a block model of lithosphere dynamics. In D. K. Chowdhury, editor, *Computational Seismology and Geodynamics*, Vol. 4, pp. 124–126, Am. Geophys. Union, Washington, D. C., 1999.

[Nar]   G. S. Narkunskaya. Personal communications.

[Nat99]   *Nature Debates, 1999.*
http://www.nature.com/nature/debates/earthquake/
equake_frameset.html

[NB87]   S. Nishenko and R. Buland. A generic recurrence interval distribution for earthquake forecasting. *Bull. Seismol. Soc. Am.*, 77(4): 1382–1399, 1987.

[NGT94]   W. Newman, A. Gabrielov, and D. L. Turcotte, editors. *Nonlinear Dynamics and Predictability of Geophysical Phenomena*, Am. Geophys. Union, Int. Union of Geodesy and Geophys., 1994.

[Nis89]   S. P. Nishenko. Circum-Pacific earthquake potential: 1989–1999. *USGS, Open File Report*: 89–86, 1989.

[NR96]   O. V. Novikova and I. M. Rotwain. Advance earthquake prediction by CN algorithm. *Trans. (Doklady) Russ. Acad. Sci.*, 348(4): 548–551, 1996.

[NS75]   P. N. Nikolaev and Yu. K. Shchyukin. Model of crust and uppermost mantle deformations for the Vrancea region. In *Deep Crustal Structure*, pp. 61–83, Nauka, Moscow, 1975 (in Russian).

[NS90]   G. S. Narkunskaya and M. G. Shnirman. Hierarchical model of defect development and seismicity. *Phys. Earth Planet. Int.*, 61: 29–35, 1990.

[NS94]   G. S. Narkunskaya and M. G. Shnirman. An algorithm of earthquake prediction. In *Computational Seismology and Geodynamics*, Vol. 1, pp. 20–24, AGU, Washington, D. C., 1994.

[NSH+00]   C. Narteau, P. Shebalin, M. Holschneider, J.-L. Le Mouël, and C. Allegre. Direct simulations of the stress redistribution in the scaling organization of fracture tectonics (SOFT) model. *Geophys. J. Int.*, 141(1): 115–135, 2000.

[NTG95]   W. I, Newman, D. L. Turcotte, and A. M. Gabrielov. Log-periodic behavior of a hierarchical failure model with application to precursory seismic activation. *Phys. Rev. E.*, 52: 4827–4835, 1995.

[NVE+95]   O. V. Novikova, I. A. Vorobieva, D. Enescu, M. Radulian, I. Kuznetzov, and G. Panza. Prediction of strong earthquakes in Vrancea, Romania, using the CN algorithm. *Pure Appl. Geophys.*, 145: 277–296, 1995.

[OB97]   M. C. Oncescu and K. P. Bonjer. A note on the depth recurrence and strain release of large Vrancea earthquakes. *Tectonophysics*, 272: 291–302, 1997.

[OBAS84]   M. C. Oncescu, V. Burlacu, M. Anghel, and V. Smalbergher. Three-dimensional P-wave velocity image under the Carpathian arc. *Tectonophysics*, 106: 305–319, 1984.

[Oga88]   Y. Ogata. Statistical models for earthquake occurrences and residual analysis for point processes. *J. Am. Stat. Assoc.*, 83(401), Applications: 9–26, 1988.

[Onc84]   M. C. Oncescu. Deep structure of the Vrancea region, Rumania, inferred from simultaneous inversion for hypocenters and 3-D velocity structure. *Ann. Geophys.*, 2: 23–28, 1984.

[OT87]   M. C. Oncescu and C.-I. Trifu. Depth variation of moment tensor principal axes in Vrancea (Rumania) seismic region. *Ann. Geophys.*, 5: 149–154, 1987.

[PA95]   F. Press and C. Allen. Patterns of seismic release in the southern California region. *J. Geophys. Res.*, 100(B4): 6421–6430, 1995.

[PB75]   F. Press and P. Briggs. Chandler wobble, earthquakes, rotation and geomagnetic changes. *Nature* (London), 256: 270–273, 1975.

[PBR98]   F. F. Pollitz, R. Burgmann, and B. Romanowicz. *Science*, 280: 1245–1249, 1998.

[PC94]   S. L. Pepke and G. M. Carlson. Predictability of self-organized systems. *Phys. Rev. E*, 50: 236–242, 1994.

[PCS94]   G. F. Pepke, J. R. Carlson and B. E. Shaw. Prediction of large events on a dynamical model of fault. *J. Geophys. Res.*, 99: 6769–6788, 1994.

[PDE]   Preliminary Determination of Epicenters (PDE): PDE monthly – FTP at the address http://ghtftp.cr.usgs.gov/pde; PDE weekly and QED – FTP at http: ghtftp.cr.usgs.gov/weekly

[Pin73]   I. Sh. Pinsker. Estimation of learning method and training set. In *Modeling and Automated Analysis of Electrocardiograms.*, pp. 13–23, Nauka, Moscow, 1973.

[PR72]   A. G. Prozorov and Y. Y. Rantsman. Earthquake statistics and morphostructures of Eastern Soviet Central Asia. *Trans. (Doklady) Akad. Sci. SSSR*, 207: 12–15, 1972.

326   References

[Pro75]   A. G. Prozorov. Changes of seismic activity connected to large earth-
           quakes. In V. I. Keilis-Borok, editor, *Interpretation of Data in Seismol-
           ogy and Neotectonics, Comput. Seismol. 8*, pp. 71–82, Nauka, Moscow,
           1975.

[Pro78]   A. G. Prozorov. A statistical analysis of P-wave residuals and the pre-
           diction of the origin times of strong earthquakes. In V. I. Keilis-Borok,
           editor, *Earthquake Prediction and the Structure of the Earth, Comput.
           Seismol.* Vol. 11, pp. 4–18, Allerton Press, New York, 1987.

[Pro91]   A. G. Prozorov. On the long range interaction among strong seismic
           events. In *Major Puzzling Problems or Paradoxes in Contemporary
           Geophysics. IUGG-U3 Abstracts*, p. 29, Vienna, 1991.

[Pro93]   A. G. Prozorov. Long range interaction of strong seismic events as a fea-
           ture of intermittent character of plate dynamics. In *1993 Spring Meet-
           ing*, May 24–28, Baltimore, Maryland, American Geophysical Union,
           Mineralogical Society of America, Geochemical Society 1993, p. 318,
           EOS, Trans., AGU, 74, 16 / Supplement.

[Pro94a]   A. G. Prozorov. An earthquake prediction algorithm for the Pamir and
           Tien Shan region based on a combination of long-range aftershocks
           and quiescent periods. In D. K. Chowdhury, editor, *Computational
           Seismology and Geodynamics*, Vol. 1, pp. 31–35, Am. Geophys. Union,
           Washington, D. C., 1994.

[Pro94b]   A. G. Prozorov. A new statistic to test the significance of long range
           interaction between large earthquakes. In V. I. Keilis-Borok, editor,
           Seismicity and Related Processes in the Environment, Vol. 1, pp. 38–
           43, Research and Coordinating Center for Seismology and Engineering,
           Moscow, 1994.

[PS84]   O. J. Perez and C. H. Scholz. Heterogeneity of the instrumental seismic-
           ity catalog (1904–1980) for strong shallow earthquakes. *Bull. Seismol.
           Soc. Am.*, 74: 669–686, 1984.

[PS90]   A. G. Prozorov and S. Yu. Schreider. Real time test of the long-range
           aftershock algorithm as a tool for mid-term earthquake prediction in
           Southern California. *Pure Appl. Geophys.*, 133: 329–347, 1990.

[PSV97]   G. F. Panza, A. A. Soloviev, and I. A. Vorobieva. Numerical modelling
           of block-structure dynamics: Application to the Vrancea region. *Pure
           Appl. Geophys.*, 149: 313–336, 1997.

[Rad67]   C. Radu, On the intermediate earthquakes in the Vrancea region. *Rev.
           Roum. Geol. Geophys. Geogr.*, 11: 113–120, 1967.

[Ran79]   E. Ia. Ranzman. *Places of Earthquakes and Morphostructures of Moun-
           tain Countries*, Nauka, Moscow, 1979 (in Russian).

[RDSS80]   Yu. V. Riznichenko, A. V. Drumya, N. Ya. Stepanenko, and N. A. Si-
           monova. Seismicity and seismic risk of the Carpathian region. In
           A. V. Drumya, editor, *The 1977 Carpathian Earthquake and its Impact*,
           pp. 46–85, Nauka, Moscow, 1980 (in Russian).

[Rea85]   P. Reasenberg. Second-order moment of Central California seismicity,
           1969–1982. *J. Geophys. Res.*, 90: 5479–5495, 1985.

[RG83]   J. Rice and J. Gu. Earthquake aftereffects and triggered seismic phe-
           nomena. *Pure Appl. Geophys.*, 121: 187-219, 1983.

[Ric58]   C. Richter. *Elementary Seismology*. Freeman, San Francisco, Calif.,
           1958.

[Ric64]   C. Richter. Comment on the paper "One Regularity in the Occurrence of Strong Earthquakes" by Keilis-Borok, V. I. and Malinovskaya, L. N. *J. Geophys. Res.*, 69: 3025, 1964.

[RJ89]   P. A. Reasenberg and L. M. Jones. Earthquake hazard after a main shock in California. *Science*, 243: 1173–1176, 1989.

[RK96]   L. L. Romashkova, and V. G.Kossobokov. Sources concentration parameters in an intermediate-term earthquake prediction algorithm. In V.I.Keilis-Borok and G.M.Molchan, editors, *Modern Problems of Seismology and Earth Dynamics, Comput. Seismol.* Vol. 28, pp. 56–66, Nauka, Moscow, 1996 (in Russian).

[RKB97]   I. Rotwain, V. I. Keilis-Borok, and L. Botvina. Premonitory transformation of steel fracturing and seismicity. *Phys. Earth Planet. Int.*, 101: 61–71, 1997.

[RKH99]   L. L. Romashkova, V. G. Kossobokov, and J. H. Healy. The 1999 Hector Mine earthquake was expected. In *AGU 1999 Fall Meeting Programme*, p. 19, 1999.

[RN99]   I. M. Rotwain and O. V. Novikova. Performance of the earthquake prediction algorithm CN in 22 regions of the world. *Phys. Earth Planet. Int.*, 111: 207–213, 1999.

[Ros70]   M. Ross. *Applied Probability Models with Optimization Applications.* San Francisco, 1970.

[Rom93]   B. Romanowicz. Spatiotemporal patterns in the energy-release of great earthquakes. *Science*, 260: 1923–1926, 1993.

[Rot]   I. M. Rotwain. Personal communications.

[RR94]   D. V. Rundkvist and I. M. Rotwain. Present-day geodynamics and seismicity of Asia Minor. In V. I. Keilis-Borok and G. M. Molchan, editors, *Theoretical Problems of Geodynamics and Seismology, Comput. Seismol.*, Vol. 27, pp. 201–244, Nauka, Moscow, 1994 (in Russian).

[RS98]   I. M. Rotwain and A. A. Soloviev. Numerical modelling of block structure dynamics: Temporal characteristics of a synthetic earthquake flow. In V. I. Keilis-Borok and G. M. Molchan, editors, *Problems in Geodynamics and Seismology, Comput. Seismol. 30*, pp. 275–288, GEOS, Moscow, 1998 (in Russian).

[RS99]   D. V. Rundquist and A. A. Soloviev. Numerical modelling of block structure dynamics: An arc subduction zone. *Phys. Earth Planet. Int.*, 111(3–4): 241–252, 1999.

[RSV98]   D. V. Rundquist, A. A. Soloviev, and G. L. Vladova. Modelling of dynamics and seismicity of arc subduction zones. In *Abstracts, XXVI General Assembly of the European Seismological Commission (ESC)*, Tel Aviv, Israel, August 23–28 1998, p. 44.

[RTK00]   J. B. Rundle, D. L. Turcotte, and W. Klein, editors. *Geocomplexity and the Physics of Earthquakes*, AGU, Washington, D. C., 2000.

[RVR98]   D. V. Rundquist, G. L. Vladova, and V. V. Rozhkova. Regularities of migration of the seismic activity along island-arcs. *Trans. (Doklady) Russ. Acad. Sci.*, 360(2): 263–266, 1998.

[Ryb79]   L. Rybach. The Suiss geotraverse from Basel to Chiaso. *Schweiz Miner. Petrogr.* 59(1/2): 199–206, 1979.

[Sad86]   M. A. Sadovsky, editor. *Long-Term Earthquake Prediction: Methodological Recommendations*, Inst. Phys. Earth, Moscow, 1986 (in Russian),

[SB98]   M. G. Shnirman and E. M. Blanter. Self-organized criticality in a mixed hierarchical system. *Phys. Rev. Lett.*, 81: 5445–5448, 1998.

[SB99]   M. G. Shnirman and E. M. Blanter. Scale invariance and invariant scaling in a mixed hierarchical system. *Phys. Rev. E*, 60: 5111–5120, 1999.

[SB01]   M. G. Shnirman and E. M. Blanter. Criticality in a dynamic mixed system. *Phys. Rev. E*, 64(5): 6123–6132, 2001.

[SBB83]   S. I. Sherman, S. A.Borniakov, V. Yu.Buddo. *Areas of Dynamic Effects of Faults*, Nauka, Novosibirsk, 1983 (in Russian).

[SBV$^+$80] O. G. Shamina, V. A. Budnikov, S. D. Vinogradov, M. P. Volarovich, I. S. Tomashevskaya. Laboratory experiments on the Physics of the earthquake source. In: *Physical Processes in Earthquake Sources* (Nauka, Moscow 1980) pp. 56–68 (in Russian).

[SC84]   D. P. Schwartz and K. J. Coppersmith. Fault behavior and characteristic earthquakes: examples from the Wasatch and San Andreas fault zones. *J. Geophys. Res.*, 89: 5681–5698, 1984.

[Sch88]   C. H. Scholz. Mechanism of seismic quiescence. *Pure Appl. Geophys.*, 126: 701–718, 1988.

[Sch90]   C. H. Scholz. *The Mechanics of Earthquakes and Faulting*, Cambridge University Press, Cambridge, UK, 1990.

[Sch97]   C. H. Scholz. Whatever happened to earthquake prediction. *Geotimes*, 42(3): 16–19, 1997.

[SCL97]   B. E. Shaw, J. M. Carlson, and J. S. Langer. Patterns of seismic activity preceding large earthquakes. *J. Geophys. Res.*, 97: 479, 1997.

[SCZ$^+$91] G. A. Sobolev, T. L. Chelidze, A. D. Zavyalov, L. B. Slavina, and V. E. Nicoladze. Maps of expected earthquakes based on cobmination of parameters. *Tectonophysics*, 193: 255–265, 1991.

[SD80]   Yu. K. Shchyukin and T. D. Dobrev. Deep geological structure, geodynamics and geophysical fields of the Vrancea region, In A. V. Drumya, editor, *The 1977 Carpathian Earthquake and its Impact*, pp. 7–40, Nauka, Moscow, 1980 (in Russian).

[SGPS84]   M. A. Sadovsky and T. V. Golubeva, V. F. Pisarenko, and M. G. Shnirman. Characteristic dimensions of rock and hierarchical properties of seismicity. *Izv. Acad. Nauk SSSR, Phys. Solid Earth.*, 20: 87–96, 1984.

[SGR+96]   P. Shebalin, N. Girardin, I. Rotwain, V. I. Keilis-Borok, and J. Dubois. Local overturn of active and non-active seismic zones as a precursor of large earthquakes in Lesser Antillean arc. *Phys. Earth Planet. Int.*, 97: 163–175, 1996.

[She87]   P. N. Shebalin. Compilation of earthquake catalogs as the task of clustering with learning. *Trans. (Doklady) Acad. Sci. SSSR*, 292: 1083–1086, 1987.

[She92]   P. N. Shebalin. Automatic duplicate identification in set of earthquake catalogues merged together. *U.S. Geol. Surv. Open-File Report 92-401, Appendix II*, 1992.

[SIGB94]   Special Issue on the Great Bolivian Earthquake of 1994. *Geophys. Res. Lett.*, 22: 2231–2280, 1994.

[SJ90]   L. R. Sykes and S. Jaumé. Seismic activity on neighboring faults as a long-term precursor to large earthquakes in the San Francisco Bay area. *Nature*, 348: 595–599, 1990.

[SK99]   P. N. Shebalin and V. I. Keilis-Borok. Phenomenon of local "seismic reversal" before strong earthquakes. *Phys. Earth Planet. Int.*, 111: 215–227, 1999.

[SKL92]   R. S. Stein, G. S. P. King, and J. Lin. Change in failure stress on the Southern San Andreas fault system caused by the 1992 magnitude=7.4 Landers earthquake. *Science*, 258: 1328–1332, 1992.

[SLT94]   J. B. Shepherd, L. L. Linch, and J. G. Tanner. A revised earthquake catalogue for the Eastern Caribbean region. In *Proceedings of the Caribbean Conference on Natural Hazards*, Trinidad and Tobago, 1994.

[Smi81]   W. Smith. The *b*-value as an earthquake precursor. *Nature*, 289: 136–139, 1981.

[Sor00]   D. Sornette. *Critical Phenomena in Natural Sciences: Chaos, Fractals, Self-organization, and Disorder. Concept and Tools*, Springer, Berlin, 2000.

[SRRV99]   A. Soloviev, D. V. Rundquist, V. V. Rozhkova, and G. L. Vladova. Application of block models to study of seismicity of arc subduction zones, H4.SMR/1150-3. *Fifth Workshop on Non-Linear Dynamics and Earthquake Prediction*, October 4–22, 1999, ICTP, Trieste, 1999.

[SS95]   D. Sornette and C. G. Sammis. Complex critical exponents from renormalization group theory of earthquakes: Implications for earthquake predictions. *J. Phys. I France*, 5: 607–619, 1995.

[SSR99]   P. O. Sobolev, A. A. Soloviev, and I. M. Rotwain. Modelling of lithosphere dynamics and seismicity for the Near East. In D. K. Chowdhury, editor, *Computational Seismology and Geodynamics*, Vol. 4, pp. 115–123, Am. Geophys. Union, Washington, D. C., 1999.

[SSS96]   C. G. Sammis, D. Sornett and H. Saleur. Complexity and earthquake forecasting. In J. B. Rundle, D. L. Turcotte, and W. Klein, editors, *SFI studies in the science of complexity*, vol. XXV, Addison-Wesley, Reading, Mass., 1996.

[SSS99]   L. R. Sykes, B. E. Shaw, and C. H. Scholz. Rethinking earthquake prediction. *Pure Appl. Geophys.*, 155: 207–232, 1999.

[STS85]   R. F. Smalley, D. L. Turcotte, and S. A. Solla. A renormalization group approach to the stick-slip behavior of faults. *J. Geophys. Res.*, 90: 1894–1900, 1985.

[SV99a]   A. A. Soloviev and Vorobieva. Study of long-range interaction between synthetic earthquakes in the model of block structure dynamics. In *Abstracts, IUGG99*, Birmingham, Week A, Monday July 19 to Saturday July 24, p. A.147, 1999.

[SV99b]   A. Soloviev and I. Vorobieva. Long-range interaction between synthetic earthquakes in the model of block structure dynamics, H4.SMR/1150-4. *Fifth Workshop on Non-Linear Dynamics and Earthquake Prediction*, October 4–22, 1999, ICTP, Trieste, 1999.

[SVP99]   A. A. Soloviev, I. A. Vorobieva, and G. F. Panza. Modelling of block-structure dynamics: Parametric study for Vrancea. *Pure Appl. Geophys.*, 156(3): 395–420, 1999.

[SVP00]   A. A. Soloviev, I. A. Vorobieva, and G. F. Panza. Modelling of block structure dynamics for the Vrancea region: Source mechanisms of the synthetic earthquakes. *Pure and Appl. Geophys.*, 157(1–2): 97–110, 2000.

[SZK00]   P. Shebalin, I. Zaliapin, and V. I. Keilis-Borok. Premonitory raise of the earthquakes' correlation range: Lesser Antilles. *Phys. Earth Planet. Int.*, 122: 241–249, 2000.

[TDRL91]   C.-I. Trifu, A. Deschamps, M. Radulian, and H. Lyon-Caen. The Vrancea earthquake of May 30, 1990: An estimate of the source parameters. In *Proceedings of the XXII General Assembly of the European Seismological Commission*, pp. 449–454, Barcelona, 1991.

[TNG00]   D. L. Turcotte, W. I. Newman, and A. Gabrielov. A statistical physics approach to earthquakes. In *Geocomplexity and the Physics of Earthquakes*, American Geophysical Union, Washington D. C., 2000.

[TR89]   C.-I. Trifu, M. Radulian. Asperity distribution and percolation as fundamentals of earthquake cycle. *Phys. Earth Planet. Int.*, 58: 277–288, 1989.

[Tra85]   V. Yu. Traskin. In *Physical and Chemical Mechanics of Natural Dispersed Systems*. Moscow University, Moscow, pp. 140–196, 1985.

[Tri90]   C.-I. Trifu. Detailed configuration of intermediate seismicity in the Vrancea region. *Rev. Geofisica*, 46: 33–40, 1990.

[Tuk77]   J. W. Tukey. *Exploratory data analysis. Series in Behavioral Science: Quantitative Methods*. Addison-Wesley, Reading, Mass., 1977.

[Tur97]   D. L. Turcotte. *Fractals and Chaos in Geology and Geophysics*, 2nd edn., Cambridge University Press, Cambridge, 1997.

[Tur99]   D. L. Turcotte. Seismicity and self-organized criticality. *Phys. Earth Planet. Int.*, 111: 275–294, 1999.

[Upd89]   R. G. Updike, editor. *Proceedings of the National Earthquake Prediction Evaluation Council, U.S. Geol. Surv. Open-File Rep. 89-114*, 1989.

[US54]   T. Utsu and A. Seki. A relation between the area of aftershock region and the energy of main shock *J. Seismol. Soc. Jpn.*, 7: 233–240, 1954.

[USGS62]   Geologic Map of North America. Scale 1:5 000 000. USGS, Washington, D. C., 1962.

[USGS65]   Tectonic Map of the United States. Scale 1:2 500 000. USGS, Washington, D. C., 1965.

[Uts61]   T. Utsu. A statistical study on the occurence of aftershocks. *Geophys. Mag.*, 30: 521–605, 1961.

[Uts77]   T. Utsu. Probabilities in earthquake prediction. *Zisin (J. Seismol. Soc. Japan.)*, 30: 179–185, 1977 (in Japanese).

[Var89]   D. J. Varnes. Predicting earthquakes by analyzing accelerating precursory seismic activity. *Pure Appl. Geophys.*, 130: 661–686, 1989.

[Ver69]   D. Vere-Jones. A note on the statistical interpretation of Bath's law. *Bull. Seismol. Soc. Am.*, 59: 1535–1541, 1969.

[Ver78]   D. Vere-Jones. Earthquake prediction – a Statistician's view. *J. Phys. Earth*, 26: 129–146, 1978.

[VGS00]   I. A. Vorobieva, A. I. Gorshkov, and A. A. Soloviev. Modelling of the block structure dynamics and seismicity of the Western Alps. In V. I. Keilis-Borok and G. M. Molchan, editors, *Problems of Dynamics and Seismicity of the Earth, Comput. Seismol. 31*, pp. 154–169. GEOS, Moscow, 2000 (in Russian).

[VL94]   I. A.Vorobieva and T. A. Levshina. Prediction of the second large earthquake based on aftershock sequence. In D. K. Chowdhury, editor, *Computational Seismology and Geodynamics*, Vol. 2, pp. 27–36, Am. Geophys. Union, Washington, D. C., 1994.

[VNK+99] I. A. Vorobieva, O. V. Novikova, I. V. Kuznetsov, D. Enescu, M. Radu-
lian, and G. Panza. Intermediate-term earthquake prediction for the
Vrancea region: Analysis of new data. In D. K. Chowdhury, editor,
Computational Seismology and Geodynamics, Vol. 4, pp. 82–93, Am.
Geophys. Union, Washington, D. C., 1999.

[Vog79] J. Vogt. Les treblements de terre en France. Bureau de Recherches
Geologiques et Minieres, Orleans (Memoire BRGM no. 96), 1979.

[Vor94] I.A.Vorobieva. Prediction of a reoccurrence of large earthquakes based
on the aftershock sequence of the first large earthquake. In V. I. Keilis-
Borok, editor, *Seismicity and Related Processes in the Environment*,
Vol. 1, pp. 33–37, Research and Coordinating Centre for Seismology
and Engineering, Moscow, 1994.

[Vor99] I. A. Vorobieva. Prediction of second large earthquake. *Phys. Earth
Planet. Int.*, 111(3-4): 197-206, 1999.

[VP93] I. A. Vorobieva and G. F. Panza. Prediction of the occurrence of related
strong earthquakes in Italy. *Pure Appl. Geophys.*, 141(1): 25–41, 1993.

[VS79] E. V. Vil'kovich and M. G. Shnirman. On an algorithm of finding out
the migration of strong earthquakes. In V. I. Keilis-Borok, editor, *Theory and Analysis of Seismological Observations, Comput. Seismol., 12*,
pp. 37–44, Nauka, Moscow, 1974 (in Russian).

[VS83] E.V. Vil'kovich and M.G. Shnirman. Epicenter migration waves: Examples and models. In V. I. Keilis-Borok and A. L. Levshin, editors,
*Mathematical Models of the Earth's Structure and the Earthquake Pre-
diction, Comput. Seismol. 14*, pp. 27–36, Allerton Press, New York,
1983.

[Weg15] A. Wegener. *Die Entstehung der Kontinente und Ozeane*, Braun-
schweig, Vieweg (Slg. Wissenschaft 23), 1915 (in German, The Origin
of Continents and Oceans).

[WGCE88] Working Group on California Earthquake Probabilities. Probabilities of
Large Earthquakes Occurring in California on the San Andreas Fault,
U.S. Geol. Surv. Open File Rep., 88–398, 1988.

[WGG+86] C. Weber, A. D. Gvishiani, P. Godefroy, A. I. Gorshkov, A. F. Kush-
nir, V. F. Pisarenko, A. Cisternas, A. V. Trusov, M. L. Tzvang, and
S. L. Tzvang. On classification of high seismicity areas in the Western
Alps. *Izv. Acad. Sci. SSSR, Phys. Earth*, 12: 3+, 1986 (in Russian).

[WH88] M. Wyss and R. Habermann. Precursory seismic quiescence. *Pure Appl.
Geophys.*, 126: 319-332, 1988.

[WLN98] F. Wenzel, D. Lungu, O. Novak, editors. *Vrancea Earthquakes: Tectonics, Hazard and Risk Mitigation*, Kluwer Academic, Dordrecht, Nether-
lands, 1998.

[WW94] S. Wiemer and M. Wyss. Seismic quiescence before the Landers
(M=7.5) and Big Bear (M=6.5) 1992 earthquakes. *Bull. Seismol. Soc.
Am.*, 84: 900–916, 1994.

[WWL98] D. A. Wiens, M. E. Wysession, and L. Lawver. Recent oceanic intraplate earthquake in Balleny Sea was largest ever detected. *EOS 79*
(July 28): 353, 1998.

[Wys86] M. Wyss. Seismic quiescence precursor to the 1983 Kaoiki (Ms=6.6),
Hawaii, earthquake. *Bull. Seismol. Soc. Am.*, 76: 785-800, 1986.

[Wys91] M. Wyss, editor. *Evaluation of Proposed Earthquake Precursors*. AGU,
Washington, D. C., 1991.

332   References

[Wys97a] M. Wyss. Cannot earthquakes be predicted? *Science*, 278: 487–488, 1997.
[Wys97b] M. Wyss. Second round of evaluation of proposed earthquake precursors. *Pure Appl. Geophys.*, 149: 3–16, 1997.
[Yam98]   T. Yamashita. Simulation of seismicity due to fluid migration in a fault zone. *Geophys. J. Int.*, 132: 674–686, 1998.
[YK92]   T. Yamashita and L. Knopoff. 1992. Model for intermediate-term precursory clustering of earthquakes. *J. Geophys. Res.*, 97: 19873–19879, 1992.
[Yod76]   H. S. Yoder, Jr. *Generation of Basaltic Magma*, Natl. Acad. Sci., Washington, D. C., 1976.
[Yos65]   K. Yosida. *Functional Analysis*, Springer-Verlag, Berlin, 1965.
[ZHK01]   G. Zoller, S. Hainzl and J. Kurths. Observation of growing correlation length as an indicator for critical point behavior prior to large earthquakes. *J. Geophys. Res.*, 106: 2167–2176, 2001.
[ZK80]   M. P. Zhidkov and V. G. Kosobokov. Identification of the sites of possible strong earthquakes. VIII: Intersections of lineaments in the east of Central Asia. In V. I. Keilis-Borok, editor, *Earthquake Prediction and the Structure of the Earth, Comput. Seismol.*, *11*, pp. 31+, Allerton Press, New York, 1980.
[ZKA02]   I. Zaliapin, V.I. Keilis-Borok, and G. Axen. Premonitory spreading of seismicity over the fault network in S. California: precursor "ACCORD". *J. Geophys. Res.*, 2002 (in press).
[ZKG01a] I. Zaliapin, V.I. Keilis-Borok and M. Ghil. A Boolean delay equation model of colliding cascades. Part I: Multiple seismic regimes, H4.SMR/1330-1. *Sixth Workshop on Non-Linear Dynamics and Earthquake Prediction*, October 15–27, 2001, ICTP, Trieste, 2001.
[ZKG01b] I. Zaliapin, V.I. Keilis-Borok and M. Ghil. A Boolean delay equation model of colliding cascades. Part II: Prediction of critical transitions, H4.SMR/1330-2. *Sixth Workshop on Non-Linear Dynamics and Earthquake Prediction*, October 15–27, 2001, ICTP, Trieste, 2001.
[ZPD+84] C. Zhang-li, L. Pu-xiong, H. De-yu, Z. Da-lin, X. Feng, and W. Zhidong. Characteristics of regional seismicity before major earthquakes. *Earthquake Prediction*, UNESCO, Paris, pp. 505–521, 1984.
[ZRS75]   M. P. Zhidkov, I. M. Rotwain, and A. M. Sadovsky. Recognition of places where strong earthquakes may occur. IV. High-seismic intersections of lineaments of the Armenian upland, the Balkans and the Aegean Sea basin. In V. I. Keilis-Borok, editor, *Interpretation of Seismology and Neotectonics Data, Comput. Seismol.*, *8*, pp. 53+, Nauka, Moscow, 1975 (in Russian).
[ZS87]   A. V. Zheligovsky and P. N. Shebalin. Worldwide earthquake catalogues in the Geophysical Data Bank for earthquake prediction. In V. I. Keilis-Borok and A. L. Levshin, editors, *Theory and Analysis of Seismological Information, Comput. Seismol.*, *18*, pp. 164–175, Allerton Press, New York, 1987.

Index